# EUREKA MATH™

## A Story of Ratios

## Grade 7, Module 5
## Statistics and Probability

**JB JOSSEY-BASS™**
A Wiley Brand

Cover design by Chris Clary

Published by Jossey-Bass
A Wiley Brand
One Montgomery Street, Suite 1200, San Francisco, CA 94104-4594—www.josseybass.com

ISBN: 978-1-118-81112-2

Printed in the United States of America
FIRST EDITION
*PB Printing*    10  9  8  7  6  5  4  3  2

When do you know you really understand something? One test is to see if you can explain it to someone else—well enough that *they* understand it. Eureka Math routinely requires students to "turn and talk" and explain the math they learned to their peers.

That is because the goal of Eureka Math (which you may know as the EngageNY math modules) is to produce students who are not merely literate, but fluent, in mathematics. By fluent, we mean not just knowing what process to use when solving a problem but understanding why that process works.

Here's an example. A student who is fluent in mathematics can do far more than just name, recite, and apply the Pythagorean theorem to problems. She can explain why $a^2 + b^2 = c^2$ is true. She not only knows the theorem can be used to find the length of a right triangle's hypotenuse, but can apply it more broadly—such as to find the distance between any two points in the coordinate plane, for example. She also can see the theorem as the glue joining seemingly disparate ideas including equations of circles, trigonometry, and vectors.

By contrast, the student who has merely memorized the Pythagorean theorem does not know why it works and can do little more than just solve right triangle problems by rote. The theorem is an abstraction—not a piece of knowledge, but just a process to use in the limited ways that she has been directed. For her, studying mathematics is a chore, a mere memorizing of disconnected processes.

Eureka Math provides much more. It offers students math knowledge that will serve them well beyond any test. This fundamental knowledge not only makes wise citizens and competent consumers, but it gives birth to budding physicists and engineers. Knowing math deeply opens vistas of opportunity.

A student becomes fluent in math—as they do in any other subject—by following a course of study that builds their knowledge of the subject, logically and thoroughly. In Eureka Math, concepts flow logically from PreKindergarten through high school. The "chapters" in the story of mathematics are "A Story of Units" for the elementary grades, followed by "A Story of Ratios" in middle school and "A Story of Functions" in high school.

This sequencing is joined with a mix of new and old methods of instruction that are proven to work. For example, we utilize an exercise called a "sprint" to develop students' fluency with standard algorithms (routines for adding, subtracting, multiplying, and dividing whole numbers and fractions). We employ many familiar models and tools such as the number line and tape diagrams (aka bar diagrams). A newer model highlighted in the curriculum is the number bond (illustrated below), which clearly shows how numbers are comprised of other numbers.

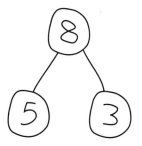

Eureka Math is designed to help accommodate different types of classrooms and serve as a resource for educators, who make decisions based on the needs of students. The "vignettes" of teacher-student interactions included in the curriculum are not scripts, but exemplars illustrating methods of instruction recommended by the teachers who have crafted our curricula.

Eureka Math has been adopted by districts from East Meadows, NY to Lafayette, LA to Chula Vista, CA. At Eureka Math we are excited to have created the most transparent math curriculum in history—every lesson, all classwork, and every problem is available online.

Many of us have less than joyful memories of learning mathematics: lots of memorization, lots of rules to follow without understanding, and problems that didn't make any sense. What if a curriculum came along that gave children a chance to avoid that math anxiety and replaced it with authentic understanding, excitement, and curiosity? Like a NY educator attending one of our trainings said: "Why didn't I learn mathematics this way when I was a kid? It is so much easier than the way I learned it!"

Eureka!

Lynne Munson
Washington DC
September 2014

GRADE

# Mathematics Curriculum

## Table of Contents[1]

# Statistics and Probability

---

[1] Each lesson is ONE day, and ONE day is considered a 45-minute period.

Grade 7 • Module 5

# Statistics and Probability

## OVERVIEW

In this module, students begin their study of probability, learning how to interpret probabilities and how to compute probabilities in simple settings. They also learn how to estimate probabilities empirically. Probability provides a foundation for the inferential reasoning developed in the second half of this module. Additionally, students build on their knowledge of data distributions that they studied in Grade 6, compare data distributions of two or more populations, and are introduced to the idea of drawing informal inferences based on data from random samples.

In Topics A and B, students learn to interpret the probability of an event as the proportion of the time that the event will occur when a chance experiment is repeated many times (**7.SP.C.5**). They learn to compute or estimate probabilities using a variety of methods, including collecting data, using tree diagrams, and using simulations. In Topic B, students move to comparing probabilities from simulations to computed probabilities that are based on theoretical models (**7.SP.C.6**, **7.SP.C.7**). They calculate probabilities of compound events using lists, tables, tree diagrams, and simulations (**7.SP.C.8**). They learn to use probabilities to make decisions and to determine whether or not a given probability model is plausible (**7.SP.C.7**). The Mid-Module Assessment follows Topic B.

In Topics C and D, students focus on using random sampling to draw informal inferences about a population (**7.SP.A.1**, **7.SP.A.2**). In Topic C, they investigate sampling from a population (**7.SP.A.2**). They learn to estimate a population mean using numerical data from a random sample (**7.SP.A.2**). They also learn how to estimate a population proportion using categorical data from a random sample. In Topic D, students learn to compare two populations with similar variability. They learn to consider sampling variability when deciding if there is evidence that the means or the proportions of two populations are actually different (**7.SP.B.3**, **7.SP.B.4**). The End-of-Module Assessment follows Topic D.

## Focus Standards

### Using random sampling to draw inferences about a population.

**7.SP.A.1**  Understand that statistics can be used to gain information about a population by examining a sample of the population; generalizations about a population from a sample are valid only if the sample is representative of that population. Understand that random sampling tends to produce representative samples and support valid inferences.

**7.SP.A.2**  Use data from a random sample to draw inferences about a population with an unknown characteristic of interest. Generate multiple samples (or simulated samples) of the same size to gauge the variation in estimates or predictions. *For example, estimate the mean*

*word length in a book by randomly sampling words from the book; predict the winner of a school election based on randomly sampled survey data. Gauge how far off the estimate or prediction might be.*

## Draw informal comparative inferences about two populations.

**7.SP.B.3**  Informally assess the degree of visual overlap of two numerical data distributions with similar variability, measuring the difference between the centers by expressing it as a multiple of a measure of variability. *For example, the mean height of players on the basketball team is 10 cm greater than the mean height of players on the soccer team, about twice the variability (mean absolute deviation) on either team; on a dot plot, the separation between the two distributions of heights is noticeable.*

**7.SP.B.4**  Use measures of center and measures of variability for numerical data from random samples to draw informal comparative inferences about two populations. *For example, decide whether the words in a chapter of a seventh-grade science book are generally longer than the words in a chapter of a fourth-grade science book.*

## Investigate chance processes and develop, use, and evaluate probability models.

**7.SP.C.5**  Understand that the probability of a chance event is a number between 0 and 1 that expresses the likelihood of the event occurring. Larger numbers indicate greater likelihood. A probability near 0 indicates an unlikely event, a probability around 1/2 indicates an event that is neither unlikely nor likely, and a probability near 1 indicates a likely event.

**7.SP.C.6**  Approximate the probability of a chance event by collecting data on the chance process that produces it and observing its long-run relative frequency, and predict the approximate relative frequency given the probability. *For example, when rolling a number cube 600 times, predict that a 3 or 6 would be rolled roughly 200 times, but probably not exactly 200 times.*

**7.SP.C.7**  Develop a probability model and use it to find probabilities of events. Compare probabilities from a model to observed frequencies; if the agreement is not good, explain possible sources of the discrepancy.

a.  Develop a uniform probability model by assigning equal probability to all outcomes, and use the model to determine probabilities of events. *For example, if a student is selected at random from a class, find the probability that Jane will be selected and the probability that a girl will be selected.*

b.  Develop a probability model (which may not be uniform) by observing frequencies in data generated from a chance process. *For example, find the approximate probability that a spinning penny will land heads up or that a tossed paper cup will land open-end down. Do the outcomes for the spinning penny appear to be equally likely based on the observed frequencies?*

**7.SP.C.8**    Find probabilities of compound events using organized lists, tables, tree diagrams, and simulation.

    a.   Understand that, just as with simple events, the probability of a compound event is the fraction of outcomes in the sample space for which the compound event occurs.

    b.   Represent sample spaces for compound events using methods such as organized lists, tables and tree diagrams.  For an event described in everyday language (e.g., "rolling double sixes"), identify the outcomes in the sample space which compose the event.

    c.   Design and use a simulation to generate frequencies for compound events.  *For example, use random digits as a simulation tool to approximate the answer to the question: If 40% of donors have type A blood, what is the probability that it will take at least 4 donors to find one with type A blood?*

## Foundational Standards

### Summarize and describe distributions.

**6.SP.B.5**    Summarize numerical data sets in relation to their context, such as by:

    a.   Reporting the number of observations.

    b.   Describing the nature of the attribute under investigation, including how it was measured and its units of measurement.

    c.   Giving quantitative measures of center (median and/or mean) and variability (interquartile range and/or mean absolute deviation), as well as describing any overall pattern and any striking deviations from the overall pattern with reference to the context in which the data were gathered.

    d.   Relating the choice of measures of center and variability to the shape of the data distribution and the context in which the data were gathered.

### Understand ratio concepts and use ratio reasoning to solve problems.

**6.RP.A.3c**    Find a percent of a quantity as a rate per 100 (e.g., 30% of a quantity means 30/100 times the quantity); solve problems involving finding the whole, given a part and the percent.

### Analyze proportional relationships and use them to solve real-world and mathematical problems.

**7.RP.A.2**    Recognize and represent proportional relationships between quantities.

# Focus Standards for Mathematical Practice

**MP.2**   **Reason abstractly and quantitatively.** Students reason quantitatively by posing statistical questions about variables and the relationship between variables. Students reason abstractly about chance experiments in analyzing possible outcomes and designing simulations to estimate probabilities.

**MP.3**   **Construct viable arguments and critique the reasoning of others.** Students construct viable arguments by using sample data to explore conjectures about a population. Students critique the reasoning of other students as part of poster or similar presentations.

**MP.4**   **Model with mathematics.** Students use probability models to describe outcomes of chance experiments. They evaluate probability models by calculating the theoretical probabilities of chance events, and by comparing these probabilities to observed relative frequencies.

**MP.5**   **Use appropriate tools strategically.** Students use simulation to approximate probabilities. Students use appropriate technology to calculate measures of center and variability. Students use graphical displays to visually represent distributions.

**MP.6**   **Attend to precision.** Students interpret and communicate conclusions in context based on graphical and numerical data summaries. Students make appropriate use of statistical terminology.

# Terminology

## New or Recently Introduced Terms

- **Probability** (A number between 0 and 1 that represents the likelihood that an outcome will occur.)
- **Probability model** (A probability model for a chance experiment specifies the set of possible outcomes of the experiment—the sample space—and the probability associated with each outcome.)
- **Uniform probability model** (A probability model in which all outcomes in the sample space of a chance experiment are equally likely.)
- **Compound event** (An event consisting of more than one outcome from the sample space of a chance experiment.)
- **Tree diagram** (A diagram consisting of a sequence of nodes and branches. Tree diagrams are sometimes used as a way of representing the outcomes of a chance experiment that consists of a sequence of steps, such as rolling two number cubes, viewed as first rolling one number cube and then rolling the second.)

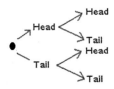

- **Simulation** (The process of generating "artificial" data that are consistent with a given probability model or with sampling from a known population.)
- **Long-run relative frequency** (The proportion of the time some outcome occurs in a very long sequence of observations.)
- **Random sample** (A sample selected in a way that gives every different possible sample of the same size an equal chance of being selected.)
- **Inference** (Using data from a sample to draw conclusions about a population.)

## Familiar Terms and Symbols[2]

- Measures of center
- Measures of variability
- Mean absolute deviation (MAD)
- Shape

# Suggested Tools and Representations

- Graphing calculator (see example below)
- Dot plots (see example below)
- Histograms (see example below)

Graphing Calculator

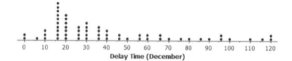

Dot Plot

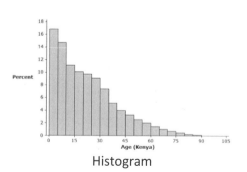

Histogram

---

[2] These are terms and symbols students have seen previously.

## Assessment Summary

| Assessment Type | Administered | Format | Standards Addressed |
|---|---|---|---|
| Mid-Module Assessment Task | After Topic B | Constructed response with rubric | 7.SP.C.5, 7.SP.C.6, 7.S.SP.7, 7.SP.C.8 |
| End-of-Module Assessment Task | After Topic D | Constructed response with rubric | 7.SP.C.1, 7.SP.C.2, 7.S.SP.3, 7.SP.C.4, 7.SP.C.5, 7.SP.C.6, 7.SP.C.7, 7.SP.C.8 |

GRADE

# Mathematics Curriculum

## Topic A:

# Calculating and Interpreting Probabilities

**7.SP.C.5, 7.SP.C.6, 7.SP.C.7, 7.SP.C.8a, 7.SP.C.8b**

| Focus Standard: | 7.SP.C.5 | Understand that the probability of a chance event is a number between 0 and 1 that expresses the likelihood of the event occurring. Larger numbers indicate greater likelihood. A probability near 0 indicates an unlikely event, a probability around 1/2 indicates an event that is neither unlikely nor likely, and a probability near 1 indicates a likely event. |
| --- | --- | --- |
| | 7.SP.C.6 | Approximate the probability of a chance event by collecting data on the chance process that produces it and observing is long-run relative frequency, and predict the approximate relative frequency given the probability. *For example, when rolling a number cube 600 times, predict that a 3 or 6 would be rolled roughly 200 times, but probably not exactly 200 times.* |
| | 7.SP.C.7 | Develop a probability model and use it to find probabilities of events. Compare probabilities from a model to observed frequencies; if the agreement is not good, explain possible sources of the discrepancy. |
| | | a. Develop a uniform probability model by assigning equal probability to all outcomes, and use the model to determine probabilities of events. *For example, if a student is selected at random from a class, find the probability that Jane will be selected and the probability that a girl will be selected.* |
| | | b. Develop a probability model (which may not be uniform) by observing frequencies in data generated from a chance process. *For example, find the approximate probability that a spinning penny will land heads up or that a tossed paper cup will land open-end down. Do the outcomes for the spinning penny appear to be equally likely based on the observed frequencies?* |
| | 7.SP.C.8a 7.SP.C.8b | Find probabilities of compound events using organized lists, tables, tree diagrams, and simulation. |
| | | a. Understand that, just as with simple events, the probability of a compound event is the fraction of outcomes in the sample space for which the compound event occurs. |

**Topic A:**     Calculating and Interpreting Probabilities

> b. Represent sample spaces for compound events using methods such as organized lists, tables and tree diagrams. For an event described in everyday language (e.g., "rolling double sixes") identify the outcomes in the sample space with compose the event.

**Instructional Days:**  7

**Lesson 1:** Chance Experiments (P)[1]

**Lesson 2:** Estimating Probabilities by Collecting Data (P)

**Lesson 3:** Chance Experiments with Equally Likely Outcomes (P)

**Lesson 4:** Calculating Probabilities for Chance Experiments with Equally Likely Outcomes (P)

**Lesson 5:** Chance Experiments with Outcomes that Are Not Equally Likely (P)

**Lesson 6:** Using Tree Diagrams to Represent a Sample Space and to Calculate Probabilities (E)

**Lesson 7:** Calculating Probabilities of Compound Events (P)

In Topic A, students begin a study of basic probability concepts (**7.SP.C.5**). They are introduced to the idea of a chance experiment and how probability is a measure of how likely it is that an event will occur. Working with spinners and other chance experiments, students estimate probabilities of outcomes (**7.SP.C.6**). In Lesson 1, students collect data they will use to estimate a probability in Lesson 2. Lesson 2 also provides additional opportunities to use data to estimate a probability. In Lesson 3, students are introduced to the terminology of probability, including *event*, *outcome*, and *sample space*. They are asked to think about chance experiments in terms of whether or not outcomes in the sample space are equally likely. In Lesson 4, they determine the sample space for a chance experiment and calculate the probabilities of events based on the sample space (**7.SP.C.7**). In this lesson, students also learn to assign probabilities to outcomes in a sample space when the outcomes are equally likely. They then calculate the probability of compound events that consist of more than a single outcome. This lesson leads students to see that when outcomes are equally likely, the probability of an event is the number of outcomes in the event divided by the number of outcomes in the sample space.

In Lesson 5, students begin to analyze chance experiments that have outcomes that are not equally likely. They calculate probabilities of various events by adding appropriate probabilities. Students learn in Lesson 6 to represent a sample space by a tree diagram and use the tree to calculate probabilities of compound events (**7.SP.C.8**). In Lesson 7, students calculate probabilities of compound events using sample spaces represented as lists of outcomes and presented as tree diagrams. This topic moves students from calculating and interpreting probabilities in simple settings into the lessons of Topic B, where students estimate probabilities empirically based on a large number of observations and by simulation.

---

[1] Lesson Structure Key: **P**-Problem Set Lesson, **M**-Modeling Cycle Lesson, **E**-Exploration Lesson, **S**-Socratic Lesson

# Lesson 1:  Chance Experiments

## Student Outcomes

- Students understand that a probability is a number between 0 and 1 that represents the likelihood that an event will occur.
- Students interpret a probability as the proportion of the time that an event occurs when a chance experiment is repeated many times.

## Classwork

> Have you ever heard a weatherman say there is a 40% chance of rain tomorrow or a football referee tell a team there is a 50/50 chance of getting a head on a coin toss to determine which team starts the game?  These are probability statements.  In this lesson, you are going to investigate probability and how likely it is that some events will occur.

### Example 1 (15 minutes):  Spinner Game

Place students into groups of 2.

Hand out a copy of the spinner and a paperclip to each group.  Read through the rules of the game and demonstrate how to use the paper clip as a spinner.

Here's how to use a paperclip and pencil to make the spinner:

1.  Unfold a paperclip to look like the paperclip pictured below.  Then, place the paperclip on the spinner so that the center of the spinner is along the edge of the big loop of the paperclip.

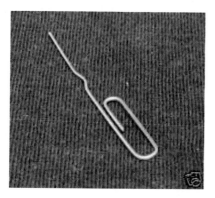

2.  Put the tip of a pencil on the center of the spinner.
3.  Flick the paperclip with your finger.  The spinner should spin around several times before coming to rest.
4.  After the paperclip has come to rest, note which color it is pointing towards.  If it lands on the line, then spin again.

---

**Example 1: Spinner Game**

Suppose you and your friend will play a game using the spinner shown here:

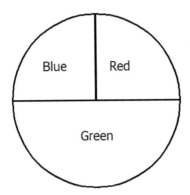

**Rules of the game:**

1.   Decide who will go first.

2.   Each person picks a color. Both players cannot pick the same color.

3.   Each person takes a turn spinning the spinner and recording what color the spinner stops on. The winner is the person whose color is the first to happen 10 times.

Play the game and remember to record the color the spinner stops on for each spin.

---

Students should try their spinners a few times before starting the game. Before students begin to play the game, discuss who should go first. Consider, for example, having the person born earliest in the year go first. If it's a tie, consider another option like tossing a coin. Discuss with students the following questions:

- Will it make a difference who goes first?
  - *The game is designed so that the spinner landing on green is more likely to occur. Therefore, if the first person selects green, this person has an advantage.*
- Who do you think will win the game?
  - *The person selecting green has an advantage.*
- Do you think this game is fair?
  - *No. The spinner is designed so that green will occur more often. As a result, the student who selects green will have an advantage.*

Play the game, and remember to record the color the spinner stops on for each spin.

**Exercises 1–4 (5 minutes)**

Allow students to work with their partners on Exercises 1–4. Then discuss and confirm as a class.

---

**Exercises 1–4**

1. Which color was the first to occur 10 times?

   *Answers will vary, but green is the most likely.*

2. Do you think it makes a difference who goes first to pick a color?

   *Yes, because the person who goes first could pick green.*

3. Which color would you pick to give you the best chance of winning the game? Why would you pick that color?

   *Green, it has the largest section on the spinner.*

4. Below are three different spinners. If you pick green for your color, which spinner would give you the best chance to win? Give a reason for your answer.

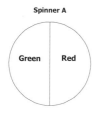 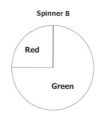 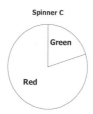

   *Spinner B, because the green section is larger for this spinner than for the other spinners.*

---

**Example 2 (10 minutes): What is Probability?**

Ask the students how they would define the word *probability,* then let them read the paragraph. After they have read the paragraph, draw the probability scale on the board. You could use the bag of balls example to emphasize the vocabulary. Present the following examples, and show how the scenario relates to the probability scale below:

Tell the students that you have a bag with four white balls.

- Ask them what would happen if you selected one ball. Discuss with students why it is certain you would draw a white ball while it would be impossible to draw a black ball.

- Under the impossible label, draw a bag with four white balls. This bag represents a bag in which it is not possible to draw a black ball. The probability of selecting a black ball would be 0. On other end, draw this same bag (four white balls). This bag represents a bag in which it is certain that you will select a white ball.

- Ask the students why "impossible" is labeled with a 0, and "certain" is labeled with a 1.

- Discuss with students that, for this example, 0 indicates that it is not possible to pick a black ball if the question is: "What is the probability of picking a black ball?" Discuss that 1 indicates that every selection would be a white ball for the question: "What is the probability of picking a white ball?"

Tell the students that you have a bag of two white and two black balls.

- Ask the students to describe what would happen if you picked a ball from that bag. Draw a model of the bag under the $\frac{1}{2}$ (or equally likely) to occur or not to occur.

- Ask the students why "equally likely" is labeled with $\frac{1}{2}$.

- Ask students what might be in a bag of balls if it was unlikely but not impossible to select a white ball.

Indicate to students that a probability is represented by a number between 0 and 1. When a probability falls in between these numbers, it can be expressed in several ways: as a fraction, a decimal, or a percent. The scale below shows the probabilities $0$, $\frac{1}{2}$, and 1, and the outcomes to the bags described above. The positions are also aligned to a description of *impossible*, *unlikely*, *equally likely*, *likely*, and *certain*. Consider providing this visual as a poster to help students interpret the value of a probability throughout this module.

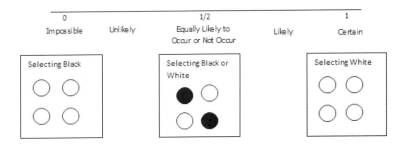

### Example 2: What is Probability?

Probability is about how likely it is that an event will happen. A probability is indicated by a number between 0 and 1. Some events are certain to happen, while others are impossible. In most cases, the probability of an event happening is somewhere between certain and impossible.

For example, consider a bag that contains only red balls. If you were to select one ball from the bag, you are certain to pick a red one. We say that an event that is certain to happen has a probability of 1. If we were to reach into the same bag of balls, it is impossible to select a yellow ball. An impossible event has a probability of 0.

| Description | Example | Explanation |
|---|---|---|
| Some events are impossible. These events have a probability of 0. | You have a bag with two green cubes, and you select one at random. Selecting a blue cube is an impossible event. | There is no way to select a blue cube if there are no blue cubes in the bag. |
| Some events are certain. These events have a probability of 1. | You have a bag with two green cubes, and you select one at random. Selecting a green cube is a certain event. | You will always get a green cube if there are only green cubes in the bag. |

| Some events are classified as equally likely to happen or to not happen. These events have a probability of $\frac{1}{2}$. | You have a bag with one blue cube and one red cube and you randomly pick one. Selecting a blue cube is equally likely to happen or to not happen. | |
|---|---|---|
| Some events are more likely to happen than to not happen. These events will have a probability that is greater than $0.5$. These events could be described as "likely" to occur. | If you have a bag that contains eight blue cubes and two red cubes, and you select one at random, it is likely that you will get a blue cube. | Even though it is not certain that you will get a blue cube, a blue cube would be selected most of the time because there are many more blue cubes than red cubes. |
| Some events are less likely to happen than to not happen. These events will have a probability that is less than $0.5$. These events could be described as "unlikely" to occur. | If you have a bag that contains eight blue cubes and two red cubes, and you select one at random, it is unlikely that you will get a red cube. | Even though it is not impossible to get a red cube, a red cube would not be selected very often because there are many more blue cubes than red cubes. |

The figure below shows the probability scale.

**Probability Scale**

| 0 | | 1/2 | | 1 |
|---|---|---|---|---|
| Impossible | Unlikely | Equally Likely to Occur or Not Occur | Likely | Certain |

## Exercises 5–8 (10 minutes)

Let the students continue to work with their partners on Questions 5–8

Then, discuss the answers. Some answers will vary. It is important for students to explain their answers. Students may disagree with one another on the exact location of the letters in Exercise 5, but the emphasis should be on student understanding of the vocabulary, not on the exact location of the letters.

P.2

> **Exercises 5–8**
>
> 5.  Decide where each event would be located on the scale below. Place the letter for each event on the appropriate place on the probability scale.
>
>     *Answers noted on probability scale below.*

Event:

A.  You will see a live dinosaur on the way home from school today.

*Probability is 0 or impossible as there are no live dinosaurs.*

B.  A solid rock dropped in the water will sink.

*Probability is 1 (or certain to occur), as rocks are typically denser than the water they displace.*

C.  A round disk with one side red and the other side yellow will land yellow side up when flipped.

*Probability is $\frac{1}{2}$, as there are two sides that are equally like to land up when the disk is flipped.*

D.  A spinner with four equal parts numbered 1–4 will land on the 4 on the next spin.

*Probability of landing on the 4 would be $\frac{1}{4}$, regardless of what spin was made. Based on the scale provided, this would indicate a probability halfway between impossible and equally likely.*

E.  Your name will be drawn when a name is selected randomly from a bag containing the names of all of the students in your class.

*Probability is between impossible and equally likely, assuming there are more than two students in the class. If there were two students, then the probability would be equally likely. If there was only one student in the class, then the probability would be certain to occur. If, however, there were two or more students, the probability would be between impossible and equally likely to occur.*

F.  A red cube will be drawn when a cube is selected from a bag that has five blue cubes and five red cubes.

*Probability would be equally likely to occur as there are an equal number of blue and red cubes.*

G.  The temperature outside tomorrow will be – 250 degrees.

*Probability is impossible (or 0) as there are no recorded temperatures at – 250 degrees Fahrenheit or Celsius.*

**Probability Scale**

6.  Design a spinner so that the probability of green is 1.

*The spinner is all green.*

7.  Design a spinner so that the probability of green is 0.

    *The spinner can include any color but green.*

8.  Design a spinner with two outcomes in which it is equally likely to land on the red and green parts.

    *The red and green areas should be equal.*

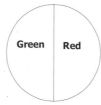

## Exercises 9–10 (5 minutes)

Have a classroom discussion about the probability values discussed in the exercises. Discuss with students that an event that is impossible has a probability of 0 and will never occur, no matter how many observations you make. This means that in a long sequence of observations, it will occur 0% of the time. An event that is certain has a probability of 1 and will always occur. This means that in a long sequence of observations, it will occur 100% of the time. Ask students to think of other examples in which the probability is 0 or 1.

---

**Exercises 9–10**

An event that is impossible has probability 0 and will never occur, no matter how many observations you make. This means that in a long sequence of observations, it will occur 0% of the time. An event that is certain, has probability 1 and will always occur. This means that in a long sequence of observations, it will occur 100% of the time.

9.  What do you think it means for an event to have a probability of $\frac{1}{2}$?

    *In a long sequence of observations, it would occur about half the time.*

10. What do you think it means for an event to have a probability of $\frac{1}{4}$?

    *In a long sequence of observations, it would occur about 25% of the time.*

---

**Closing**

> ### Lesson Summary
>
> - <u>Probability</u> is a measure of how likely it is that an event will happen.
>
> - A probability is a number between 0 and 1.
>
> - The probability scale is:
>
>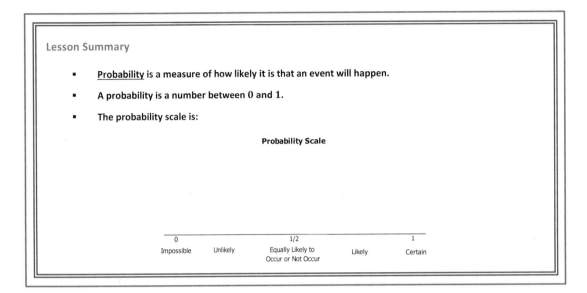

**Exit Ticket (5 minutes)**

Name _____     Date_____

# Lesson 1:  Chance Experiments

Exit Ticket

Decide where each of the following events would be located on the scale below.  Place the letter for each event on the appropriate place on the probability scale.

## Probability Scale

| 0 | | 1/2 | | 1 |
|---|---|---|---|---|
| Impossible | Unlikely | Equally Likely to Occur or Not Occur | Likely | Certain |

The numbers from 1 to 10 are written on small pieces of paper and placed in a bag.  A piece of paper will be drawn from the bag.

    A.   A piece of paper with a 5 is drawn from the bag.

    B.   A piece of paper with an even number is drawn.

    C.   A piece of paper with a 12 is drawn.

    D.   A piece of paper with a number other than 1 is drawn.

    E.   A piece of paper with a number divisible by 5 is drawn.

## Exit Ticket Sample Solutions

Decide where each of the following events would be located on the scale below.  Place the letter for each event on the appropriate place on the probability scale.

**Probability Scale**

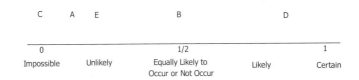

The numbers from 1 to 10 are written on small pieces of paper and placed in a bag.  A piece of paper will be drawn from the bag.

    A.    A piece of paper with a 5 is drawn from the bag.

    B.    A piece of paper with an even number is drawn.

    C.    A piece of paper with a 12 is drawn.

    D.    A piece of paper with a number other than 1 is drawn.

    E.    A piece of paper with a number divisible by 5 is drawn.

## Problem Set Sample Solutions

1.    Match each spinner below with the words Impossible, Unlikely, Equally likely to occur or not occur, Likely, and Certain to describe the chance of the spinner landing on black.

Spinner A: *Unlikely*        Spinner B: *Likely*        Spinner C: *Impossible*

Spinner A        Spinner B        Spinner C

    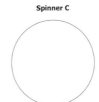

Spinner D: *Equally Likely*        Spinner E: *Certain*

Spinner D        Spinner E

2.  Decide if each of the following events is Impossible, Unlikely, Equally likely to occur or not occur, Likely, or Certain to occur.

   a.   A vowel will be picked when a letter is randomly selected from the word "lieu."

   *Likely; most of the letters of the word lieu are vowels.*

   b.   A vowel will be picked when a letter is randomly selected from the word "math."

   *Unlikely; most of the letters of the word math are not vowels.*

   c.   A blue cube will be drawn from a bag containing only five blue and five black cubes.

   *Equally likely to occur or not occur; the number of blue and black cubes in the bag is the same.*

   d.   A red cube will be drawn from a bag of 100 red cubes.

   *Certain; the only cubes in the bag are red.*

   e.   A red cube will be drawn from a bag of 10 red and 90 blue cubes.

   *Unlikely; most of the cubes in the bag are blue.*

3.  A shape will be randomly drawn from the box shown below.  Decide where each event would be located on the probability scale.  Then, place the letter for each event on the appropriate place on the probability scale.

Event:

   A.   A circle is drawn.

   B.   A square is drawn.

   C.   A star is drawn.

   D.   A shape that is not a square is drawn.

**Probability Scale**

| A | C | B D | | |
|---|---|---|---|---|
| 0 | | 1/2 | | 1 |
| Impossible | Unlikely | Equally Likely to Occur or Not Occur | Likely | Certain |

4.  Color the cubes below so that it would be equally likely to choose a blue or yellow cube.

    *Color 5 five blue and 5 five yellow.*

5.  Color the cubes below so that it would be likely but not certain to choose a blue cube from the bag.

    *7, 8, or 9 blue, and the rest any other color.*

6.  Color the cubes below so that it would be unlikely but not impossible to choose a blue cube from the bag.

    *1, 2, or 3 blue, and the others any other color.*

7. Color the cubes below so that it would be impossible to choose a blue cube from the bag.

   *Any color but blue.*

# Lesson 2:  Estimating Probabilities by Collecting Data

## Student Outcomes

- Students estimate probabilities by collecting data on an outcome of a chance experiment.
- Students use given data to estimate probabilities.

## Lesson Overview

This lesson builds on students' beginning understanding of probability.  In Lesson 1, students were introduced to an informal idea of probability and the vocabulary:  impossible, unlikely, equally likely, likely, certain to describe the chance of an event occurring.  In this lesson, students begin by playing a game similar to the game they played in Lesson 1.  Now, we use the results of the game to introduce a method for finding an estimate for the probability of an event occurring.  Then, students use data given in a table to estimate the probability of an event.

## Classwork

### Example 1 (10 minutes):  Carnival Game

Place students into groups of two.  Hand out a copy of the spinner and a paperclip to each group.  Read through the rules of the game and demonstrate how to use the paper clip as a spinner.

 Before playing the game, display the probability scale from Lesson 1 and ask the students where they would place the probability of winning the game.

Remind students to carefully record the results of each spin.

---

**Example 1:  Carnival Game**

At the school carnival, there is a game in which students spin a large spinner.  The spinner has four equal sections numbered 1–4 as shown below.  To play the game, a student spins the spinner twice and adds the two numbers that the spinner lands on.  If the sum is greater than or equal to 5, the student wins a prize.

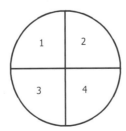

---

## Exercises 1–8 (15 minutes)

Allow students to work with their partner on Exercises 1–6. Then discuss and confirm as a class.

Sample responses to the questions should be based on the outcomes recorded by students. The following outcomes were generated by two students. They are used to provide sample responses to the questions that follow:

---

**Exercises 1–8**

You and your partner will play this game 15 times. Record the outcome of each spin in the table below.

| Turn | 1st Spin Results | 2nd Spin Results | Sum |
|------|------------------|------------------|-----|
| 1 | 4 | 1 | 5 |
| 2 | 1 | 3 | 4 |
| 3 | 3 | 2 | 5 |
| 4 | 1 | 1 | 2 |
| 5 | 2 | 1 | 3 |
| 6 | 1 | 4 | 5 |
| 7 | 4 | 1 | 5 |
| 8 | 3 | 1 | 4 |
| 9 | 2 | 4 | 6 |
| 10 | 4 | 4 | 8 |
| 11 | 1 | 1 | 2 |
| 12 | 4 | 3 | 7 |
| 13 | 3 | 4 | 7 |
| 14 | 3 | 1 | 4 |
| 15 | 1 | 2 | 3 |

1. Out of the 15 turns how many times was the sum greater than or equal to 5?

   *Answers will vary and should reflect the results from students playing the game 15 times. In the example above, eight outcomes had a sum greater than or equal to 5.*

2. What sum occurred most often?

   *5 occurred the most.*

3. What sum occurred least often?

   *6 and 8 occurred the least. (Anticipate a range of answers as this was only done 15 times. We anticipate that 2 and 8 will not occur as often.)*

4. If students played a lot of games, what proportion of the games played will they win? Explain your answer.

   *Based on the above outcomes, $\frac{8}{15}$ represents the proportion of outcomes with a sum of 5 or more.*

5. Name a sum that would be impossible to get while playing the game.

   *Answers will vary. One possibility is getting a sum of 100.*

6. What event is certain to occur while playing the game?

   *Answers will vary. One possibility is getting a sum between 2 and 8 as all possible sums are between 2 and 8.*

---

Before students work on Exercises 7 and 8, discuss the definition of a *chance experiment*. A *chance experiment* is the process of making an observation when the outcome is not certain (that is, when there is more than one possible outcome). If students struggle with this idea, present some examples of a chance experiment, such as: flipping a coin 15 times or selecting a cube from a bag of 20 cubes. Then, display the formula for finding an estimate for the probability of an event. Using the game that the students just played, explain that the denominator is the total number of times they played the game, and the numerator is the number of times they recorded a sum greater than or equal to 5.

> When you were spinning the spinner and recording the outcomes, you were performing a <u>chance experiment</u>. You can use the results from a chance experiment to estimate the probability of an event. In the example above, you spun the spinner 15 times and counted how many times the sum was greater than or equal to 5. An estimate for the probability of a sum greater than or equal to 5 is:
>
> $$P(sum \geq 5) = \frac{Number\ of\ observed\ occurrences\ of\ the\ event}{Total\ number\ of\ observations}$$

Give the students a few minutes to answer Exercises 7 and 8, and then ask each group to share their results. After students have shared their results, point out that not every group had exactly the same answer.

- Ask the students to explain why their answers are estimates of the probability of getting a sum of 5 or more.

> 7. Based on your experiment of playing the game, what is your estimate for the probability of getting a sum of 5 or more?
>
> *Answers will vary. Students should answer this question based on their results. For the results indicated above, $\frac{8}{15}$ or approximately $0.53$ or $53\%$ would estimate the probability of getting a sum of 5 or more.*
>
> 8. Based on your experiment of playing the game, what is your estimate for the probability of getting a sum of exactly 5?
>
> *Answers will vary. Students should answer this question based on their results. Using the above 15 outcomes, $\frac{4}{15}$ or approximately $0.27$ or $27\%$ of the time represents an estimate for the probability of getting a sum of exactly 5.*

Students will learn how to determine a theoretical probability for problems similar to this game. Before they begin determining the theoretical probability, however, summarize how an estimated probability is based on the proportion of the number of specific outcomes to the total number of outcomes. Students may also begin to realize that the more outcomes they determine, the more confident they are that the proportion of winning the game is providing an accurate estimate of the probability. These ideas will be developed more fully in the following lessons.

## Example 2 (10 minutes): Animal Crackers

Have students read the example. You may want to show a box of animal crackers and demonstrate how a student can take a sample from the box. Explain that the data presented result from a student taking a sample of 20 crackers from a very large jar of Animal Crackers and recording the results for each draw.

Display the table of data.

Ask students:

- What was the total number of observations?
- If we want to estimate the probability of selecting a zebra, how many zebras were chosen?
- What is the estimate for the probability of selecting a zebra?

The main point of this example is for students to estimate the probability of selecting a certain type of animal cracker. Use the data collected to make this estimate.

---

**Example 2**

A student brought a very large jar of animal crackers to share with students in class. Rather than count and sort all the different types of crackers, the student randomly chose 20 crackers and found the following counts for the different types of animal crackers.

| Lion | 2 |
|---|---|
| Camel | 1 |
| Monkey | 4 |
| Elephant | 5 |
| Zebra | 3 |
| Penguin | 3 |
| Tortoise | 2 |
| | Total 20 |

The student can now use that data to find an estimate for the probability of choosing a zebra from the jar by dividing the observed number of zebras by the total number of crackers selected. The estimated probability of picking a zebra is $\frac{3}{20}$, or $0.15$ or $15\%$. This means that an estimate of the proportion of the time a zebra will be selected is $0.15$, or $15\%$ of the time. This could be written as $P(zebra) = 0.15$, or the probability of selecting a zebra is $0.15$.

---

## Exercises 9–15 (5 minutes)

Place the students in groups of 2, and allow them time to answer each question. You may wish to specify in which form they should answer. For this exercise, it is acceptable for students to write answers in fraction form to emphasis the formula. As a class, briefly discuss students' answers. Specifically, discuss the answers for Exercises 11 and 15. Each of these questions involve "or." For these questions, students should indicate that they would add the outcomes as indicated in the question to form their proportion.

**Exercises 9–15**

If a student were to randomly select a cracker from the large jar:

9.   What is your estimate for the probability of selecting a lion?

$$\frac{2}{20} = \frac{1}{10} = 0.1$$

10.  What is the estimate for the probability of selecting a monkey?

$$\frac{4}{20} = \frac{1}{5} = 0.2$$

11.  What is the estimate for the probability of selecting a penguin *or* a camel?

$$\frac{(3+1)}{20} = \frac{4}{20} = \frac{1}{5} = 0.2$$

12.  What is the estimate for the probability of selecting a rabbit?

$$\frac{0}{20} = 0$$

13.  Is there the same number of each animal cracker in the large jar?  Explain your answer.

*No, there appears to be more elephants than other types of crackers.*

14.  If the student were to randomly select another 20 animal crackers, would the same results occur?  Why or why not?

*Probably not.  Results may be similar, but it is very unlikely they would be exactly the same.*

15.  If there are 500 animal crackers in the jar, how many elephants are in the jar?  Explain your answer.

$$\frac{5}{20} = \frac{1}{4} = 0.25, \text{ *hence an estimate for number of elephants would be* } 125.$$

## Closing

Discuss with the students the Lesson Summary.  Ask students to summarize how they would find the probability of an event.

> **Lesson Summary**
>
> An estimate for finding the probability of an event occurring is
>
> $$P(event\ occurring) = \frac{Number\ of\ observed\ occurrences\ of\ the\ event}{Total\ number\ of\ observations}$$

## Exit Ticket (5 minutes)

Name _____ Date_____

# Lesson 2: Estimating Probabilities by Collecting Data

Exit Ticket

In the following problems, round all of your decimal answers to 3 decimal places. Round all of your percents to the nearest tenth of a percent.

A student randomly selected crayons from a large bag of crayons. Below is the number of each color the student selected. Now, suppose the student were to randomly select one crayon from the bag.

| Color | Number |
|-------|--------|
| Brown | 10 |
| Blue | 5 |
| Yellow | 3 |
| Green | 3 |
| Orange | 3 |
| Red | 6 |

1.  What is the estimate for the probability of selecting a blue crayon from the bag? Express your answer as a fraction, decimal or percent.

2.  What is the estimate for the probability of selecting a brown crayon from the bag?

3.  What is the estimate for the probability of selecting a red crayon *or* a yellow crayon from the bag?

4.  What is the estimate for the probability of selecting a pink crayon from the bag?

5.  Which color is most likely to be selected?

6.  If there are 300 crayons in the bag, how many will be red? Justify your answer.

## Exit Ticket Sample Solutions

In the following problems, round all of your decimal answers to 3 decimal places. Round all of your percents to the nearest tenth of a percent.

A student randomly selected crayons from a large bag of crayons. Below is the number of each color the student selected. Now, suppose the student were to randomly select one crayon from the bag.

| Color | Number |
|--------|--------|
| Brown | 10 |
| Blue | 5 |
| Yellow | 3 |
| Green | 3 |
| Orange | 3 |
| Red | 6 |

1. What is the estimate for the probability of selecting a blue crayon from the bag? Express your answer as a fraction, decimal or percent.

$$\frac{5}{30} = \frac{1}{6} = 0.167 \ or \ 16.7\%$$

2. What is the estimate for the probability of selecting a brown crayon from the bag?

$$\frac{10}{30} = \frac{1}{3} = 0.33$$

3. What is the estimate for the probability of selecting a red crayon *or* a yellow crayon from the bag?

$$\frac{9}{30} = \frac{3}{10} = 0.3$$

4. What is the estimate for the probability of selecting a pink crayon from the bag?

$$\frac{0}{30} = 0$$

5. Which color is most likely to be selected?

*Brown.*

6. If there are 300 crayons in the bag, how many will be red? Justify your answer.

*There are 6 out of 30 crayons that are red, or $\frac{1}{5}$ or 0.2. Anticipate $\frac{1}{5}$ of 300 crayons are red, or approximately 60 crayons.*

## Problem Set Sample Solutions

1.  Play a game using the two spinners below. Spin each spinner once, and then multiply the outcomes together. If the result is less than or equal to 8, you win the game. Play the game 15 times, and record your results in the table below.

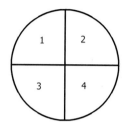
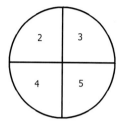

| Turn | 1st Spin Results | 2nd Spin Results | Product |
|------|------------------|------------------|---------|
| 1 | | | |
| 2 | | | |
| 3 | | | |
| 4 | | | |
| 5 | | | |
| 6 | | | |
| 7 | | | |
| 8 | | | |
| 9 | | | |
| 10 | | | |
| 11 | | | |
| 12 | | | |
| 13 | | | |
| 14 | | | |
| 15 | | | |

2.  a.  What is your estimate for the probability of getting a product of 8 or less?

    *Students should find the number of times the product was 8 or less and divide by 15. Answers should be approximately 7, 8, or 9, divided by 15.*

    b.  What is your estimate for the probability of getting a product more than 8?

    *Subtract the answer to part (a) from 1, or 1 − the answer from part (a). Approximately 8, 7, or 6, divided by 15.*

    c.  What is your estimate for the probability of getting a product of exactly 8?

    *Students should find the number of times 8 occurred and divide by 15. Approximately 1 or 2 divided by 15.*

    d.  What is the most likely product for this game?

    *Students should record the product that occurred most often. Possibilities are 4, 6, 8, and 12.*

    e.  If you played this game another 15 times, will you get the exact same results? Explain.

    *No, since this is a chance experiment, results could change for each time the game is played.*

3.  A seventh grade student surveyed students at her school. She asked them to name their favorite pet. Below is a bar graph showing the results of the survey.

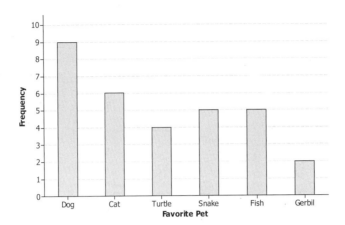

Use the results from the survey to answer the following questions.

a.  How many students answered the survey question?

31

b.  How many students said that a snake was their favorite pet?

5

Now, suppose a student will be randomly selected and asked what his or her favorite pet is.

c.  What is your estimate for the probability of that student saying that a dog is his or her favorite pet?

*(Allow any form.)* $\dfrac{9}{31}$, or approximately $0.29$, or approximately $29\%$.

d.  What is your estimate for the probability of that student saying that a gerbil is his or her favorite pet?

*(Allow any form.)* $\dfrac{2}{31}$, or approximately $0.06$, or approximately $6\%$.

e.  What is your estimate for the probability of that student saying that a frog is his or her favorite pet?

$\dfrac{0}{31}$ *or* $0$ *or* $0\%$.

4.  A seventh grade student surveyed 25 students at her school.  She asked them how many hours a week they spend playing a sport or game outdoors.  The results are listed in the table below.

| Number of hours | Tally | Frequency |
|---|---|---|
| 0 | I I I | 3 |
| 1 | I I I I | 4 |
| 2 | ┼┼┼┼ | 5 |
| 3 | ┼┼┼┼ I I | 7 |
| 4 | I I I | 3 |
| 5 | | 0 |
| 6 | I I | 2 |
| 7 | | 0 |
| 8 | I | 1 |

a.  Draw a dot plot of the results.

Number of Hours

Suppose a student will be randomly selected.

b.  What is your estimate for the probability of that student answering 3 hours?

$\frac{7}{25}$

c.  What is your estimate for the probability of that student answering 8 hours?

$\frac{1}{25}$

d.  What is your estimate for the probability of that student answering 6 or more hours?

$\frac{3}{25}$

e.  What is your estimate for the probability of that student answering 3 or less hours?

$\frac{19}{25}$

f.  If another 25 students were surveyed, do you think they will give the exact same results?  Explain your answer.

*No, each group of 25 students could answer the question differently.*

g.  If there are 200 students at the school, what is your estimate for the number of students who would say they play a sport or game outdoors 3 hours per week?  Explain your answer.

$200 \cdot \left(\frac{7}{25}\right) = 28$ *students. This is based on estimating that of the 200 students, $\frac{7}{25}$ would play a sport or game outdoors 3 hours per week, as $\frac{7}{25}$ represented the probability of playing a sport or game outdoors 3 hours per week from the 7th grade class surveyed.*

5.  A student played a game using one of the spinners below.  The table shows the results of 15 spins.  Which spinner did the student use?  Give a reason for your answer.

*Spinner B.  Tallying the results:  1 occurred 6 times, 2 occurred 6 times and 3 occurred 3 times.  In Spinner B, the section labeled 1 and 2 are equal and larger than section 3.*

| Spin | Results |
|------|---------|
| 1 | 1 |
| 2 | 1 |
| 3 | 2 |
| 4 | 3 |
| 5 | 1 |
| 6 | 2 |
| 7 | 3 |
| 8 | 2 |
| 9 | 2 |
| 10 | 1 |
| 11 | 2 |
| 12 | 2 |
| 13 | 1 |
| 14 | 3 |
| 15 | 1 |

Spinner A

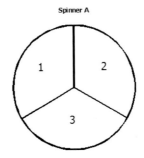

Spinner B

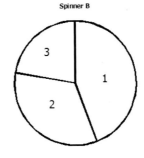

Spinner C

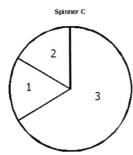

# Lesson 3: Chance Experiments with Equally Likely Outcomes

## Student Outcomes

- Students determine the possible outcomes for simple chance experiments.
- Given a description of a simple chance experiment, students determine the sample space for the experiment.
- Given a description of a chance experiment and an event, students determine for which outcomes in the sample space the event will occur.
- Students distinguish between chance experiments with equally likely outcomes and chance experiments for which the outcomes are not equally likely.

## Lesson Overview

This lesson continues to build students' understanding of probability. This lesson begins to formalize concepts explored in the first two lessons of the module. Students are presented with several descriptions of simple chance experiments such as spinning a spinner and drawing objects from a bag. Note that students will be working on a chance experiment in class that utilizes paper cups. Teachers will need to provide cups for the experiment.

## Classwork

### Example 1 (5 minutes)

Begin this example by showing the students a paper cup and asking them how it might land when tossed. Explain that each possibility is called an *outcome* of the experiment. Then list the three outcomes on the board and explain that the three outcomes form the *sample space* of the experiment.

Provide other examples of an experiment, and ask students to describe the sample space:

- Flipping a coin: heads, tails
- Drawing a colored cube from a bag that has 3 red, 2 blue, 1 yellow, and 4 green: red, blue, yellow, and green
- Picking a letter from the word "classroom": c, l, a, s, r, o, m

Note: Some students will want to list all of the letters but explain that, when listing the sample space, you only need to list the possibilities, not how many times the letter appears.

**Example 1**

Jamal, a 7<sup>th</sup> grader, wants to design a game that involves tossing paper cups. Jamal tosses a paper cup five times and records the outcome of each toss. An <u>outcome</u> is the result of a single trial of an experiment.

Here are the results of each toss:

Jamal noted that the paper cup could land in one of three ways: on its side, right side up, or upside down. The collection of these three outcomes is called the *sample space* of the experiment. The <u>sample space</u> of an experiment is the set of all possible outcomes of that experiment.

For example, the sample space when flipping a coin is Heads, Tails.

The sample space when drawing a colored cube from a bag that has 3 red, 2 blue, 1 yellow, and 4 green cubes is red, blue, yellow, green.

## Exercises 1–6 (15 minutes)

Allow students to work with a partner on Exercises 1–6. Then discuss and confirm as a class.

**Exercise 1–6**

For each of the following chance experiments, list the sample space (i.e., all the possible outcomes).

1. Drawing a colored cube from a bag with 2 green, 1 red, 10 blue, and 3 black.

   *Green, Red, Blue, Black.*

2. Tossing an empty soup can to see how it lands.

   *Right side up, upside down, on its side.*

3. Shooting a free-throw in a basketball game.

   *Made shot, missed shot.*

4. Rolling a number cube with the numbers 1– 6 on its faces.

   *1, 2, 3, 4, 5, or 6.*

5. Selecting a letter from the word: probability

   *p, r, o, b, a, i, l, t, y.*

6. Spinning the spinner:

   *1, 2, 3, 4, 5, 6, 7, or 8.*

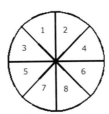

MP.2

### Example 2 (15 minutes):  Equally Likely Outcomes

In this example, students will carry out the experiment of tossing a paper cup.  Small Dixie cups work well.  Styrofoam cups tend not to work as well.  You will need one cup for each pair of students.  Before the students begin this experiment ask:

- Do you think each of the possible outcomes of tossing the paper cup has the same chance of occurring?

Explain to the students that if the outcomes have the same chance of occurring, then they are *equally likely* to occur.

The students should toss the paper cup 30 times, and record the results of each toss.

---

**Example 2:  Equally Likely Outcomes**

The sample space for the paper cup toss was on its side, right side up, and upside down.  Do you think each of these outcomes has the same chance of occurring?  If they do, then they are equally likely to occur.

The outcomes of an experiment are equally likely to occur when the probability of each outcome is equal.

You and your partner toss the paper cup 30 times and record in a table the results of each toss.

| Toss | Outcome |
|:----:|:-------:|
| 1 | |
| 2 | |
| 3 | |
| 4 | |
| 5 | |
| 6 | |
| 7 | |
| 8 | |
| 9 | |
| 10 | |
| 11 | |
| 12 | |
| 13 | |
| 14 | |
| 15 | |
| 16 | |
| 17 | |
| 18 | |
| 19 | |
| 20 | |
| 21 | |
| 22 | |
| 23 | |
| 24 | |
| 25 | |
| 26 | |
| 27 | |
| 28 | |
| 29 | |
| 30 | |

---

## Exercises 7–12 (10 minutes)

Allow students to work with a partner on Exercises 7–12. Then, discuss and confirm as a class.

---

**Exercises 7–12**

7. Using the results of your experiment, what is your estimate for the probability of a paper cup landing on its side?

   *Answers will vary. Students should write their answer as a fraction with a denominator of 30 and a numerator of the number of times cup landed on its side.*

8. Using the results of your experiment, what is your estimate for the probability of a paper cup landing upside down?

   *Answers will vary. Students should write their answer as a fraction with a denominator of 30 and a numerator of the number of times cup landed upside down.*

9. Using the results of your experiment, what is your estimate for the probability of a paper cup landing right side up?

   *Answers will vary. Students should write their answer as a fraction with a denominator of 30 and a numerator of the number of times cup landed right side up.*

MP.6

10. Based on your results, do you think the three outcomes are equally likely to occur?

    *Answers will vary but, generally, the results are not equally likely.*

---

Based on their results of tossing the cup 30 times, ask students to predict how many times the cup will land on its side, right side up, or upside down for approximately 120 tosses. If time permits, allow students to carry out the experiment for a total of 120 tosses, or combine results of students to examine the number of outcomes for approximately 120 tosses. Compare the predicted numbers and the actual numbers. It is likely the results from approximately 120 tosses will not match the predicted numbers. Discuss with students why they generally do not agree.

---

11. Using the spinner below, answer the following questions.

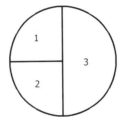

a. Are the events spinning and landing on 1 or a 2 equally likely?

   *Yes, the area of sections 1 and 2 are equal.*

b. Are the events spinning and landing on 2 or 3 equally likely?

   *No, the area of sections 2 and 3 are not equal.*

c. How many times do you predict the spinner to land on each section after 100 spins?

   *Based on the area of the sections, approximately 25 times each for sections 1 and 2, and 50 times for section 3.*

---

12.   Draw a spinner that has 3 sections that are equally likely to occur when the spinner is spun.  How many times do you think the spinner will land on each section after 100 spins?

*The three sectors should be equal in area.  Expect the spinner to land on each section approximately 33 times (30–35 times).*

## Closing (5 minutes)

Remind students of the new vocabulary:  *outcome, sample space, and equally likely.*

Use the following example to summarize the main ideas of the lesson:

Suppose a bag contains 10 Green, 10 Red, 10 yellow, 10 Orange, and 10 Purple crayons.  If one crayon is selected from the bag and the color is noted, the *outcome* is the color that will will be chosen.  The *sample space* will be the colors: green, red, yellow, orange, and purple.  Each color is *equally likely* to be selected because each color has the same chance of being chosen.

---

**Lesson Summary**

An <u>outcome</u> is the result of a single observation of an experiment.

The <u>sample space</u> of an experiment is the set of all possible outcomes of that experiment.

The outcomes of an experiment are <u>equally likely</u> to occur when the probability of each outcome is equal.

Suppose a bag of crayons contains 10 green, 10 red, 10 yellow, 10 orange, and 10 purple pieces of crayons.  If one crayon is selected from the bag and the color is noted, the *outcome* is the color that will be chosen.  The *sample space* will be the colors:  green, red, yellow, orange, and purple.  Each color is *equally likely* to be selected because each color has the same chance of being chosen.

---

## Exit Ticket (10 minutes)

Name _____     Date_____

# Lesson 3:  Chance Experiments with Equally Likely Outcomes

Exit Ticket

The numbers from 1–10 are written on note cards and placed in a bag.  One card will be drawn from the bag at random.

1.  List the sample space for this experiment.

2.  Are the events selecting an even number and selecting an odd number equally likely?  Explain your answer.

3.  Are the events selecting a number divisible by 3 and selecting a number divisible by 5 equally likely?  Explain your answer.

## Exit Ticket Sample Solutions

The numbers from $1-10$ are written on note cards and placed in a bag.  One card will be drawn from the bag at random.

1.  List the sample space for this experiment.

    *1, 2, 3, 4, 5, 6, 7, 8, 9, 10.*

2.  Are the events selecting an even number and selecting an odd number equally likely?  Explain your answer.

    *Yes, each has the same chance of occurring.  There are 5 even and 5 odd numbers in the bag.*

3.  Are the events selecting a number divisible by 3 and selecting a number divisible by 5 equally likely?  Explain your answer.

    *No.  There are 3 numbers divisible by 3 (3, 6, and 9), but only 2 numbers divisible by 5 (5 and 10).  So the chance of selecting a number divisible by 3 is slightly greater than selecting a number divisible by 5.*

## Problem Set Sample Solutions

1.  For each of the following chance experiments, list the sample space (all the possible outcomes).

    a.  Rolling a 4-sided die with the numbers $1-4$ on the faces of the die.

        *1, 2, 3, or 4.*

    b.  Selecting a letter from the word:  mathematics.

        *m, a, t, h, e, i, c, or s.*

    c.  Selecting a marble from a bag containing 50 black marbles and 45 orange marbles.

        *Black or Orange.*

    d.  Selecting a number from the even numbers from $2-14$, inclusive.

        *2, 4, 6, 8, 10, 12, or 14.*

    e.  Spinning the spinner below:

        *1, 2, 3, or 4.*

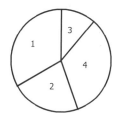

2. For each of the following decide if the two outcomes listed are equally likely to occur. Give a reason for your answer.

    a. Rolling a 1 or a 2 when a 6-sided number cube with the numbers 1– 6 on the faces of the cube is rolled.

      *Yes, each has the same chance of occurring.*

    b. Selecting the letter *a* or *k* from the word: take.

      *Yes, each has the same chance of occurring.*

    c. Selecting a black or an orange marble from a bag containing 50 black and 45 orange marbles.

      *No, Black has a slightly greater chance of being chosen.*

    d. Selecting a 4 or an 8 from the even numbers from 2– 14, inclusive.

      *Yes, each has the same chance of being chosen.*

    e. Landing on a 1 or 3 when spinning the spinner below.

      *No, 1 has a larger area, so it has a greater chance of occurring.*

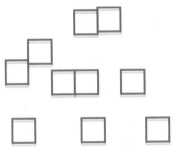

3. Color the cubes below so that it would be equally likely to choose a blue or yellow cube.

    *Answers will vary, but the students should have 5 colored blue and 5 colored yellow.*

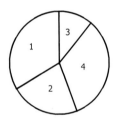

4.  Color the cubes below so that it would be more likely to choose a blue than a yellow cube.

    *Answers will vary. Students should have more cubes colored blue than yellow.*

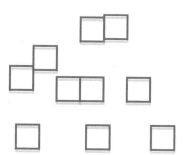

5.  You are playing a game using the spinner below. The game requires that you spin the spinner twice. For example, one outcome could be Yellow on 1$^{st}$ spin and Red on 2$^{nd}$ spin. List the sample space (all the possible outcomes) for the two spins.

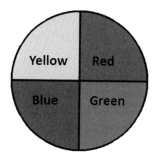

    *There are 16 possibilities:*

    | 1$^{st}$ spin | 2$^{nd}$ spin |
    |---|---|
    | Y | Y |
    | Y | R |
    | Y | G |
    | Y | B |
    | R | Y |
    | R | R |
    | R | G |
    | R | B |
    | G | Y |
    | G | R |
    | G | G |
    | G | B |
    | B | Y |
    | B | R |
    | B | G |
    | B | B |

6.  List the sample space for the chance experiment of flipping a coin twice.

    *There are four possibilities:*

    | 1$^{st}$ toss | 2$^{nd}$ toss |
    |---|---|
    | H | H |
    | H | T |
    | T | H |
    | T | T |

# Lesson 4: Calculating Probabilities for Chance Experiments with Equally Likely Outcomes

## Student Outcomes

- Students will calculate probabilities of events for chance experiments that have equally likely outcomes.

## Classwork

### Example 1 (8 minutes): Theoretical Probability

This example is a chance experiment similar to those conducted in Lesson 2. The experiment requires a brown paper bag that contains 10 yellow, 10 green, 10 red, and 10 blue cubes. Unifix cubes work well for this experiment. In the experiment, 20 cubes are drawn at random and *with replacement*. After each cube is drawn, have students record the outcome in the table.

Before starting the experiment ask students:

- What does it mean to draw a cube out at random?
  - *Random means that cube has an equal chance in being selected.*
- What does it mean to draw a cube with replacement?
  - *The cube is put back before you pick again.*

---

**Example 1: Theoretical Probability**

In a previous lesson, you saw that to find an estimate of the probability of an event for a chance experiment you divide:

$$P(event) = \frac{Number\ of\ observed\ occurrences\ of\ the\ event}{Total\ number\ of\ observations}$$

Your teacher has a bag with some cubes colored yellow, green, blue, and red. The cubes are identical except for their color. Your teacher will conduct a chance experiment by randomly drawing a cube with replacement from the bag. Record the outcome of each draw in the table below.

| Trial | Outcome |
|-------|---------|
| 1 | |
| 2 | |
| 3 | |
| 4 | |
| 5 | |
| 6 | |
| 7 | |
| 8 | |

---

| 9 | |
|----|----|
| 10 | |
| 11 | |
| 12 | |
| 13 | |
| 14 | |
| 15 | |
| 16 | |
| 17 | |
| 18 | |
| 19 | |
| 20 | |

## Exercises 1–6 (20 minutes)

Allow students to work with their partners on Exercises 1–3. After students have completed the three questions, discuss the answers.

**Exercises 1–6**

1.    Based on the 20 trials, estimate for the probability of

a.    choosing a yellow cube.

*Answers vary – but should be approximately $\frac{5}{20}$ or $\frac{1}{4}$.*

b.    choosing a green cube.

*Answers vary – but should be approximately $\frac{5}{20}$ or $\frac{1}{4}$.*

c.    choosing a red cube.

*Answers vary – but should be approximately $\frac{5}{20}$ or $\frac{1}{4}$.*

d.    choosing a blue cube.

*Answers vary – but should be approximately $\frac{5}{20}$ or $\frac{1}{4}$.*

2.    If there are 40 cubes in the bag, how many cubes of each color are in the bag? Explain.

*Answers will vary. Because the estimated probabilities are about the same for each color, we can predict that there are approximately the same number of each color of cubes in the bag. Since an equal number of each color is estimated, approximately 10 of each color are predicted.*

3. If your teacher were to randomly draw another 20 cubes one at a time and with replacement from the bag, would you see exactly the same results? Explain.

*No, this is an example of a chance experiment, so the results will vary.*

Now tell the students what is in the bag (10 each of yellow, green, red, and blue cubes). Allow students to work with a partner on Exercise 4. Then discuss and confirm the answers.

4. Find the fraction of each color of cubes in the bag.

Yellow    $\frac{10}{40}$ or $\frac{1}{4}$

Green    $\frac{10}{40}$ or $\frac{1}{4}$

Red    $\frac{10}{40}$ or $\frac{1}{4}$

Blue    $\frac{10}{40}$ or $\frac{1}{4}$

Present the formal definition of the *theoretical probability* of an outcome when outcomes are equally likely. Then ask:

▪ Why is the numerator of the fraction just 1?

Define the word *event* as "a collection of outcomes." Then, present that definition to students and ask:

▪ Why is the numerator of the fraction not always 1?

Use the cube example to explain the difference between an *outcome* and an *event*. Explain that each cube is equally likely to be chosen (an outcome) while the probability of drawing a blue cube (an event) is $\frac{10}{40}$.

Each fraction is the <u>theoretical probability</u> of choosing a particular color of cube when a cube is randomly drawn from the bag.

When all the possible outcomes of an experiment are equally likely, the probability of each outcome is

$$P(outcome) = \frac{1}{Number\ of\ possible\ outcomes}.$$

An event is a collection of outcomes, and when the outcomes are equally likely, the theoretical probability of an event can be expressed as

$$P(event) = \frac{Number\ of\ favorable\ outcomes}{Number\ of\ possible\ outcomes}.$$

The theoretical probability of drawing a blue cube is

$$P(blue) = \frac{Number\ of\ blue\ cubes}{Total\ number\ of\ cubes} = \frac{10}{40}.$$

Allow students to work with a partner to answer Exercises 5 and 6.  Then discuss and confirm the answers.

---

5.   **Is each color equally likely to be chosen?  Explain your answer.**

*Yes, there are the same numbers of cubes for each color.*

6.   **How do the theoretical probabilities of choosing each color from Exercise 4 compare to the experimental probabilities you found in Exercise 1?**

*Answers will vary.*

---

## Example 2 (10 minutes)

This example connects the concept of sample space from Lesson 3 to finding probability.  Present the example of flipping a nickel and then a dime.  List the sample space representing the outcomes of a head or tail on the nickel and a head or tail on the dime (HH, HT, TH, and TT).  Discuss how each outcome is equally likely to occur.  Then ask students:

- What is the probability of getting two heads?

  - *Probability is $\frac{1}{4}$ or $0.25$ or $25\%$.*

- What is the probability of getting exactly one head of either the nickel or the dime?  (This is an example of an event with two outcomes.)

  - *Probability of the outcomes of HT and TH, or $\frac{2}{4}$ or $\frac{1}{2}$ or $0.5$ or $50\%$.*

---

**Example 2**

An experiment consisted of flipping a nickel and a dime.  The first step in finding the theoretical probability of obtaining a head on the nickel and a head on the dime is to list the sample space.  For this experiment, the sample space is shown below.

| Nickel | Dime |
|--------|------|
| H | H |
| H | T |
| T | H |
| T | T |

If the counts are fair, these outcomes are equally likely, so the probability of each outcome is $\frac{1}{4}$.

| Nickel | Dime | Probability |
|--------|------|-------------|
| H | H | $\frac{1}{4}$ |
| H | T | $\frac{1}{4}$ |
| T | H | $\frac{1}{4}$ |
| T | T | $\frac{1}{4}$ |

The probability of two heads is $\frac{1}{4}$ or $P(two\ heads) = \frac{1}{4}$.

---

## Exercises 7–10 (10 minutes)

Allow students to work with a partner on Exercises 7–10.

---

**Exercises 7–10**

7.  Consider a chance experiment of rolling a number cube.

    a.   What is the sample space?  List the probability of each outcome in the sample space.

        *Sample Space:* $1, 2, 3, 4, 5,$ *and* $6$

        *Probability of each outcome is* $\frac{1}{6}$.

    b.   What is the probability of rolling an odd number?

        $\frac{3}{6}$ *or* $\frac{1}{2}$

    c.   What is the probability of rolling a number less than 5?

        $\frac{4}{6}$ *or* $\frac{2}{3}$

8.  Consider an experiment of randomly selecting a letter from the word:  number.

    a.   What is the sample space?  List the probability of each outcome in the sample space.

        *Sample space:* $n, u, m, b, e,$ *and* $r.$

        *Probability of each outcome is* $\frac{1}{6}$.

    b.   What is the probability of selecting a vowel?

        $\frac{2}{6}$ *or* $\frac{1}{3}$

    c.   What is the probability of selecting the letter z?

        $\frac{0}{6}$ *or* $0$

---

MP.2
&
MP.6

9.  Consider an experiment of randomly selecting a cube from a bag of 10 cubes.

    a.  Color the cubes below so that the probability of selecting a blue cube is $\frac{1}{2}$.

        *Answers will vary; 5 of the cubes should be colored blue.*

        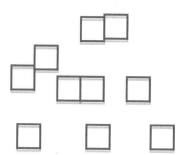

    b.  Color the cubes below so that the probability of selecting a blue cube is $\frac{4}{5}$.

        *Answers will vary; 8 of the cubes will be colored blue.*

        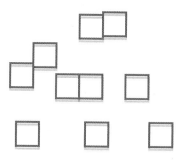

10. Students are playing a game that requires spinning the two spinners shown below.  A student wins the game if both spins land on Red.  What is the probability of winning the game?  Remember to first list the sample space and the probability of each outcome in the sample space.  There are eight possible outcomes to this chance experiment.

    *Sample Space:  R1 R2, R1 B2, R1 G2, R1 Y2, B1 R2, B1 B2, B1 G2, and B1 Y2.*

    *Each outcome has a probability of $\frac{1}{8}$.*

    *Probability of a win (both red) is $\frac{1}{8}$.*

    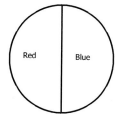    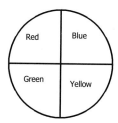

## Closing (5 minutes)

Summarize the two formal definitions of theoretical probability. The first is the probability of an outcome when all of the possible outcomes are equally likely, and the second is the probability of an event when the possible outcomes are equally likely. Remind students that an event is a collection of outcomes.

For example, in the experiment of rolling two number cubes, obtaining a sum of 7 is an event.

---

**Lesson Summary**

When all the possible outcomes of an experiment are equally likely, the probability of each outcome is

$$P(outcome) = \frac{1}{Number\ of\ possible\ outcomes}.$$

An event is a collection of outcomes, and when all outcomes are equally likely, the theoretical probability of an event can be expressed as

$$P(event) = \frac{Number\ of\ favorable\ outcomes}{Number\ of\ possible\ outcomes}.$$

---

## Exit Ticket (10 minutes)

Name _____        Date_____

# Lesson 4:  Calculating Probabilities for Chance Experiments with Equally Likely Outcomes

Exit Ticket

An experiment consists of randomly drawing a cube from a bag containing three red and two blue cubes.

1.   What is the sample space of this experiment?

2.   List the probability of each outcome in the sample space.

3.   Is the probability of selecting a red cube equal to the probability of selecting a blue cube?  Explain.

## Exit Ticket Sample Solutions

An experiment consists of randomly drawing a cube from a bag containing three red and two blue cubes.

1.  What is the sample space of this experiment?

    *R, R, R, B, and B are representing the five cubes.*

2.  List the probability of each outcome in the sample space.

    *Each outcome has a probability of $\frac{1}{5}$.*

3.  Is the probability of selecting a red cube equal to the probability of selecting a blue cube?  Explain.

    *No, there are more red cubes than blue cubes, so red has a greater probability of being chosen.*

## Problem Set Sample Solutions

1.  In a seventh grade class of 28 students, there are 16 girls and 12 boys.  If one student is randomly chosen to win a prize, what is the probability that a girl is chosen?

    $\frac{16}{28}$ *or* $\frac{4}{7}$

2.  An experiment consists of spinning the spinner once.

    a.  Find the probability of landing on a 2.

        $\frac{2}{8}$ *or* $\frac{1}{4}$

    b.  Find the probability of landing on a 1.

        $\frac{3}{8}$

    c.  Is landing in each section of the spinner equally likely to occur?  Explain.

        *Yes, each section is the same size.*

3.  An experiment consists of randomly picking a square section from the board shown below.

    a.   **Find the probability of choosing a triangle.**

         $\frac{8}{16}$ *or* $\frac{1}{2}$

    b.   **Find the probability of choosing a star.**

         $\frac{4}{16}$ *or* $\frac{1}{4}$

    c.   **Find the probability of choosing an empty square.**

         $\frac{4}{16}$ *or* $\frac{1}{4}$

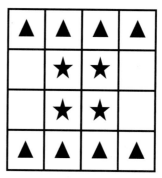

    d.   **Find the probability of choosing a circle.**

         $\frac{0}{16}$ *or* 0

4.  Seventh graders are playing a game where they randomly select two integers from $0-9$, inclusive, to form a two-digit number. The same integer might be selected twice.

    a.   **List the sample space for this chance experiment. List the probability of each outcome in the sample space.**

         *Sample Space: numbers from $00-99$. Probability of each outcome is $1/100$.*

    b.   **What is the probability that the number formed is between 90 and 99, inclusive?**

         $\frac{10}{100}$ *or* $\frac{1}{10}$

    c.   **What is the probability that the number formed is evenly divisible by 5?**

         $\frac{20}{100}$ *or* $\frac{1}{5}$

    d.   **What is the probability that the number formed is a factor of 64?**

         $\frac{7}{100}$ *(Factors of 64 are 1, 2, 4, 8, 16, 32, and 64.)*

5.  A chance experiment consists of flipping a coin and rolling a number cube with the numbers $1-6$ on the faces of the cube.

    a.   **List the sample space of this chance experiment. List the probability of each outcome in the sample space.**

         $h1, h2, h3, h4, h5, h6, t1, t2, t3, t4, t5,$ *and* $t6$. *The probability of each outcome is* $\frac{1}{12}$.

    b.   **What is the probability of getting a head on the coin and the number 3 on the number cube?**

         $\frac{1}{12}$

    c.   **What is the probability of getting a tail on the coin and an even number on the number cube?**

         $\frac{3}{12}$ *or* $\frac{1}{4}$

6.   A chance experiment consists of spinning the two spinners below.

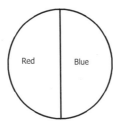   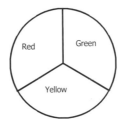

a.   List the sample space and the probability of each outcome.

*Sample Space:  R1 R2, R1 G2, R1 Y2, B1 R2, B1 G2, and B1 Y2.  Each outcome has a probability of* $\frac{1}{6}$.

b.   Find the probability of the event of getting a red on the first spinner and a red on the second spinner.

$\frac{1}{6}$

c.   Find the probability of a red on at least one of the spinners.

$\frac{4}{6}$ *or* $\frac{2}{3}$

# Lesson 5:  Chance Experiments with Outcomes that are Not Equally Likely

## Student Outcomes

- Students calculate probabilities for chance experiments that do not have equally likely outcomes.

## Classwork

It is important that this lesson address the common student misconception that outcomes are always equally likely. Take the opportunity to remind them that this is not always the case. To make this point you can use a spinner that has sections with areas that are not equal as an example. Just because there are four sections on the spinner does not mean that each of the four possible outcomes will have a probability of $\frac{1}{4}$. A weighted die is another example you can use to show a situation where the six possible outcomes would not be equally likely to occur and so it is not reasonable to think that each of the six outcomes has a probability of $\frac{1}{6}$.

> In previous lessons, you learned that when the outcomes in a sample space are equally likely, the probability of an event is the number of outcomes in the event divided by the number of outcomes in the sample space. However, when the outcomes in the sample space are *not* equally likely, we need to take a different approach.

## Example 1 (5 minutes)

This straightforward example lets students experience probabilities given in a table, outcomes with unequal probabilities, adding probabilities, and finding the probability of a complement (i.e., the probability that an event does *not* happen).

Ask students:

- What might be an alternative way of answering part (e)?  Add the probabilities for $0, 1, 2, 4,$ and $5$ bananas.
  - $0.1 + 0.1 + 0.1 + 0.2 + 0.3 = 0.8$

---

**Example 1**

When Jenna goes to the farmer's market she usually buys bananas.  The numbers of bananas she might buy and their probabilities are shown in the table below.

| Number of Bananas | 0 | 1 | 2 | 3 | 4 | 5 |
|---|---|---|---|---|---|---|
| Probability | 0.1 | 0.1 | 0.1 | 0.2 | 0.2 | 0.3 |

a.    What is the probability that Jenna buys exactly 3 bananas?

*You can see from the table that the probability that Jenna buys exactly 3 bananas is 0.2.*

---

b.  What is the probability that Jenna doesn't buy any bananas?

*The probability that Jenna buys 0 bananas is 0.1.*

c.  What is the probability that Jenna buys more than 3 bananas?

*The probability that Jenna buys 4 or 5 bananas is 0.2 + 0.3 = 0.5.*

d.  What is the probability that Jenna buys at least 3 bananas?

*The probability that Jenna buys , 4, or 5 bananas is 0.2 + 0.2 + 0.3 = 0.7.*

e.  What is the probability that Jenna doesn't buy exactly 3 bananas?

*Remember that the probability that an event does __not__ happen is*

$1 - (the\ probability\ that\ the\ event\ does\ happen).$

*So, the probability that Jenna does not buy exactly 3 bananas is*

$1 - (the\ probability\ that\ she\ does\ buy\ exactly\ 3\ bananas)$

$1 - 0.2 = 0.8.$

Notice that the probabilities in the table add to 1. $(0.1 + 0.1 + 0.1 + 0.2 + 0.2 + 0.3 = 1)$ This is always true; when we add up the probabilities of all the possible outcomes, the result is always 1. So, taking 1 and subtracting the probability of the event gives us the probability of something NOT occurring.

## Exercises 1–2 (6 minutes)

Decide if you want to let students use calculators on this exercise. Doing the exercise using only pencil and paper would provide useful practice with adding decimals. However, you might decide that the time saved by allowing calculators on this exercise will give you more time to devote to working with fractions in the examples and exercises that follow. Let students work independently and confirm their answers with a neighbor.

### Exercises 1–2

Jenna's husband, Rick, is concerned about his diet. On any given day, he eats 0, 1, 2, 3, or 4 servings of fruit and vegetables. The probabilities are given in the table below.

| Number of Servings of Fruit and Vegetables | 0 | 1 | 2 | 3 | 4 |
|---|---|---|---|---|---|
| Probability | 0.08 | 0.13 | 0.28 | 0.39 | 0.12 |

1.  On a given day, find the probability that Rick eats:
    a.  Two servings of fruit and vegetables.

    0.28

    b.  More than two servings of fruit and vegetables.

    $0.39 + 0.12 = 0.51$

    c.  At least two servings of fruit and vegetables.

    $0.28 + 0.39 + 0.12 = 0.79$

2.  Find the probability that Rick does not eat exactly two servings of fruit and vegetables.

$1 - 0.28 = 0.72$

## Example 2 (8 minutes)

Here, the concepts in Example 1 are repeated, this time using fractions.  This provides an opportunity to remind students how to add and subtract fractions using common denominators, and how to reduce fractions.

---

**Example 2**

Luis works in an office, and the phone rings occasionally.  The possible numbers of phone calls he receives in an afternoon and their probabilities are given in the table below.

| Number of Phone Calls | 0 | 1 | 2 | 3 | 4 |
|---|---|---|---|---|---|
| Probability | $\frac{1}{6}$ | $\frac{1}{6}$ | $\frac{2}{9}$ | $\frac{1}{3}$ | $\frac{1}{9}$ |

a.  Find the probability that Luis receives 3 or 4 phone calls.

*The probability that Luis receives 3 or 4 phone calls is* $\frac{1}{3} + \frac{1}{9} = \frac{3}{9} + \frac{1}{9} = \frac{4}{9}$.

b.  Find the probability that Luis receives fewer than 2 phone calls.

*The probability that Luis receives fewer than 2 phone calls is* $\frac{1}{6} + \frac{1}{6} = \frac{2}{6} = \frac{1}{3}$.

c.  Find the probability that Luis receives 2 or fewer phone calls.

*The probability that Luis receives 2 or fewer phone calls is* $\frac{2}{9} + \frac{1}{6} + \frac{1}{6} = \frac{4}{18} + \frac{3}{18} + \frac{3}{18} = \frac{10}{18} = \frac{5}{9}$.

d.  Find the probability that Luis does not receive 4 phone calls.

*The probability that Luis does not receive 4 phone calls is* $1 - \frac{1}{9} = \frac{9}{9} - \frac{1}{9} = \frac{8}{9}$.

---

If there is time available, ask students:

▪  How would you calculate the probability that Luis receives at least one call?  What might be a quicker way of doing this?

  ▫  *The straightforward answer is to add the probabilities for* 1, 2, 3, *and* 4:

$$\frac{1}{6} + \frac{2}{9} + \frac{1}{3} + \frac{1}{9} = \frac{3}{18} + \frac{4}{18} + \frac{2}{18} = \frac{15}{18} = \frac{5}{6}$$

  *The quicker way is to say that the probability he gets at least one call is the probability that he doesn't get zero calls, and so the required probability is* $1 - \frac{1}{6} = \frac{5}{6}$.

### Exercises 3–6 (8 minutes)

The main challenge this question presents is adding and subtracting fractions, so it is important that students do not use calculators on this exercise or on the other questions that involve fractions. Students can either work independently or with a partner. Teachers should use discretion on how important it is to complete this exercise. This exercise provides students an opportunity to review their work with fractions.

---

**Exercises 3–6**

When Jenna goes to the farmer's market, she also usually buys some broccoli. The possible number of heads of broccoli that she buys and the probabilities are given in the table below.

| Number of Heads of Broccoli | 0 | 1 | 2 | 3 | 4 |
|---|---|---|---|---|---|
| Probability | $\frac{1}{12}$ | $\frac{1}{6}$ | $\frac{5}{12}$ | $\frac{1}{4}$ | $\frac{1}{12}$ |

Find the probability that Jenna:

3. buys exactly 3 heads of broccoli. $\frac{1}{4}$

4. does not buy exactly 3 heads of broccoli. $1 - \frac{1}{4} = \frac{3}{4}$

5. buys more than 1 head of broccoli. $\frac{5}{12} + \frac{1}{4} + \frac{1}{12} = \frac{3}{4}$

6. buys at least 3 heads of broccoli. $\frac{1}{4} + \frac{1}{12} = \frac{1}{3}$

---

### Exercises 7–10 (8 minutes)

Here, students calculate the probabilities of a given scenario. Because this is the first time this is covered in this class, some students might need assistance.

---

**Exercises 7–10**

The diagram below shows a spinner designed like the face of a clock. The sectors of the spinner are colored red (R), blue (B), green (G), and yellow (Y).

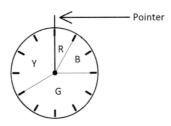

Spin the pointer, and award the player a prize according to the color on which the pointer stops.

---

7. Writing your answers as fractions in lowest terms, find the probability that the pointer stops on

   a. red: $\frac{1}{12}$

   b. blue: $\frac{2}{12} = \frac{1}{6}$

   c. green: $\frac{5}{12}$

   d. yellow: $\frac{4}{12} = \frac{1}{3}$

8. Complete the table of probabilities below.

| Color | Red | Blue | Green | Yellow |
|---|---|---|---|---|
| Probability | $\frac{1}{12}$ | $\frac{1}{6}$ | $\frac{5}{12}$ | $\frac{1}{3}$ |

**MP.4**

9. Find the probability that the pointer stops in either the blue region or the green region.

   $\frac{1}{6} + \frac{5}{12} = \frac{7}{12}$

10. Find the probability that the pointer does not stop in the green region.

   $1 - \frac{5}{12} = \frac{7}{12}$

## Closing

Discuss with students the Lesson Summary.

### Lesson Summary

In a probability experiment where the outcomes are not known to be equally likely, the formula for the probability of an event does not necessarily apply:

$$P(event) = \frac{Number\ of\ outcomes\ in\ the\ event}{Number\ of\ outcomes\ in\ the\ sample\ space}.$$

For example:

- To find the probability that the score is greater than 3, add the probabilities of all the scores that are greater than 3.

- To find the probability of not getting a score of 3, calculate $1 - (the\ probability\ of\ getting\ a\ 3)$.

## Exit Ticket (5 minutes)

Name _____   Date_____

# Lesson 5: Chance Experiments with Outcomes that are Not Equally Likely

Exit Ticket

Carol is sitting on the bus on the way home from school and is thinking about the fact that she has three homework assignments to do tonight. The table below shows her estimated probabilities of completing 0, 1, 2, or all 3 of the assignments.

| Number of Homework Assignments Completed | 0 | 1 | 2 | 3 |
|---|---|---|---|---|
| Probability | $\frac{1}{6}$ | $\frac{2}{9}$ | $\frac{5}{18}$ | $\frac{1}{3}$ |

1. Writing your answers as fractions in lowest terms, find the probability that Carol completes
   a. exactly one assignment.


   b. more than one assignment.


   c. at least one assignment.



2. Find the probability that the number of homework assignments Carol completes is not exactly 2.


3. Carol has a bag of colored chips. She has 3 red chips, 10 blue chips, and 7 green chips in the bag. Estimate the probability (as a fraction or decimal) of Carol reaching into her bag and pulling out a green chip.

## Exit Ticket Sample Solutions

Carol is sitting on the bus on the way home from school and is thinking about the fact that she has three homework assignments to do tonight. The table below shows her estimated probabilities of completing 0, 1, 2, or all 3 of the assignments.

| Number of Homework Assignments Completed | 0 | 1 | 2 | 3 |
|---|---|---|---|---|
| Probability | $\frac{1}{6}$ | $\frac{2}{9}$ | $\frac{5}{18}$ | $\frac{1}{3}$ |

1. Writing your answers as fractions in lowest terms, find the probability that Carol completes

   a. exactly one assignment.

   $\frac{2}{9}$

   b. more than one assignment.

   $\frac{5}{18} + \frac{1}{3} = \frac{5}{18} + \frac{6}{18} = \frac{11}{18}$

   c. at least one assignment.

   $\frac{2}{9} + \frac{5}{18} + \frac{1}{3} = \frac{4}{18} + \frac{5}{18} + \frac{6}{18} = \frac{15}{18} = \frac{5}{6}$

2. Find the probability that the number of homework assignments Carol completes is not exactly 2.

   $1 - \frac{5}{18} = \frac{13}{18}$

3. Carol has a bag of colored chips. She has 3 red chips, 10 blue chips, and 7 green chips in the bag. Estimate the probability (as a fraction or decimal) of Carol reaching into her bag and pulling out a green chip.

   *An estimate of the probability would be 7 out of 20, or* $\frac{7}{20}$*, or* $0.35$*.*

## Problem Set Sample Solutions

1. The "Gator Girls" are a soccer team. The possible number of goals the Gator Girls will score in a game and their probabilities are shown in the table below.

| Number of Goals | 0 | 1 | 2 | 3 | 4 |
|---|---|---|---|---|---|
| Probability | 0.22 | 0.31 | 0.33 | 0.11 | 0.03 |

   Find the probability that the Gator Girls

   a. score more than two goals:      $0.11 + 0.03 = 0.14$

   b. score at least two goals:      $0.33 + 0.11 + 0.03 = 0.47$

   c. do not score exactly 3 goals:      $1 - 0.11 = 0.89$

2. The diagram below shows a spinner. The pointer is spun, and the player is awarded a prize according to the color on which the pointer stops.

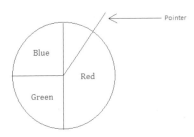

a. What is the probability that the pointer stops in the red region?

$\frac{1}{2}$

b. Complete the table below showing the probabilities of the three possible results.

| Color | Red | Green | Blue |
|---|---|---|---|
| Probability | $\frac{1}{2}$ | $\frac{1}{4}$ | $\frac{1}{4}$ |

c. Find the probability that the pointer stops on green or blue.

$\frac{1}{4} + \frac{1}{4} = \frac{1}{2}$

d. Find the probability that the pointer does not stop on green.

$1 - \frac{1}{4} = \frac{3}{4}$

3. Wayne asked every student in his class how many siblings (brothers and sisters) they had. Survey results are shown in the table below. (Wayne included himself in the results.)

| Number of Siblings | 0 | 1 | 2 | 3 | 4 |
|---|---|---|---|---|---|
| Number of Students | 4 | 5 | 14 | 6 | 3 |

(Note: The table tells us that 4 students had no siblings, 5 students had one sibling, 14 students had two siblings, and so on.)

a. How many students are in Wayne's class?

$4 + 5 + 14 + 6 + 3 = 32$

b. What is the probability that a randomly selected student does not have any siblings? Write your answer as a fraction in lowest terms.

$\frac{4}{32} = \frac{1}{8}$

c.  The table below shows the possible number of siblings and the probabilities of each number.  Complete the table by writing the probabilities as fractions in lowest terms.

| Number of Siblings | 0 | 1 | 2 | 3 | 4 |
|---|---|---|---|---|---|
| Probability | $\dfrac{1}{8}$ | $\dfrac{5}{32}$ | $\dfrac{7}{16}$ | $\dfrac{3}{16}$ | $\dfrac{3}{32}$ |

d.  Writing your answers as fractions in lowest terms, find the probability that the student:

i.  has fewer than two siblings.

$$\frac{1}{8} + \frac{5}{32} = \frac{9}{32}$$

ii.  has two or fewer siblings.

$$\frac{1}{8} + \frac{5}{32} + \frac{7}{16} = \frac{23}{32}$$

iii.  does not have exactly one sibling.

$$1 - \frac{5}{32} = \frac{27}{32}$$

**Lesson 6:  Using Tree Diagrams to Represent a Sample Space and to Calculate Probabilities**

### Student Outcomes

- Given a description of a chance experiment that can be thought of as being performed in two or more stages, students use tree diagrams to organize and represent the outcomes in the sample space.
- Students calculate probabilities of compound events.

> Suppose a girl attends a preschool where the students are studying primary colors.  To help teach calendar skills, the teacher has each student maintain a calendar in his or her cubby.  For each of the four days that the students are covering primary colors in class, each student gets to place a colored dot on his/her calendar:  blue, yellow, or red.  When the four days of the school week have passed (Monday–Thursday), what might the young girl's calendar look like?
>
> One outcome would be four blue dots if the student chose blue each day.  But consider that the first day (Monday) could be blue, and the next day (Tuesday) could be yellow, and Wednesday could be blue, and Thursday could be red.  Or, maybe Monday and Tuesday could be yellow, Wednesday could be blue, and Thursday could be red.  Or, maybe Monday, Tuesday, and Wednesday could be blue, and Thursday could be red …
>
> As hard to follow as this seems now, we have only mentioned 3 of the 81 possible outcomes in terms of the four days of colors!  Listing the other 78 outcomes would take several pages!  Rather than listing outcomes in the manner described above (particularly when the situation has multiple stages, such as the multiple days in the case above), we often use a *tree diagram* to display all possible outcomes visually.  Additionally, when the outcomes of each stage are the result of a chance experiment, tree diagrams are helpful for computing probabilities.

### Classwork

**Example 1 (10 minutes):  Two Nights of Games**

- The tree diagram is an important way of organizing and visualizing outcomes.
- The tree diagram is a particularly useful device when the experiment can be thought of as occurring in stages.
- When the information about probabilities associated with each branch is included, the tree diagram facilitates the computation of the probabilities of the different possible outcomes.

Example 1: Two Nights of Games

Imagine that a family decides to play a game each night. They all agree to use a tetrahedral die (i.e., a four-sided pyramidal die where each of four possible outcomes is equally likely—see image on page 9) each night to randomly determine if they will play a board game ($B$) or a card game ($C$). The tree diagram mapping the possible overall outcomes over two consecutive nights will be developed below.

To make a tree diagram, first present all possibilities for the first stage. (In this case, Monday.)

Monday        Tuesday        Outcome


B



C

Then, from *each* branch of the first stage, attach all possibilities for the second stage (Tuesday).

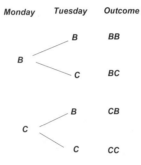

Monday        Tuesday        Outcome

                  B          BB
            B
                  C          BC

                  B          CB
            C
                  C          CC

Note: If the situation has more than two stages, this process would be repeated until all stages have been presented.

a.   If "$BB$" represents two straight nights of board games, what does "$CB$" represent?

     *"CB" would represent a card game on the first night and a board game on the second night.*

b.   List the outcomes where exactly one board game is played over two days. How many outcomes were there?

     *BC and CB—there are two outcomes.*

## Example 2 (10 minutes): Two Nights of Games (with Probabilities)

Now include probabilities on the tree diagram from Example 1. Explain that the probability for each "branch of the tree" can be found by multiplying the probabilities of the outcomes from each stage.

Pose each question in the example to the class. Give students a moment to think about the problem.

**Example 2: Two Nights of Games (with Probabilities)**

In the example above, each night's outcome is the result of a chance experiment (rolling the tetrahedral die). Thus, there is a probability associated with each night's outcome.

By multiplying the probabilities of the outcomes from each stage, we can obtain the probability for each "branch of the tree." In this case, we can figure out the probability of each of our four outcomes: $BB$, $BC$, $CB$, and $CC$.

For this family, a card game will be played if the die lands showing a value of $1$ and a board game will be played if the die lands showing a value of $2$, $3$, or $4$. This makes the probability of a board game ($B$) on a given night $0.75$.

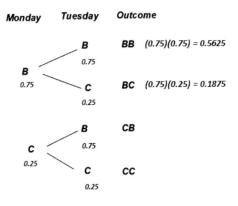

a.   The probabilities for two of the four outcomes are shown. Now, compute the probabilities for the two remaining outcomes.

*CB:* $(0.75)(0.25) = 0.1875$

*CC:* $(0.25)(0.25) = 0.0625$

b.   What is the probability that there will be exactly one night of board games over the two nights?

*The two outcomes which contain exactly one night of board games are BC and CB (see Example 2). The probability of exactly one night of board games would be the sum of the probabilities of these outcomes (since the outcomes are disjoint).* $0.1875 + 0.1875 = 0.375$.

## Exercises 1–3 (15 minutes): Two Children

 After developing the tree diagram, pose the questions to students one at a time. Allow for more than one student to offer an answer for each question, encouraging a brief (2 minute) discussion.

**Exercises 1–3: Two Children**

Two friends meet at a grocery store and remark that a neighboring family just welcomed their second child. It turns out that both children in this family are girls, and they are not twins. One of the friends is curious about what the chances are of having 2 girls in a family's first 2 births. Suppose that for each birth the probability of a "boy" birth is $0.5$ and the probability of a "girl" birth is also $0.5$.

1.  Draw a tree diagram demonstrating the four possible birth outcomes for a family with 2 children (no twins). Use the symbol "*B*" for the outcome of "boy" and "*G*" for the outcome of "girl." Consider the first birth to be the "first stage." (Refer to Example 1 if you need help getting started.)

First     Second
Child     Child     Outcome

B       BB   (0.5)(0.5) = 0.25
                    0.5
          B
          0.5
                  G       BG   (0.5)(0.5) = 0.25
                    0.5

                  B       GB   (0.5)(0.5) = 0.25
                    0.5
          G
          0.5
                  G       GG   (0.5)(0.5) = 0.25
                    0.5

2.  Write in the probabilities of each stage's outcome to the tree diagram you developed above, and determine the probabilities for each of the 4 possible birth outcomes for a family with 2 children (no twins).

    *In this case, since the probability of a boy is 0.5 and the probability of a girl is 0.5, all four outcomes will have a probability of (0.5)(0.5), or 0.25 probability of occurring.*

3.  What is the probability of a family having 2 girls in this situation? Is that greater than or less than the probability of having exactly 1 girl in 2 births?

    *The probability of a family having 2 girls is 0.25. This is less than the probability of having exactly 1 girl in 2 births, which is 0.5 (the sum of the probabilities of BG and GB).*

## Closing (5 minutes)

Consider posing the following question; discuss with students:

- Can you think of any situations where the first stage of a tree diagram might have two possibilities but the second stage might have more than two possibilities attached to each first-stage "branch"?

  □ *Answers will vary, but an example will be shown in Lesson 7 where males and females are then split into Democrat, Republican, and Other.*

Lesson Summary

Tree diagrams can be used to organize outcomes in the sample space for chance experiments that can be thought of as being performed in multiple stages. Tree diagrams are also useful for computing probabilities of events with more than one outcome.

## Exit Ticket (5 minutes)

Name _____    Date_____

# Lesson 6:  Using Tree Diagrams to Represent a Sample Space and to Calculate Probabilities

Exit Ticket

In a laboratory experiment, two mice will be placed in a simple maze with one decision point where a mouse can turn either left ($L$) or right ($R$).  When the first mouse arrives at the decision point, the direction it chooses is recorded.  Then, the process is repeated for the second mouse.

1.  Draw a tree diagram where the first stage represents the decision made by the first mouse, and the second stage represents the decision made by the second mouse.  Determine all four possible decision outcomes for the two mice.

2.  If the probability of turning left is 0.5, and the probability of turning right is 0.5 for each mouse, what is the probability that only one of the two mice will turn left?

3.  If the researchers add food in the simple maze such that the probability of each mouse turning left is now 0.7, what is the probability that only one of the two mice will turn left?

## Exit Ticket Sample Solutions

In a laboratory experiment, two mice will be placed in a simple maze with one decision point where a mouse can turn either left ($L$) or right ($R$). When the first mouse arrives at the decision point, the direction it chooses is recorded. Then, the process is repeated for the second mouse

1. Draw a tree diagram where the first stage represents the decision made by the first mouse, and the second stage represents the decision made by the second mouse. Determine all four possible decision outcomes for the two mice.

First Mouse / Second Mouse / Outcome

L (0.5) → L (0.5) → LL  (0.5)(0.5) = 0.25
L (0.5) → R (0.5) → LR  (0.5)(0.5) = 0.25
R (0.5) → L (0.5) → RL  (0.5)(0.5) = 0.25
R (0.5) → R (0.5) → RR  (0.5)(0.5) = 0.25

2. If the probability of turning left is $0.5$, and the probability of turning right is $0.5$ for each mouse, what is the probability that only one of the two mice will turn left?

   *There are two outcomes that have exactly one mouse turning left: $LR$ and $RL$. Each has a probability of $0.25$, so the probability of only one of the two mice turning left is $0.25 + 0.25 = 0.5$.*

3. If the researchers add food in the simple maze such that the probability of each mouse turning left is now $0.7$, what is the probability that only one of the two mice will turn left?

First Mouse / Second Mouse / Outcome

L (0.7) → L (0.7) → LL  (0.7)(0.7) = 0.49
L (0.7) → R (0.3) → LR  (0.7)(0.3) = 0.21
R (0.3) → L (0.7) → RL  (0.3)(0.7) = 0.21
R (0.3) → R (0.3) → RR  (0.3)(0.3) = 0.09

   *As in Question 2, there are two outcomes that have exactly one mouse turning left: $LR$ and $RL$. However, with the adjustment made by the researcher, each of these outcomes now has a probability of $0.21$. So now, the probability of only one of the two mice turning left is $0.21 + 0.21 = 0.42$.*

## Problem Set Sample Solutions

1. Imagine that a family of three (Alice, Bill, and Chester) plays bingo at home every night. Each night, the chance that any one of the three players will win is $\frac{1}{3}$.

    a. Using "$A$" for Alice wins, "$B$" for Bill wins, and "$C$" for Chester wins, develop a tree diagram that shows the nine possible outcomes for two consecutive nights of play.

    | First Night | Second Night | Outcome |
    |---|---|---|
    | | A | AA |
    | A | B | AB |
    | | C | AC |
    | | A | BA |
    | B | B | BB |
    | | C | BC |
    | | A | CA |
    | C | B | CB |
    | | C | CC |

    b. Is the probability that "Bill wins both nights" the same as the probability that "Alice wins the first night and Chester wins the second night"? Explain.

    *Yes. The probability of Bill winning both nights is $\frac{1}{3} \cdot \frac{1}{3} = \frac{1}{9}$, which is the same as the probability of Alice winning the first night and Chester winning the second night $(\frac{1}{3} \cdot \frac{1}{3} = \frac{1}{9})$.*

2. According to the Washington, DC Lottery's website for its "Cherry Blossom Doubler" instant scratch game, the chance of winning a prize on a given ticket is about $17\%$. Imagine that a person stops at a convenience store on the way home from work every Monday and Tuesday to buy a "scratcher" ticket to play the game.

    (Source: http://dclottery.com/games/scratchers/1223/cherry-blossom-doubler.aspx accessed May 27, 2013).

    a. Develop a tree diagram showing the four possible outcomes of playing over these two days. Call stage 1 "Monday", and use the symbols "$W$" for a winning ticket and "$L$" for a non-winning ticket.

    | Monday | Tuesday | Outcome |
    |---|---|---|
    | | W | WW |
    | W | | |
    | | L | WL |
    | | W | LW |
    | L | | |
    | | L | LL |

b.  What is the chance that the player will not win on Monday but will win on Tuesday?

*LW outcome:* $(0.83)(0.17) = 0.1411$

c.  What is the chance that the player will win at least once during the two-day period?

*"Winning <u>at least</u> once" would include all outcomes except LL (which has a $0.6889$ probability). The probabilities of these outcomes would sum to $0.3111$.*

**Image of Tetrahedral Die**

Source: <u>http://commons.wikimedia.org/wiki/File:4-sided_dice_250.jpg</u>

# Lesson 7: Calculating Probabilities of Compound Events

## Student Outcomes

- Students will calculate probabilities of compound events.

## Lesson Overview

The content of this lesson is deliberately similar to the material in Lesson 6. For most questions, an extra stage has been added to the corresponding original Lesson 6 question.

Students are likely to notice this relation to previous material. When they do, consider reinforcing the idea that tree diagrams can be used to effectively map out a sample space, and to compute probabilities, no matter how many stages or outcomes a situation has.

> A previous lesson introduced *tree diagrams* as an effective method of displaying the possible outcomes of certain multistage chance experiments. Additionally, in such situations, tree diagrams were shown to be helpful for computing probabilities.
>
> In those previous examples, diagrams primarily focused on cases with two stages. However, the basic principles of tree diagrams can apply to situations with more than two stages.

## Classwork

### Example 1 (4–6 minutes): Three Nights of Games

Discuss examples in which a tree diagram could organize and visualize outcomes. Pose the following discussion problems to students, and have them sketch trees as possible ways to organize outcomes simply. Emphasize the ways in which the trees help to organize the outcomes in these two examples.

- Decorating a House

  A designer has three different paints to use for a client's house—red, white, and blue. The designer is responsible for decorating two rooms of the house—the kitchen and a home office. If each room is expected to be a different color, how could a tree diagram be used to organize the possible outcomes of what color each room is painted? (Organize the tree by setting up kitchen to office. Under each room, use a tree to organize the possible paint colors that could be used for each room.) What are the different ways the two rooms could be painted?

- Organizing a School Schedule

  Consider the following possible scheduling problem. Ralph needs to select courses for two hours of his school day. Possible courses he can take are music, science, and mathematics. Each course is offered during the first and second hours of a school day. How could a tree diagram be used to organize the possible outcomes of Ralph's schedule for the two hours? (Organize the tree by setting up Hour 1 to Hour 2. Under each period, use a tree to organize the possible courses Ralph could take.) What are the different ways Ralph's schedule could look?

**Example 1: Three Nights of Games**

Recall a previous example where a family decides to play a game each night, and they all agree to use a four-sided die in the shape of a pyramid (where each of four possible outcomes is equally likely) each night to randomly determine if the game will be a board ($B$) or a card ($C$) game. The tree diagram mapping the possible overall outcomes over two consecutive nights was as follows:

| Monday | Tuesday | Outcome |
|--------|---------|---------|
| B | B | BB |
| | C | BC |
| C | B | CB |
| | C | CC |

But how would the diagram change if you were interested in mapping the possible overall outcomes over three consecutive nights? To accommodate this additional "third stage," you would take steps similar to what you did before. You would attach all possibilities for the third stage (Wednesday) to each branch of the previous stage (Tuesday).

| Monday | Tuesday | Wednesday | Outcome |
|--------|---------|-----------|---------|
| B | B | B | BBB |
| | | C | BBC |
| | C | B | BCB |
| | | C | BCC |
| C | B | B | CBB |
| | | C | CBC |
| | C | B | CCB |
| | | C | CCC |

Read through the example in the student lesson as a class. Convey the following important points about the tree diagram for this example:

- The tree diagram is an important way of organizing and visualizing outcomes.
- The tree diagram is particularly useful when the experiment can be thought of as occurring in stages.
- When the information about probabilities associated with each branch is included, the tree diagram facilitates the computation of the probabilities of possible outcomes.
- The basic principles of tree diagrams can apply to situations with more than two stages.

### Exercises 1–3 (6 minutes)

Let students work independently, and then discuss and confirm answers as a class.

---

Exercises 1–3

1.  If "*BBB*" represents three straight nights of board games, what does "*CBB*" represent?

    *"CBB" would represent a card game on the first night and a board game on each of the second and third nights.*

2.  List all outcomes where exactly two board games were played over three days.  How many outcomes were there?

    *BBC, BCB, and CBB—there are 3 outcomes.*

3.  There are eight possible outcomes representing the three nights.  Are the eight outcomes representing the three nights equally likely?  Why or why not?

    *As in the exercises of the previous lesson, the probability of C and B are not the same.  As a result, the probability of the outcome CCC (all three nights involve card games) is not the same as BBB (all three nights playing board games).  The probability of playing cards was only $\frac{1}{4}$ in the previous lesson.*

---

### Example 2 (5 minutes):  Three Nights of Games (with Probabilities)

---

Example 2:  Three Nights of Games (with Probabilities)

In the example above, each night's outcome is the result of a chance experiment (rolling the four-sided die).  Thus, there is a probability associated with each night's outcome.

By multiplying the probabilities of the outcomes from each stage, you can obtain the probability for each "branch of the tree."  In this case, you can figure out the probability of each of our eight outcomes.

For this family, a card game will be played if the die lands showing a value of 1, and a board game will be played if the die lands showing a value of 2, 3, or 4.  This makes the probability of a board game (*B*) on a given night 0.75.

Let's use a tree to examine the probabilities of the outcomes for the three days.

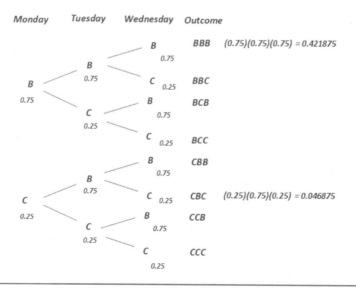

---

Read through the example with students and explain how to calculate the probabilities for the outcomes $BBB$ and $CBC$ as shown on the tree diagram. Then ask students:

MP.4

- What does "$CBC$" represent?
  - "$CBC$" represents the sequence of card game the first night, board game the second night, and card came the third night.
- The probability of "$CBC$" is approximately 0.046785. What does this probability mean?
  - *The probability of "$CBC$" is very small. This means that the outcome of $CBC$ is not expected to happen very often.*


### Exercises 4–6 (5 minutes)

Let students work independently on Exercises 4–5, and confirm their answers with a neighbor.

---

**Exercises 4–6**

4. Probabilities for two of the eight outcomes are shown. Calculate the approximate probabilities for the remaining six outcomes.

   *BBC:* $0.75(0.75)(0.25) = 0.140625$

   *BCB:* $0.75(0.25)(0.75) = 0.140625$

   *BCC:* $0.75(0.25)(0.25) = 0.046875$

   *CBB:* $0.25(0.75)(0.27) = 0.140625$

   *CCB:* $0.25(0.25)(0.75) = 0.046875$

   *CCC:* $0.25(0.25)(0.25) = 0.015625$

5. What is the probability that there will be exactly two nights of board games over the three nights?

   *The three outcomes that contain exactly two nights of board games are BBC, BCB, and CBB. The probability of exactly two nights of board games would be the sum of the probabilities of these outcomes, or:*

   $0.140625 + 0.140625 + 0.140625 = 0.421875$

---

Point out to students that the sum of probabilities of these three outcomes is not $\frac{3}{8}$, or 0.375. Why? Students previously discussed that outcomes for the three nights are not equally likely.

---

6. What is the probability that the family will play at least one night of card games?

   *This "at least" question could be answered by subtracting the probability of no card games (BBB) from 1. Discuss with students that when you remove the BBB from the list of eight outcomes, the remaining outcomes include at least one night of card games. The probability of three nights of board games (BBB) is 0.421875. Therefore, the probability of at least one night of card games would be the probability of all outcomes (or 1) minus the probability of all board games, or: $1 - 0.421875 = 0.578125$.*

---

## Exercises 7–10 (15 minutes):  Three Children

After developing the tree diagram, pose the questions to students one at a time.  Allow for more than one student to offer an answer for each question, encouraging a brief (2 minute) discussion.

---

**Exercises 7–10**

A neighboring family just welcomed their third child.  It turns out that all 3 of the children in this family are girls, and they are not twins.  Suppose that for each birth the probability of a "boy" birth is $0.5$, and the probability of a "girl" birth is also $0.5$.  What are the chances of having 3 girls in a family's first 3 births?

7.  Draw a tree diagram showing the eight possible birth outcomes for a family with 3 children (no twins).  Use the symbol "$B$" for the outcome of "boy" and "$G$" for the outcome of "girl."  Consider the first birth to be the "first stage."  (Refer to Example 1 if you need help getting started.)

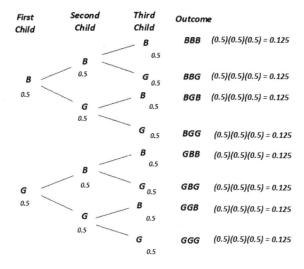

8.  Write in the probabilities of each stage's outcomes in the tree diagram you developed above, and determine the probabilities for each of the eight possible birth outcomes for a family with 3 children (no twins).

*In this case, since the probability of a boy is $0.5$, and the probability of a girl is $0.5$, all eight outcomes will have a probability of $0.125$ of occurring.*

9.  What is the probability of a family having 3 girls in this situation?  Is that greater than or less than the probability of having exactly 2 girls in 3 births?

*The probability of a family having 3 girls is $0.125$. This is less than the probability of having exactly 2 girls in 3 births, which is $0.375$ (the sum of the probabilities of $GGB$, $GBG$, and $BGG$).  Another way of explaining this to students is to point out that the probability of each outcome is the same, or $0.125$.  Therefore, the probability of 3 girls in 3 births is the probability of just one possible outcome.  The probability of having exactly 2 girls in 3 births is sum of the probabilities of three outcomes.  Therefore, the probability of one of the outcomes is less than the probability of the sum of three outcomes (and again, emphasize that the probability of each outcome is the same).*

10.  What is the probability of a family of 3 children having at least 1 girl?

*The probability of having at least 1 girl would be found by subtracting the probability of no girls (or all boys, $BBB$) from 1, or $1 - 0.125 = 0.875$.*

---

Ask students if they know any families with exactly 3 children.  Consider organizing the possible examples students provide ($BBB$, $GGB$, etc.).  Compare and discuss the relative frequencies of the examples to the above probabilities.

## Closing (5 minutes)

Consider posing the following questions, and allow a few student responses for each:

- If you knew that a certain game where the only outcomes are "win" or "lose" offered a probability of winning each time of 0.35, given what has been discussed so far, can you calculate the probability of losing such a game five consecutive times?
    - *If the chance of winning a game is 0.35, the chance of losing a game is 0.65. So the chance of losing five times would be $(0.65)(0.65)(0.65)(0.65)(0.65)$ or approximately 0.116.*

- Did you need to draw a tree diagram for that question? Why or why not?
    - *No, it is not necessary for this particular outcome. When the outcome represents the same event occurring several times, the probability of that outcome is the probability of the event multiplied by itself for each time the event occurred. In this case, it was $(0.65)$ times itself 5 times.*

- What are the chances of winning *at least* once in those five games?
    - *If you win at least once, the only outcome you avoided would be "losing five times." So all other outcomes would be valid, and the sum of all of their probabilities would be $1.000 - 0.116 = 0.884$, so there's actually a good chance you would win at least once over five games.*

---

Lesson Summary

The use of tree diagrams is not limited to cases of just two stages. For more complicated experiments, tree diagrams are used to organize outcomes and to assign probabilities. The tree diagram is a visual representation of outcomes that involve more than one event.

---

## Exit Ticket (5–7 minutes)

Lesson 7:    Calculating Probabilities of Compound Events

Name _____    Date_____

# Lesson 7:  Calculating Probabilities of Compound Events

In a laboratory experiment, three mice will be placed in a simple maze that has just one decision point where a mouse can turn either left ($L$) or right ($R$).  When the first mouse arrives at the decision point, the direction he chooses is recorded.  The same is done for the second and the third mice.

1.  Draw a tree diagram where the first stage represents the decision made by the first mouse, and the second stage represents the decision made by the second mouse, and so on.  Determine all eight possible outcomes of the decisions for the three mice.

2. Use the tree diagram from Question 1 to help answer the following question. If, for each mouse, the probability of turning left is 0.5 and the probability of turning right is 0.5, what is the probability that only one of the three mice will turn left?

3. If the researchers conducting the experiment add food in the simple maze such that the probability of each mouse turning left is now 0.7, what is the probability that only one of the three mice will turn left? To answer question, use the tree diagram from Question 1.

## Exit Ticket Sample Solutions

In a laboratory experiment, three mice will be placed in a simple maze that has just one decision point where a mouse can turn either left ($L$) or right ($R$). When the first mouse arrives at the decision point, the direction he chooses is recorded. The same is done for the second and the third mice.

1.  Draw a tree diagram where the first stage represents the decision made by the first mouse, and the second stage represents the decision made by the second mouse, and so on. Determine all eight possible outcomes of the decisions for the three mice.

| First Mouse | Second Mouse | Third Mouse | Outcome | |
|---|---|---|---|---|
| | | L | LLL | $(0.5)(0.5)(0.5) = 0.125$ |
| | L 0.5 | R | LLR | $(0.5)(0.5)(0.5) = 0.125$ |
| L 0.5 | | L | LRL | $(0.5)(0.5)(0.5) = 0.125$ |
| | R 0.5 | R | LRR | $(0.5)(0.5)(0.5) = 0.125$ |
| | | L | RLL | $(0.5)(0.5)(0.5) = 0.125$ |
| | L 0.5 | R | RLR | $(0.5)(0.5)(0.5) = 0.125$ |
| R 0.5 | | L | RRL | $(0.5)(0.5)(0.5) = 0.125$ |
| | R 0.5 | R | RRR | $(0.5)(0.5)(0.5) = 0.125$ |

2.  If the probability of turning left is $0.5$ and the probability of turning right is $0.5$ for each mouse, what is the probability that only one of the three mice will turn left?

    *There are three outcomes that have exactly one mouse turning left: $LRR$, $RLR$, and $RRL$. Each has a probability of $0.125$, so the probability of having only one of the three mice turn left is $0.375$.*

3. If the researchers conducting the experiment add food in the simple maze such that the probability of each mouse turning left is now $0.7$, what is the probability that only one of the three mice will turn left?

*As in Question 2, there are three outcomes that have exactly one mouse turning left: LRR, RLR, and RRL. However, with the adjustment made by the researcher, each of the three outcomes now has a probability of $0.063$. So now, the probability of having only one of the three mice turn left is the sum of three equally likely outcomes of $0.63$, or $0.63(0.06) = 0.189$. The tree provides a way to organize the outcomes and the probabilities.*

| First Mouse | Second Mouse | Third Mouse | Outcome | |
|---|---|---|---|---|
| | | L 0.7 | LLL | $(0.7)(0.7)(0.7) = 0.343$ |
| | L 0.7 | R 0.3 | LLR | $(0.7)(0.7)(0.3) = 0.147$ |
| L 0.7 | | L 0.7 | LRL | $(0.7)(0.3)(0.7) = 0.147$ |
| | R 0.3 | R 0.3 | LRR | $(0.7)(0.3)(0.3) = 0.063$ |
| | L 0.7 | L 0.7 | RLL | $(0.3)(0.7)(0.7) = 0.147$ |
| | | R 0.3 | RLR | $(0.3)(0.7)(0.3) = 0.063$ |
| R 0.3 | R 0.3 | L 0.7 | RRL | $(0.3)(0.3)(0.7) = 0.063$ |
| | | R 0.3 | RRR | $(0.3)(0.3)(0.3) = 0.027$ |

## Problem Set Sample Solutions

1. According to the Washington, DC Lottery's website for its "Cherry Blossom Doubler" instant scratch game, the chance of winning a prize on a given ticket is about $17\%$. Imagine that a person stops at a convenience store on the way home from work every Monday, Tuesday, and Wednesday to buy a "scratcher" ticket and plays the game.

(Source: http://dclottery.com/games/scratchers/1223/cherry-blossom-doubler.aspx accessed May 27, 2013)

a. Develop a tree diagram showing the eight possible outcomes of playing over these three days. Call stage one "Monday," and use the symbols "$W$" for a winning ticket and "$L$" for a non-winning ticket.

| Monday | Tuesday | Wednesday | Outcome |
|---|---|---|---|
| | | W | WWW |
| | W | L | WWL |
| W | | W | WLW |
| | L | L | WLL |
| | | W | LWW |
| | W | L | LWL |
| L | | W | LLW |
| | L | L | LLL |

b.  What is the probability that the player will not win on Monday but will win on Tuesday and Wednesday?

    *LWW outcome:* $0.83(0.17)(0.17) = 0.024$

  c.  What is the probability that the player will win at least once during the 3-day period?

    *"Winning at least once" would include all outcomes except LLL (which has a $0.5718$ probability). The probabilities of these outcomes would sum to about $0.4282$. This is also equal to $1 - 0.5718$.*

**2.**  A survey company is interested in conducting a statewide poll prior to an upcoming election. They are only interested in talking to registered voters.

  Imagine that $55\%$ of the registered voters in the state are male, and $45\%$ are female. Also, consider that the distribution of ages may be different for each group. In this state, $30\%$ of male registered voters are age 18–24, $37\%$ are age 25–64, and $33\%$ are 65 or older. $32\%$ of female registered voters are age 18–24, $26\%$ are age 25–64, and $42\%$ are 65 or older.

  The following tree diagram describes the distribution of registered voters. The probability of selecting a male registered voter age 18–24 is $0.165$.

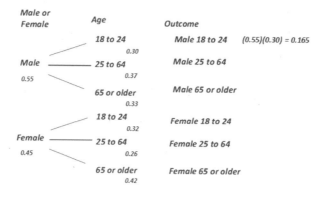

  a.  What is the chance that the polling company will select a registered female voter age 65 or older?

    *Female 65 or Older:* $(0.45)(0.42) = 0.189.$

  b.  What is the chance that the polling company will select any registered voter age 18–24?

    *The probability of selecting any registered voter age 18–24 would be the sum of the probability of selecting a male registered voter age 18–24 and the probability of selecting a female registered voter age 18–24. Those values are $(0.55)(0.30) = 0.165$ and $(0.45)(0.32) = 0.144$ respectively. So the sum is $0.165 + 0.144 = 0.309.$*

# Mathematics Curriculum

## Topic B:

# Estimating Probabilities

### 7.SP.C.6, 7.SP.C.7, 7.SP.C.8c

| Focus Standard: | 7.SP.C.6 | Approximate the probability of a chance event by collecting data on the chance process that produces it and observing its long-run relative frequency, and predict the approximate relative frequency given the probability. *For example, when rolling a number cube 600 times, predict that a 3 or 6 would be rolled roughly 200 times, but probably not exactly 200 times.* |
|---|---|---|
| | 7.SP.C.7 | Develop a probability model and use it to find probabilities of events. Compare probabilities from a model to observed frequencies; if the agreement is not good, explain possible sources of the discrepancy. |
| | | a. Develop a uniform probability model by assigning equal probability to all outcomes, and use the model to determine probabilities of events. *For example, if a student is selected at random from a class, find the probability that Jane will be selected and the probability that a girl will be selected.* |
| | | b. Develop a probability model (which may not be uniform) by observing frequencies in data generated from a chance process. *For example, find the approximate probability that a spinning penny will land heads up or that a tossed paper cup will land open-end down. Do the outcomes for the spinning penny appear to be equally likely based on the observed frequencies?* |
| | 7.SP.C.8c | Find probabilities of compound events using organized lists, tables, tree diagrams, and simulation. |
| | | c. Design and use a simulation to generate frequencies for compound events. *For example, use random digits as a simulation tool to approximate the answer to the question: If 40% of donors have type A blood, what is the probability that it will take at least 4 donors to find one with type A blood?* |

| **Instructional Days:** | 5 |
|---|---|
| **Lesson 8:** | The Difference Between Theoretical Probabilities and Estimated Probabilities (P)[1] |
| **Lesson 9:** | Comparing Estimated Probabilities to Probabilities Predicted by a Model (E) |
| **Lessons 10–11:** | Using Simulation to Estimate a Probability (P,P) |
| **Lesson 12:** | Using Probability to Make Decisions (P) |

In Topic B, students estimate probabilities empirically and by using simulation. In Lesson 8, students make the distinction between a theoretical probability and an estimated probability. For a simple chance experiment, students carry out the experiment many times and use observed frequencies to estimate known theoretical probabilities. Students also consider chance experiments for which they cannot compute theoretical probabilities. In Lesson 9, students continue to collect data from a chance experiment and use it to estimate probabilities. Students compare these probabilities to theoretical probabilities from a model and then assess the plausibility of the model. In Lessons 10 and 11, students work with simulations. They are either given results from a simulation to approximate a probability (Lesson 10), or they design their own simulation, carry out the simulation, and use the simulation results to approximate a probability (Lesson 11). Lesson 12 concludes this topic by providing students with opportunities to use probabilities to make decisions.

---

[1] Lesson Structure Key: **P**-Problem Set Lesson, **M**-Modeling Cycle Lesson, **E**-Exploration Lesson, **S**-Socratic Lesson

**Lesson 8: The Difference Between Theoretical Probabilities and Estimated Probabilities**

## Student Outcomes

- Given theoretical probabilities based on a chance experiment, students describe what they expect to see when they observe many outcomes of the experiment.
- Students distinguish between theoretical probabilities and estimated probabilities.
- Students understand that probabilities can be estimated based on observing outcomes of a chance experiment.

> Did you ever watch the beginning of a Super Bowl game? After the traditional handshakes, a coin is tossed to determine which team gets to kick-off first. Whether or not you are a football fan, the toss of a fair coin is often used to make decisions between two groups.

## Classwork

### Example 1 (5 minutes): Why a Coin?

> **Example 1: Why a Coin?**
>
> Coins were discussed in previous lessons of this module. What is special about a coin? In most cases, a coin has two different sides: a head side ("heads") and a tail side ("tails"). The sample space for tossing a coin is {heads, tails}. If each outcome has an equal chance of occurring when the coin is tossed, then the probability of getting heads is $\frac{1}{2}$, or $0.5$. The probability of getting tails is also $0.5$. Note that the sum of these probabilities is $1$.
>
> The probabilities formed using the sample space and what we know about coins are called the theoretical probabilities. Using observed relative frequencies is another method to estimate the probabilities of heads or tails. A relative frequency is the proportion derived from the number of the observed outcomes of an event divided by the total number of outcomes. Recall from earlier lessons that a relative frequency can be expressed as a fraction, a decimal, or a percent. Is the estimate of a probability from this method close to the theoretical probability? The following exercise investigates how relative frequencies can be used to estimate probabilities.

This lesson focuses on the chance experiment of tossing a coin. The outcomes are simple, and in most cases, students understand the theoretical probabilities of the outcomes. It is also a good example to build on their understanding of estimated probabilities. This example then sets up the situation of estimating these same probabilities using relative frequencies. The term *relative frequency* is introduced and defined in this example and the following exercise.

Have students read through the example. Then, use the following questions to guide the discussion:

- Are there other situations where a coin toss would be used?
  - *As a part of this discussion, you might indicate that in several state constitutions, if two candidates receive the same number of votes, the winning candidate is determined by a coin toss.*
- Is it possible to toss a fair coin and get 3 heads in a row? How about 5 heads?
  - *Make sure students understand that it is possible to get several heads or tails in a row, and that evaluating how likely it would be to get three, five, or even 20 heads in a row, are examples of probability problems.*

## Exercises 1–9 (15 minutes)

The following Exercises are designed to have students develop an estimate of the probability of getting heads by collecting data. In this example, students are provided with data from actual tosses of a fair coin. Students calculate the relative frequencies of getting heads from the data, and then use the relative frequencies to estimate the probability of getting a head.

Let students work in small groups to complete Exercises 1–9.

---

**Exercises 1–9**

Beth tosses a coin 10 times and records her results. Here are the results from the 10 tosses:

| Toss | 1 | 2 | 3 | 4 | 5 | 6 | 7 | 8 | 9 | 10 |
|---|---|---|---|---|---|---|---|---|---|---|
| Result | H | H | T | H | H | H | T | T | T | H |

The total number of heads divided by the total number of tosses is the relative frequency of heads. It is the proportion of the time that heads occurred on these tosses. The total number of tails divided by the total number of tosses is the relative frequency of tails.

1. Beth started to complete the following table as a way to investigate the relative frequencies. For each outcome, the total number of tosses increased. The total number of heads or tails observed so far depends on the outcome of the current toss. Complete this table for the 10 tosses recorded above.

| Toss | Outcome | Total number of heads so far | Relative frequency of heads so far (to the nearest hundredth) | Total number of tails so far | Relative frequency of tails so far (to the nearest hundredth) |
|---|---|---|---|---|---|
| 1 | H | 1 | $\frac{1}{1} = 1$ | 0 | $\frac{0}{1} = 0$ |
| 2 | H | 2 | $\frac{2}{2} = 1$ | 0 | $\frac{0}{2} = 0$ |
| 3 | T | 2 | $\frac{2}{3} = 0.67$ | 1 | $\frac{1}{3} = 0.33$ |
| 4 | H | 3 | $\frac{3}{4} = 0.75$ | 1 | $\frac{1}{4} = 0.25$ |
| 5 | H | 4 | $\frac{4}{5} = 0.80$ | 1 | $\frac{1}{5} = 0.20$ |
| 6 | H | 5 | $\frac{5}{6} = 0.83$ | 1 | $\frac{1}{6} = 0.17$ |
| 7 | T | 5 | $\frac{5}{7} = 0.71$ | 2 | $\frac{2}{7} = 0.29$ |
| 8 | T | 5 | $\frac{5}{8} = 0.63$ | 3 | $\frac{3}{8} = 0.37$ |
| 9 | T | 5 | $\frac{5}{9} = 0.56$ | 4 | $\frac{4}{9} = 0.44$ |
| 10 | H | 6 | $\frac{6}{10} = 0.60$ | 4 | $\frac{4}{10} = 0.40$ |

---

2. What is the sum of the relative frequency of heads and the relative frequency of tails for each row of the table?

   *The sum of the relative frequency of heads and the relative frequency of tails for each row is 1.00.*

3. Beth's results can also be displayed using a graph. Complete this graph using the values of relative frequency of heads so far from the table above:

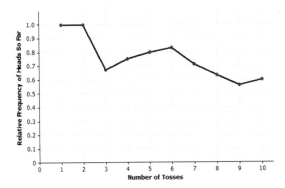

4. Beth continued tossing the coin and recording results for a total of 40 tosses. Here are the results of the next 30 tosses:

| Toss | 11 | 12 | 13 | 14 | 15 | 16 | 17 | 18 | 19 | 20 |
|------|----|----|----|----|----|----|----|----|----|----|
| Result | T | H | T | H | T | H | H | T | H | T |

| Toss | 21 | 22 | 23 | 24 | 25 | 26 | 27 | 28 | 29 | 30 |
|------|----|----|----|----|----|----|----|----|----|----|
| Result | H | T | T | H | T | T | T | T | H | T |

| Toss | 31 | 32 | 33 | 34 | 35 | 36 | 37 | 38 | 39 | 40 |
|------|----|----|----|----|----|----|----|----|----|----|
| Result | H | T | H | T | H | T | H | H | T | T |

As the number of tosses increases, the relative frequency of heads changes. Complete the following table for the 40 coin tosses:

| Number of tosses | Total number of heads so far | Relative frequency of heads so far (to the nearest hundredth) |
|------------------|------------------------------|---------------------------------------------------------------|
| 1 | 1 | 1.00 |
| 5 | 4 | 0.80 |
| 10 | 6 | 0.60 |
| 15 | 8 | 0.53 |
| 20 | 11 | 0.55 |
| 25 | 13 | 0.52 |
| 30 | 14 | 0.47 |
| 35 | 17 | 0.49 |
| 40 | 19 | 0.48 |

5.  Complete the graph below using the relative frequency of heads so far from the table above for total number of tosses of 1, 5, 10, 15, 20, 25, 30, 35, and 40:

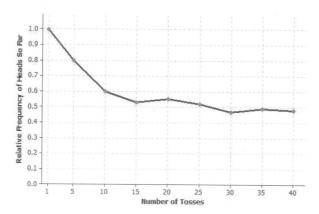

6.  What do you notice about the changes in the relative frequency of number of heads so far as the number of tosses increases?

    *The relative frequencies seem to change less as the number of tosses increases. The line drawn to connect the relative frequencies seems to be leveling off.*

7.  If you tossed the coin 100 times, what do you think the relative frequency of heads would be? Explain your answer.

    *Answers will vary. Anticipate most students will indicate 50 heads result in 100 tosses, for a relative frequency of 0.50. This is a good time to indicate that the value of 0.50 is where the graph of the relative frequencies seems to be approaching. However, the relative frequencies will vary. For example, if the relative frequency for 100 tosses were 0.50 (and it could be), what would the relative frequency for 101 tosses be? Point out to students that no matter the outcome on the 101st toss, the relative frequency of heads would not be exactly 0.50.*

8.  Based on the graph and the relative frequencies, what would you estimate the probability of getting heads to be? Explain your answer.

    *Answers will vary. Anticipate that students will estimate the probability to be 0.50, as that is what they determined in the opening discussion, and that is the value that the relative frequencies appear to be approaching. Some students may estimate the probability as 0.48, as that was the last relative frequency obtained after 40 tosses. That estimate is also a good estimate of the probability.*

9.  How close is your estimate in Exercise 8 to the theoretical probability of 0.5? Would the estimate of this probability have been as good if Beth had only tossed the coin a few times instead of 40?

    *In the beginning, the relative frequencies jump around. The estimated probabilities and the theoretical probabilities should be nearly the same as the number of observations increase. The estimated probabilities would likely not be as good after just a few coin tosses. Direct students to draw a horizontal line representing their estimate of the probability (or, in most cases, 0.5).*

The value you gave in Exercise 8 is an estimate of the theoretical probability and is called an experimental or estimated probability.

If time permits, you might point out some history on people who wanted to observe long-run relative frequencies. Share with students how, for each of these cases, the relative frequencies were close to 0.5. Students may also find it interesting that the relative frequencies were not exactly 0.5. Ask students, "If they were closer to 0.5 than in our example, why do you think that was the case?"

- The French naturalist Count Buffon (1707–1788) tossed a coin 4,040 times.

   Result: 2,048 heads, or proportion 2,048/4,040 = 0.5069 for heads.

- Around 1900, the English statistician, Karl Pearson, tossed a coin 24,000 times.

   Result: 12,012 heads, a proportion of 0.5005.

- While imprisoned by the Germans during World War II, the South African mathematician, John Kerrich, tossed a coin 10,000 times.

   Result: 5067 heads, a proportion of 0.5067.

## Example 2 (5 minutes): More Pennies!

> **Example 2:  More Pennies!**
>
> Beth received nine more pennies.  She securely taped them together to form a small stack.  The top penny of her stack showed heads, and the bottom penny showed tails.  If Beth tosses the stack, what outcomes could she observe?

This example moves the discussion to a chance experiment in which the theoretical probability is not known.  Prepare several stacks of the pennies as described in this example.  Make sure the 10 pennies are stacked with one end showing heads and the other end tails.  It is suggested you use scotch tape to wrap the entire stack.  Because constructing the stacks might result in pennies flying around, it is suggested you prepare the stacks before this exercise is started.

Introduce students to the following exercise by tossing the stack a few times (and testing that it did not fall apart!).  Then ask:

- What are the possible outcomes?

   *Head, Tail, and on the Side.  These three outcomes represent the sample space.*

## Exercises 10–17 (15 minutes)

Let students continue to work in small groups.

> **Exercises 10–17**
>
> 10. Beth wanted to determine the probability of getting heads when she tosses the stack.  Do you think this probability is the same as the probability of getting heads with just one coin?  Explain your answer.
>
>    *The outcomes when tossing this stack would be: {Head, Tail, Side}.  This changes the probability of getting heads, as there are three outcomes.*

11. Make a sturdy stack of 10 pennies in which one end of the stack has a penny showing heads and the other end tails. Make sure the pennies are taped securely, or you may have a mess when you toss the stack. Toss the stack to observe possible outcomes. What is the sample space for tossing a stack of 10 pennies taped together? Do you think the probability of each outcome of the sample space is equal? Explain your answer.

*The sample space is {Head, Tail, Side}. A couple of tosses should clearly indicate to students that the stack lands often on its side. As a result, the probabilities of heads, tails, and on the side do not appear to be the same.*

12. Record the results of 10 tosses. Complete the following table of the relative frequencies of heads for your 10 tosses:

*Answers will vary; the results of an actual toss are shown below.*

| Toss | 1 | 2 | 3 | 4 | 5 | 6 | 7 | 8 | 9 | 10 |
|---|---|---|---|---|---|---|---|---|---|---|
| Result | Head | Head | Side | Side | Side | Tail | Side | Side | Tail | Side |
| Relative frequency of heads so far | 1.00 | 1.00 | 0.67 | 0.50 | 0.40 | 0.33 | 0.29 | 0.25 | 0.22 | 0.20 |

13. Based on the value of the relative frequencies of heads so far, what would you estimate the probability of getting heads to be?

*If students had a sample similar to the above, they would estimate the probability of tossing a head as .20 (or something close to that last relative frequency).*

14. Toss the stack of 10 pennies another 20 times. Complete the following table:

*Answers will vary; student data will be different.*

| Toss | 11 | 12 | 13 | 14 | 15 | 16 | 17 | 18 | 19 | 20 |
|---|---|---|---|---|---|---|---|---|---|---|
| Result | Head | Head | Tail | Side | Side | Tail | Tail | Side | Tail | Side |

| Toss | 21 | 22 | 23 | 24 | 25 | 26 | 27 | 28 | 29 | 30 |
|---|---|---|---|---|---|---|---|---|---|---|
| Result | Side | Head | Side | Side | Head | Tail | Tail | Head | Head | Side |

15. Summarize the relative frequencies of heads so far by completing the following table:

*Sample table is provided using data from Exercise 14.*

| Number of tosses | Total number of heads so far | Relative frequency of heads so far (to the nearest hundredth) |
|---|---|---|
| 1 | 1 | 1.00 |
| 5 | 2 | 0.4 |
| 10 | 2 | 0.2 |
| 15 | 4 | 0.27 |
| 20 | 4 | 0.2 |
| 25 | 6 | 0.24 |
| 30 | 8 | 0.27 |

**16.** Based on the relative frequencies for the 30 tosses, what is your estimate of the probability of getting heads? Can you compare this estimate to a theoretical probability like you did in the first example? Explain your answer.

*Answers will vary. Students are anticipated to indicate an estimated probability equal or close to the last value in the relative frequency column. For this example, that would be $0.27$. An estimate of $0.25$ for this sample would have also been a good estimate. Students would indicate that they could not compare this to a theoretical probability, as the theoretical probability is not known for this example. Allow for a range of estimated probabilities. Factors that might affect the results for the long-run frequencies include how much tape is used to create the stack and how sturdy the stack is. Discussing these points with students is a good summary of this lesson.*

**17.** Create another stack of pennies. Consider creating a stack using 5 pennies, 15 pennies, or 20 pennies taped together in the same way you taped the pennies to form a stack of 10 pennies. Again, make sure the pennies are taped securely, or you might have a mess!

Toss the stack you made 30 times. Record the outcome for each toss:

| Toss | 1 | 2 | 3 | 4 | 5 | 6 | 7 | 8 | 9 | 10 |
|------|---|---|---|---|---|---|---|---|---|----|
| Result | | | | | | | | | | |

| Toss | 11 | 12 | 13 | 14 | 15 | 16 | 17 | 18 | 19 | 20 |
|------|----|----|----|----|----|----|----|----|----|----|
| Result | | | | | | | | | | |

| Toss | 21 | 22 | 23 | 24 | 25 | 26 | 27 | 28 | 29 | 30 |
|------|----|----|----|----|----|----|----|----|----|----|
| Result | | | | | | | | | | |

*The problem set involves another example of obtaining results from a stack of pennies. Suggestions include stacks of 5, 15, and 20 (or a number of your choice). The problem set includes questions based on the results from tossing one of these stacks. Provide students in small groups one of these stacks. Each group should collect data for 30 tosses to use for the problem set.*

## Closing (5 minutes)

When students finish collecting data for the problem set, ask the following:

- When you toss the stack and calculate a relative frequency, are you getting an estimated probability or a theoretical probability?
  - *You are getting an estimated probability.*
- Is there an exact number of times you should toss the stack to estimate the probability of getting heads?
  - *There is no exact number of times you should toss the stack; however, the larger the number of tosses, the closer the estimated probability will approach to the probability of the event.*

> **Lesson Summary**
>
> - Observing the long-run relative frequency of an event from a chance experiment (or the proportion of an event derived from a long sequence of observations) approximates the theoretical probability of the event.
>
> - After a long sequence of observations, the observed relative frequencies get close to the probability of the event occurring.
>
> - When it is not possible to compute the theoretical probabilities of chance experiments, then the long-run relative frequencies (or the proportion of events derived from a long sequence of observations) can be used as estimated probabilities of events.

**Exit Ticket (5–8 minutes)**

Name _____ Date_____

# Lesson 8: The Difference Between Theoretical Probabilities and Estimated Probabilities

Exit Ticket

1. Which of the following graphs would NOT represent the relative frequencies of heads when tossing 1 penny? Explain your answer.

**Graph A**                                    **Graph B**

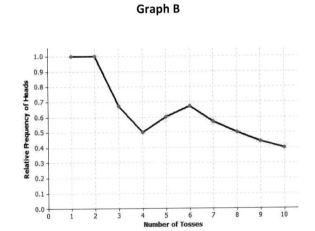

2. Jerry indicated that after tossing a penny 30 times, the relative frequency of heads was 0.47 (to the nearest hundredth). He indicated that after 31 times, the relative frequency of heads was 0.55. Are Jerry's summaries correct? Why or why not?

3. Jerry observed 5 heads in 100 tosses of his coin. Do you think this was a fair coin? Why or why not?

Exit Ticket Sample

1. Which of the following graphs would NOT represent the relative frequencies of heads when tossing 1 penny? Explain your answer.

Graph A

Graph B

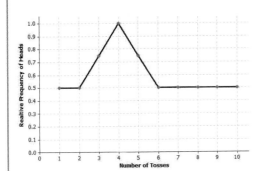

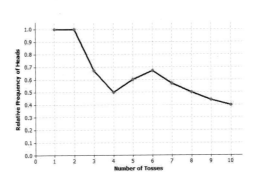

*Graph A would not represent a possible graph of the relative frequencies. Students could point out a couple of reasons. The first problem is the way Graph A starts. After the first toss, the probability would either be a 0 or a 1. Also, it seems to settle exactly to the theoretical probability without showing the slight changes from toss to toss.*

2. Jerry indicated that after tossing a penny 30 times, the relative frequency of heads was 0.47 (to the nearest hundredth). He indicated that after 31 times, the relative frequency of heads was 0.55. Are Jerry's summaries correct? Why or why not?

*Something is wrong with Jerry's information. If he tossed the penny 30 times, and the relative frequency of heads was 0.47, then he had 14 heads. If his next toss were heads, then the relative frequency would be $\frac{15}{31}$ or 0.48 (to the nearest hundredth). If his next toss were tails, then the relative frequency would be $\frac{14}{31}$ or 0.45 (to the nearest hundredth).*

3. Jerry observed 5 heads in 100 tosses of his coin? Do you think this was a fair coin? Why or why not?

*Students should indicate Jerry's coin is probably not a fair coin. The relative frequency of heads for a rather large number of tosses should be close to the theoretical probability. For this problem, the relative frequency of 0.05 is quite different from 0.5, and probably indicates that the coin is not fair.*

## Problem Set Sample Solutions

1.  If you created a stack of 15 pennies taped together, do you think the probability of getting a head on a toss of the stack would be different than for a stack of 10 pennies? Explain your answer.

    *The estimated probability of getting a head for a stack of 15 pennies would be different than for a stack of 10 pennies. A few tosses indicate that it is very unlikely that the outcome of heads or tails would result as the stack almost always lands on its side. (The possibility of a head or a tail is noted, but it has a small probability of being observed.)*

2.  If you created a stack of 20 pennies taped together, what do you think the probability of getting a head on a toss of the stack would be? Explain your answer.

    *The estimated probability of getting a head for a stack of 20 pennies is very small. The toss of a stack of this number of pennies almost always lands on its side. Students might indicate there is a possibility but with this example, the observed outcomes are almost all on their side.*

    *Note: If students selected a stack of 5 coins, the outcomes are nearly the same as if it was only 1 coin. The probability of landing on its side is small (close to 0). As more pennies are added, the probability of the stack landing on its side increases, until it is nearly $1.00$ (or $100\%$).*

3.  Based on your work in this lesson, complete the following table of the relative frequencies of heads for the stack you created:

    *Answers will vary based on the outcomes of tossing the stack. As previously stated, as more pennies are added to the stack, the probability that the stack will land on its side increases. Anticipate results of 0 for a stack of 20 pennies. Samples involving 15 pennies will have a very small probability of showing heads.*

    | Number of tosses | Total number of heads so far | Relative frequency of heads so far (to the nearest hundredth) |
    |---|---|---|
    | 1 | | |
    | 5 | | |
    | 10 | | |
    | 15 | | |
    | 20 | | |
    | 25 | | |
    | 30 | | |

4.  What is your estimate of the probability that your stack of pennies will land heads up when tossed? Explain your answer.

    *Answers will vary based on the relative frequencies.*

5.  Is there a theoretical probability you could use to compare to the estimated probability? Explain you answer.

    *There is no theoretical probability that could be calculated to compare to the estimated probability.*

# Lesson 9: Comparing Estimated Probabilities to Probabilities Predicted by a Model

## Student Outcomes

- Students compare estimated probabilities to those predicted by a probability model.

## Classwork

This lesson continues students' exploration in collecting data from chance experiments to estimate probabilities. Students will be presented a scenario of a pretend quiz show (Picking Blue!). A summary of the game is provided to students and should be read as a group before the activity of this lesson begins. The game is simple, but it is important that students understand its rules.

Prepare bags of chips, labeled as Bag A and Bag B, for students to use in this lesson. As a suggestion, make Bag A the bag with the same number of blue and red chips (e.g., five red and five blue), and Bag B the bag with a different number of blue to red (e.g., three blue and nine red). Students will ultimately decide which of two bags gives them a better chance of picking a blue chip. When the game begins, a chip is picked from one of the bags. If a blue chip is picked, students who selected that bag win a Blue Token. The chip is then placed back in the bag, and another chip is selected. If this chip is blue, they win another Blue Token, and the process continues; but, if the chip is red, the game moves to the other bag. If a chip picked from that bag is blue, the students who picked that bag win a Blue Token. The process continues until a red chip is selected that ends the game.

Students are provided a 20-minute research time. It is the research that allows students to estimate the probability of picking a blue chip and to use their estimated probability to decide which bag to select when they play the game. They are faced with two bags. They know the probability of picking a blue chip is 0.5 in one of the bags, although they do not know which bag it is. Nor do they know the composition of red and blue chips in the other bag. The process of pulling out a chip, recording its color, putting it back in the bag, and then repeating this process with each bag provides data to estimate the probability. Students examine the relative frequencies of picking blue from each bag and continue to make selections until they are confident that they have a good estimate of the probability of picking blue based on the relative frequencies.

The exploration is presented to students in stages. Questions are posed at each stage that offer opportunities to help students think about the process and the goals of this game.

## Exploratory Challenge (35 minutes)

---

**Game Show—Picking Blue!**

Imagine, for a moment, the following situation: You and your classmates are contestants on a quiz show called *Picking Blue!* There are two bags in front of you, Bag A and Bag B. Each bag contains red and blue chips. You are told that one of the bags has exactly the same number of blue chips as red chips. But you are told nothing about the number of blue and red chips in the second bag.

Each student in your class will be asked to select either Bag A or Bag B. Starting with Bag A, a chip is randomly selected from the bag. If a blue chip is drawn, all of the students in your class who selected Bag A win a Blue Token. The chip is put back in the bag. After mixing up the chips in the bag, another chip is randomly selected from the bag. If the chip is blue, the students who picked Bag A win another Blue Token. After the chip is placed back into the bag, the process continues until a red chip is picked. When a red chip is picked, the game moves to Bag B. A chip from the Bag B is then randomly selected. If it is blue, all of the students who selected Bag B win a Blue Token. But if the chip is red, the game is over. Just like for Bag A, if the chip is blue, the process repeats until a red chip is picked from the bag. When the game is over, the students with the greatest number of Blue Tokens are considered the winning team.

Without any information about the bags, you would probably select a bag simply by guessing. But surprisingly, the show's producers are going to allow you to do some research before you select a bag. For the next 20 minutes, you can pull a chip from either one of the two bags, look at the chip, and then put the chip back in the bag. You can repeat this process as many times as you want within the 20 minutes. At the end of 20 minutes, you must make your final decision and select which of the bags you want to use in the game.

**Getting Started**

Assume that the producers of the show do not want give away a lot of their Blue Tokens. As a result, if one bag has the same number of red and blue chips, do you think the other bag would have more, or fewer, blue chips than red chips? Explain your answer.

> *Anticipate answers that indicate the producers would likely want the second bag to have fewer blue chips. If a participant selects that bag, it would mean the participant is more likely to lose this game.*

---

**P.2** The question posed in this stage asks students to think about the number of blue and red chips in the bag that does not have the same number of blue and red chips. During the discussion of this question, ask students to consider their options. What if there were more blue chips than red chips? What if there was the same number of blue and red chips in this bag also? What if there were fewer blue chips than red chips? Is it likely the producers of the game would create a second bag in which the probability of picking a blue chip is less than 0.5? The purpose of this question, however, is to have students think about what they will find from their research.

**Planning the Research**

Your teacher will provide you with two bags labeled A and B.  You have 20 minutes to experiment with pulling chips one at a time from the bags.  After you examine a chip, you must put it back in the bag.  Remember, no fair peeking in the bags as that will disqualify you from the game.  You can pick chips from just one bag, or you can pick chips from one bag and then the other bag.

Use the results from 20 minutes of research to determine which bag you will choose for the game.

Provide a description outlining how you will carry out your research:

Organize the class in small groups.  Before class, prepare Bag A and Bag B for each group so that they can carry out their research.  As previously suggested, make Bag A the bag with same number of blue and red chips (e.g., five red and five blue), and Bag B the bag with a different number of blue to red (e.g., three blue and nine red).  These two bag compositions were tested, and at about 40 picks from each bag, the estimated probabilities were able to provide a good indication of the probability of picking a blue chip.  (Remember, however, there is still a possibility that you or the students may run into a situation that is not clear after 40 picks!)

Students are asked to present their plans for developing this research.  They are expected to outline the steps they will implement during the next 20 minutes to estimate the probability of picking blue for each bag.  They are expected to carry out their plans as a small group.  The plans provide you with an indication of whether or not students understand the game.

Students should organize the results of the data collection on their own.  Most students will likely summarize the outcomes in a table.  Provide students a ruler to assist in making a table.  A table was intentionally not provided in the student lesson so that you can see if they can organize the data and calculate the relative frequencies.  For students who struggle with this process, however, you are encouraged to review the tables used in the previous lesson or to provide them with a blank table similar to the table used in Lesson 8.

**Carrying Out the Research**

Share your plan with your teacher.  Your teacher will verify whether your plan is within the rules of the quiz show.  Approving your plan does not mean, however, that your teacher is indicating that your research method offers the most accurate way to determine which bag to select.  If your teacher approves your research, carry out your plan as outlined.  Record the results from your research, as directed by your teacher.

Students implement their plan in small groups.  Ask students what the data indicate about the probability of picking a blue chip.  If time permits, have students produce a brief report of their research that summarizes their findings.  Their report should include what choice they are making and why.  Allow students to explain to you their reports, or encourage students to briefly summarize their findings to the class.

> **Playing the Game**
>
> After the research has been conducted, the competition begins.  First, your teacher will shake up Bag A.  A chip is selected.  If the chip is blue, all students who selected Bag A win an imaginary Blue Token.  The chip is put back in the bag, and the process continues.  When a red chip is picked from Bag A, students selecting Bag A have completed the competition.  Your teacher will now shake up Bag B.  A chip is selected.  If it is blue, all students who selected Bag B win an imaginary Blue Token.  The process continues until a red chip is picked.  At that point, the game is over.
>
> How many Blue Tokens did you win?

After each group has obtained an estimated probability of picking blue from each bag, play the game.  Divide the class into two groups based on which bag they will pick when the game starts.  It is very possible the entire class will pick the same bag.  If that happens, then for the purposes of a wrap-up discussion, you select the other bag and tell students that when you pick a chip from that bag, you will get the Blue Token if a blue chip is selected.

Play the game a couple of times by picking a chip and determining if a Blue Token is awarded.  It is possible no tokens were given out if the first chip picked from each bag was red.  Play the game several times, so that students can discuss the possible outcomes of winning.

> **Examining Your Results**
>
> At the end of the game, your teacher will open the bags and reveal how many blue and red chips were in each bag.  Answer the following questions.  (*See:  Closing Questions*.)  After you have answered these questions, discuss them with your class.

After playing the game several times, open the bags and show students the contents.  Discuss whether the estimated probabilities provided an indication of the contents of the bags.  Expect students to comment that they were looking for an estimated probability of 0.5 and that the relative frequencies provided them with an indication of the bag with the same number of red and blue chips.

Direct students to read and answer the five questions provided in the student lesson as a summary of the process.  After 5 minutes of reading and writing their responses, discuss the questions as a group.  Use their answers as an indication of student outcomes. These questions serve as the Exit Ticket for this lesson.

## Closing (10 minutes)

> 1.  **Before you played the game, what were you trying to learn about the bags from your research?**
>
>     *Students should indicate they were trying to learn the estimated probability of picking a blue chip, without knowing the theoretical probability.*
>
> 2.  **What did you expect to happen when you pulled chips from the bag with the same number of blue and red chips?  Did the bag that you thought had the same number of blue and red chips yield the results you expected?**
>
>     *Students should indicate that they were looking for an estimated probability that was close to 0.5.  They should connect that estimated probability of picking blue with the bag that had the same number of red and blue chips.*
>
> 3.  **How confident were you in predicting which bag had the same number of blue and red chips?  Explain.**
>
>     *Answers will vary.  Students' confidence will be based on the data collected.  The more data collected, the closer the estimates are likely to be to the actual probabilities.*

4.  **What bag did you select to use in the competition and why?**

    *Answers will vary. It is anticipated that they would pick the bag with the larger estimated probability of picking a blue chip based on the many chips they selected during the research stage. Ask students to provide evidence of their choice. For example, "I picked 50 chips from Bag A, and 24 were blue and 26 were red. The results are very close to 0.5 probability of picking each color. I think this indicates that there were likely an equal number of each color in this bag."*

5.  **If you were the show's producers, how would make up the second bag? (Remember, one bag has the same number of red and blue chips.)**

    *Answers will vary. Allow students to speculate on a number of possible scenarios. Most students should indicate a second bag that had fewer blue chips, but some may speculate on a second bag that is nearly the same (to make the game more of a guessing game), or a bag with very few blue, (thus providing a clearer indication which bag had the same number of red and blue chips).*

6.  **If you picked a chip from Bag B 100 times and found that you picked each color exactly 50 times, would you know for sure that that bag was the one with equal numbers of each color?**

    *Students indicate that they are quite confident they have the right bag, as getting that result is likely to occur from the bag with equal numbers of each colored chips (or close to equal numbers). However, students should understand that even a bag with a different number of colored chips could have that outcome. But the probability of that happening from a bag with noticeably different numbers of chips is low but not impossible.*

---

Lesson Summary

- The long-run relative frequencies can be used as estimated probabilities of events.
- Collecting data on a game or chance experiment is one way to estimate the probability of an outcome.
- The more data collected on the outcomes from a game or chance experiment, the closer the estimates of the probabilities are likely to be the actual probabilities.

---

## Exit Ticket

As this was an exploratory lesson, the exit problem is incorporated into the closing questions of Examining Your Results.

## Problem Set Sample Solutions

Jerry and Michael played a game similar to *Picking Blue!*. The following results are from their research using the same two bags:

Jerry's research:

| | Number of Red chips picked | Number of Blue chips picked |
|---|---|---|
| Bag A | 2 | 8 |
| Bag B | 3 | 7 |

Michael's research:

| | Number of Red chips picked | Number of Blue chips picked |
|---|---|---|
| Bag A | 28 | 12 |
| Bag B | 22 | 18 |

1. If all you knew about the bags were the results of Jerry's research, which bag would you select for the game? Explain your answer.

   *Using only Jerry's research, the greater relative frequency of picking a blue chip would be Bag A, or 0.8. There were 10 selections. Eight of the selections resulted in picking blue.*

2. If all you knew about the bags were the results of Michael's research, which bag would you select for the game? Explain your answer.

   *Using Michael's research, the greater relative frequency of picking a blue chip would be Bag B, or 0.45. There were 40 selections. Eighteen of the selections resulted in picking blue.*

3. Does Jerry's research or Michael's research give you a better indication of the make-up of the blue and red chips in each bag? Explain why you selected this research.

   *Michael's research would provide a better indication of the probability of picking a blue chip as it was carried out 40 times compared to 10 times for Jerry's research. The more outcomes that are carried out, the closer the relative frequencies approach the estimated probability of picking a blue chip.*

4. Assume there are 12 chips in each bag. Use either Jerry's or Michael's research to estimate the number of red and blue chips in each bag. Then explain how you made your estimates.

   Bag A

   Number of red chips:

   Number of blue chips:

   Bag B

   Number of red chips:

   Number of blue chips:

   *Answers will vary. Anticipate that students see Bag B as the bag with nearly the same number of blue and red chips, and Bag A as possibly having a third of the chips blue. Answers provided by students should be based on the relative frequencies. A sample answer based on this reasoning is*

   *Bag A:      4 blue chips and 8 red chips.*

   *Bag B:      6 blue chips and 6 red chips.*

5. In a different game of *Picking Blue!*, two bags each contain red, blue, green, and yellow chips. One bag contains the same number of red, blue, green, and yellow chips. In the second bag, half the chips are blue. Describe a plan for determining which bag has more blue chips than any of the other colors.

   *Students should describe a plan similar to the plan implemented in the lesson. Students would collect data by selecting chips from each bag. After several selections, the estimated probabilities of selecting blue from each bag would suggest which bag has more blue chips than chips of the other colors.*

**Lesson 10: Using Simulation to Estimate a Probability**

## Student Outcomes

- Students learn simulation as a method for estimating probabilities that can be used for problems in which it is difficult to collect data by experimentation or by developing theoretical probability models.
- Students learn how to perform simulations to estimate probabilities.
- Students use various devices to perform simulations (e.g., coin, number cube, cards).
- Students compare estimated probabilities from simulations to theoretical probabilities.

## Lesson Overview

Simulation uses devices such as coins, number cubes, and cards to generate outcomes that represent real outcomes. Students may find it difficult to make the connection between device outcomes and the real outcomes of the experiment.

This lesson eases students into simulation by defining the real outcomes of an experiment and specifying a device to simulate the outcome. Then we will carefully lead students to define how an outcome of the device represents the real outcome, define a trial for the simulation, and identify what is meant by a trial resulting in a "success" or "failure." Be sure that students see how a device that may have many outcomes can be used to simulate a situation that has only two outcomes. For example, how a number cube can be used to represent a boy birth (e.g., even outcome, prime number outcome, or any three of its digits.)

## Classwork

In previous lessons, you estimated probabilities of events by collecting data empirically or by establishing a theoretical probability model. There are real problems for which those methods may be difficult or not practical to use. Simulation is a procedure that will allow you to answer questions about real problems by running experiments that closely resemble the real situation.

It is often important to know the probabilities of real-life events that may not have known theoretical probabilities. Scientists, engineers, and mathematicians design simulations to answer questions that involve topics such as diseases, water flow, climate changes, or functions of an engine. Results from the simulations are used to estimate probabilities that help researchers understand problems and provide possible solutions to these problems.

## Example 1 (10 minutes): Families

This first example begins with an equally likely model that simulates boy and girl births in a family of three children. Note that in human populations, the probabilities of a boy birth and of a girl birth are not actually equal, but we treat them as equal here. The example uses a coin in the simulation.

There are five steps in the simulation:

- The first is to define the basic outcome of the real experiment, e.g., a birth.

- The second is to choose a device and define which possible outcomes of the device will represent an outcome of the real experiment, (e.g., toss of a coin, head represents boy; roll of a number cube, prime number (P) represents boy; choice of a card, black card represents boy.)

- The third is to define what is meant by a trial in the simulation that represents an outcome in the real experiment (e.g., three tosses of the coin represents three births; three rolls of a number cube represents three births; three cards chosen with replacement represents three births.)

- The fourth is to define what is meant by a success in the performance of a trial (e.g., using a coin, HHT represents exactly two boys in a family of three children; using a number cube, NPP represents exactly two boys in a family of three children; using cards, BRB represents exactly two boys in a family of three children.) Be sure that your students realize that in using a coin, HHT, HTH, and THH all represent exactly two boys in a family of three children whereas HHH is the only way to represent three boys in a family of three children.

- The fifth step is to perform $n$ trials (the more the better), count the number of successes in the $n$ trials, and divide the number of successes by $n$, which produces the estimate of the probability based on the simulation.

You may find it useful to reiterate the five steps for every problem in this lesson so that your students gain complete understanding of the simulation procedure.

---

**Example 1:  Families**

How likely is it that a family with three children has all boys or all girls?

Let's assume that a child is equally likely to be a boy or a girl. Instead of observing the result of actual births, a toss of a fair coin could be used to simulate a birth. If the toss results in heads (H), then we could say a boy was born; if the toss results in tails (T), then we could say a girl was born. If the coin is fair (i.e., heads and tails are equally likely), then getting a boy or a girl is equally likely.

---

Pose the following questions to the class one at a time and allow for multiple responses:

- How could a number cube be used to simulate getting a boy or a girl birth?
  - *An even-number outcome represents boy, and an odd-number outcome represents girl; a prime-number outcome represents boy, and a non-prime outcome represents girl; or, any three-number cube digits can be chosen to represent boy with the rest of them, girl.*

- How could a deck of cards be used to simulate getting a boy or a girl birth?
  - *The most natural one would be black card for one gender, and red for the other.*

## Exercises 1–2 (5 minutes)

> **Exercises 1–2**
>
> Suppose that a family has three children. To simulate the genders of the three children, the coin or number cube or a card would need to be used three times, once for each child. For example, three tosses of the coin resulted in HHT, representing a family with two boys and one girl. Note that HTH and THH also represent two boys and one girl.

Let students work with a partner. Then discuss and confirm answers as a class.

> 1. Suppose a prime number (P) result of a rolled number cube simulates a boy birth, and a non-prime (N) simulates a girl birth. Using such a number cube, list the outcomes that would simulate a boy birth, and those that simulate a girl birth. Are the boy and girl birth outcomes equally likely?
>
>    *The outcomes are* 2, 3, 5 *for a boy birth and* 1, 4, 6 *for a girl birth. The boy and girl births are thereby equally likely.*
>
> 2. Suppose that one card is drawn from a regular deck of cards, a red card (R) simulates a boy birth and a black card (B) simulates a girl birth. Describe how a family of three children could be simulated.
>
>    *The key response has to include the drawing of three cards with replacement. If a card is not replaced and the deck shuffled before the next card is drawn, then the probabilities of the genders have changed (ever so slightly, but they are not* 50 − 50 *from draw to draw). Simulating the genders of three children requires three cards to be drawn with replacement.*

## Example 2 (5 minutes)

> **Example 2**
>
> Simulation provides an estimate for the probability that a family of three children would have three boys or three girls by performing three tosses of a fair coin many times. Each sequence of three tosses is called a trial. If a trial results in either HHH or TTT, then the trial represents all boys or all girls, which is the event that we are interested in. These trials would be called a "success." If a trial results in any other order of H's and T's, then it is called a "failure."
>
> The estimate for the probability that a family has either three boys or three girls based on the simulation is the number of successes divided by the number of trials. Suppose 100 trials are performed, and that in those 100 trials, 28 resulted in either HHH or TTT. Then the estimated probability that a family of three children has either three boys or three girls would be $\frac{28}{100} = 0.28$.

This example describes what is meant by a trial, a success, and how to estimate the probability of the desired event (i.e., that a family has three boys or three girls). It uses the coin device, 100 trials, in which 28 of them were either HHH or TTT. Hence the estimated probability that a family of three children has three boys or three girls is $\frac{28}{100}$. Then ask:

- What is the estimated probability that the three children are not all the same gender?
    - $1 − 0.28 = 0.72$.

## Exercises 3–5 (15 minutes)

Let students continue to work with their partners on Exercises 3–5.  Discuss the answer to Exercise 5 (c) as a class.

---

**Exercises 3–5**

3.  Find an estimate of the probability that a family with three children will have exactly one girl using the following outcomes of 50 trials of tossing a fair coin three times per trial.  Use H to represent a boy birth, and T to represent a girl birth.

| | | | | | | | | | |
|---|---|---|---|---|---|---|---|---|---|
| HHT | HTH | HHH | TTH | THT | THT | HTT | HHH | TTH | HHH |
| HHT | TTT | HHH | TTH | HHH | HTH | THH | TTT | THT | THT |
| THT | HHH | THH | HTT | HTH | TTT | HTT | HHH | TTH | THT |
| THH | HHT | TTT | TTH | HTT | THH | HTT | HTH | TTT | HHH |
| HTH | HTH | THT | TTH | TTT | HHT | HHT | THT | TTT | HTT |

*T represents a girl.  Students should go through the list, count the number of times that they find HHT, HTH, or THH and divide that number of successes by 50.  They should find the simulated probability to be $\frac{16}{50} = 0.32$. If time permits, ask students, "Based on what you found, is it likely that a family with three children will have exactly one girl? From your own experiences, how many families do you know who have three children and exactly one girl?"*

4.  Perform a simulation of 50 trials by rolling a fair number cube in order to find an estimate of the probability that a family with three children will have exactly one girl.

   a.   Specify what outcomes of one roll of a fair number cube will represent a boy and what outcomes will represent a girl.

   b.   Simulate 50 trials, keeping in mind that one trial requires three rolls of the number cube.  List the results of your 50 trials.

   c.   Calculate the estimated probability.

   *Answers will vary. For example, they could identify a girl birth as 1, 2, 3 outcome on one roll of the number cube, and roll the number cube three times to simulate three children (one trial).  They need to list their 50 trials. Note that an outcome of 412 would represent two girls, 123 would represent three girls, and 366 would represent one girl, as would 636 and 663. Be sure that they are clear about how to do all five steps of the simulation process.*

5.  Calculate the theoretical probability that a family with three children will have exactly one girl.

   a.   List the possible outcomes for a family with three children.  For example, one possible outcome is BBB (all three children are boys).

   *The sample space is:  BBB, BBG, BGB, GBB, BGG, GBG, GGB, GGG.*

   b.   Assume that having a boy and having a girl are equally likely.  Calculate the theoretical probability that a family with three children will have exactly one girl.

   *Each is equally likely, so the theoretical probability of getting exactly one girl is $\frac{3}{8} = 0.375$ (BBG, BGB, GBB.)*

   c.   Compare it to the estimated probabilities found in parts (a) and (b) above.

   *Answers will vary; the estimated probabilities from the first two parts of this exercise should be around 0.375.  If not, you may suggest that they conduct more trials.*

---

P.4

## Example 3 (5 minutes):  Basketball Player

**Example 3:  Basketball Player**

Suppose that, on average, a basketball player makes about three out of every four foul shots.  In other words, she has a 75% chance of making each foul shot she takes.  Since a coin toss produces equally likely outcomes, it could not be used in a simulation for this problem.

Instead, a number cube could be used by specifying that the numbers 1, 2, or 3 represent a hit, the number 4 represents a miss, and the numbers 5 and 6 would be ignored.  Based on the following 50 trials of rolling a fair number cube, find an estimate of the probability that she makes five or six of the six foul shots she takes.

| | | | | |
|---|---|---|---|---|
| 441323 | 342124 | 442123 | 422313 | 441243 |
| 124144 | 333434 | 243122 | 232323 | 224341 |
| 121411 | 321341 | 111422 | 114232 | 414411 |
| 344221 | 222442 | 343123 | 122111 | 322131 |
| 131224 | 213344 | 321241 | 311214 | 241131 |
| 143143 | 243224 | 323443 | 324243 | 214322 |
| 214411 | 423221 | 311423 | 142141 | 411312 |
| 343214 | 123131 | 242124 | 141132 | 343122 |
| 121142 | 321442 | 121423 | 443431 | 214433 |
| 331113 | 311313 | 211411 | 433434 | 323314 |

Present the beginning of the example.  This example has a real outcome probability of 0.75.  Ask students:

- What device could be used to generate a probability of 3 out of 4?
    - *If they have studied Platonic solids, they may suggest a tetrahedron.  Note that the outcome would be the result that is face down.*

Continue reading through the text of the example as a class.  Explain how a number cube could be used in which a 1, 2, or 3 would represent a hit, the number 4 would represent a miss, and 5 and 6 would be ignored.  The simulation is designed to fit the probabilities of 25% and 75%. Outcomes of 1, 2, 3, and 4 provide those probabilities.  Adding the outcomes of 5 or 6 would change the probabilities; therefore, they are ignored (or simply not counted as outcomes).

Ask students:

- Is there another way to assign the numbers on the cube to the outcomes?
    - *Yes,* 1 *could represent a miss and* 2, 3, 4 *could represent a hit.*

50 trials of six numbers each are shown.  Students are to estimate the probability that the player makes five or six of the six shots she takes.  So, they should count how many of the trials have five or six of the numbers 1, 2, 3 in them as successes.  They should find the estimated probability to be $\frac{27}{50} = 0.54$.

As an aside, the theoretical probability is calculated by considering the possible outcomes for six foul shots that have either 6 successes (SSSSSS) or 5 successes (FSSSSS, SFSSSS, SSFSSS, SSSFSS, SSSSFS, and SSSSSF).  The outcome that consists of six successes has probability $(0.75)^6$ and each of the 6 outcomes with five successes has probability $(0.75)^5(0.25)$.  Based on the binominal probability distribution, the theoretical probability is $6(0.75)^5(0.25) + (0.75)^6$, or approximately 0.5339.

## Closing (5 minutes)

> **Lesson Summary**
>
> In previous lessons, you estimated probabilities by collecting data and found theoretical probabilities by creating a model. In this lesson you used simulation to estimate probabilities in real problems and in situations for which empirical or theoretical procedures are not easily calculated.
>
> Simulation is a method that uses an artificial process (like tossing a coin or rolling a number cube) to represent the outcomes of a real process that provides information about the probability of events. In several cases, simulations are needed to both understand the process as well as provide estimated probabilities.

## Exit Ticket (5–10 minutes)

Name _____   Date_____

# Lesson 10: Using Simulation to Estimate a Probability

1. Nathan is your school's star soccer player. When he takes a shot on goal, he scores half of the time on average. Suppose that he takes six shots in a game. To estimate probabilities of the number of goals Nathan makes, use simulation with a number cube. One roll of a number cube represents one shot.

   a. Specify what outcome of a number cube you want to represent a goal scored by Nathan in one shot.

   b. For this problem, what represents a trial of taking six shots?

   c. Perform and list the results of ten trials of this simulation.

   d. Identify the number of goals Nathan made in each of the ten trials you did in part (c).

   e. Based on your ten trials, what is your estimate of the probability that Nathan scores three goals if he takes six shots in a game?

2. Suppose that Pat scores 40% of the shots he takes in a soccer game. If he takes six shots in a game, what would one simulated trial look like using a number cube in your simulation?

Exit Ticket Sample Solutions

---

1.  Nathan is your school's star soccer player. When he takes a shot on goal, he scores half of the time on average. Suppose that he takes six shots in a game. To estimate probabilities of the number of goals Nathan makes, use simulation with a number cube. One roll of a number cube represents one shot.

    a.  Specify what outcome of a number cube you want to represent a goal scored by Nathan in one shot.

        *Answers will vary; students need to determine which three numbers on the number cube represent scoring a goal.*

    b.  For this problem, what represents a trial of taking six shots?

        *Rolling the cube six times represents taking six shots on goal or 1 simulated trial.*

    c.  Perform and list the results of ten trials of this simulation.

        *Answers will vary; students working in pairs works well for these problems. Performing only ten trials is a function of time. Ideally many more trials should be done. If there is time, have your class pool their results.*

    d.  Identify the number of goals Nathan made in each of the ten trials you did in part (c).

        *Answers will vary.*

    e.  Based on your ten trials, what is your estimate of the probability that Nathan scores three goals if he takes six shots in a game?

        *Answers will vary; the probability of scoring per shot is $\frac{1}{2}$.*

2.  Suppose that Pat scores $40\%$ of the shots he takes in a soccer game. If he takes six shots in a game, what would one simulated trial look like using a number cube in your simulation?

    *Students need to realize that $40\%$ is 2 out of 5. In order to use the number cube as the device, 1 and 2 could represent goals, while 3, 4, and 5 could represent missed shots, and 6 is ignored. Rolling the number cube six times creates 1 simulated trial.*

## Problem Set Sample Solutions

1.  A mouse is placed at the start of the maze shown below.  If it reaches station B, it is given a reward.  At each point where the mouse has to decide which direction to go, assume that it is equally likely to go in either direction.  At each decision point $1, 2, 3$, it must decide whether to go left (L) or right (R).  It cannot go backwards.

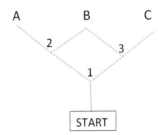

a.  Create a theoretical model of probabilities for the mouse to arrive at terminal points A, B, and C.

   i.  List the possible paths of a sample space for the paths the mouse can take.  For example, if the mouse goes left at decision point 1, and then right at decision point 2, then the path would be denoted LR.

   *The possible paths in the sample space are {LL, LR, RL, RR}.*

   ii.  Are the paths in your sample space equally likely?  Explain.

   *Each of these outcomes has an equal probability of $\frac{1}{4}$, since at each decision point there are only two possible choices, which are equally likely.*

   iii.  What are the theoretical probabilities that a mouse reaches terminal points A, B, and C?  Explain.

   *The probability of reaching terminal point A is $\frac{1}{4}$, since it is accomplished by path LL.  Similarly, reaching terminal point C is $\frac{1}{4}$, since it is found by path RR.  However, reaching terminal point B is $\frac{1}{2}$, since it is reached via LR or RL.*

b.  Based on the following set of simulated paths, estimate the probabilities that the mouse arrives at points A, B, and C.

| RR | RR | RL | LL | LR | RL | LR | LL | LR | RR |
|----|----|----|----|----|----|----|----|----|----|
| LR | RL | LR | RR | RL | LR | RR | LL | RL | RL |
| LL | LR | LR | LL | RR | RR | RL | LL | RR | LR |
| RR | LR | RR | LR | LR | LL | LR | RL | RL | LL |

*Students need to go through the list and count the number of paths that go to A, B, and C.  They should find the estimated probabilities to be $\frac{8}{40} = 0.2$ for A, $\frac{22}{40} = 0.55$ for B, and $\frac{10}{40} = 0.25$ for C.*

c.  How do the simulated probabilities in part (b) compare to the theoretical probabilities of part (a)?

*The probabilities are reasonably close for parts (a) and (b).  Students should realize that probabilities based on taking 400 trials should be closer than those based on 40, but that the probabilities based on 40 are in the ballpark.*

2.  Suppose that a dartboard is made up of the $8 \times 8$ grid of squares shown below.  Also, suppose that when a dart is thrown, it is equally likely to land on any one of the 64 squares.  A point is won if the dart lands on one of the 16 black squares.  Zero points are earned if the dart lands in a white square.

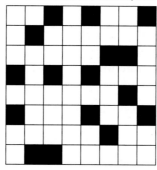

a.  For one throw of a dart, what is the probability of winning a point?  Note that a point is won if the dart lands on a black square.

*The probability of winning a point is $\frac{16}{64} = 0.25$.*

b.  Lin wants to use a number cube to simulate the result of one dart.  She suggests that 1 on the number cube could represent a win.  Getting 2, 3, or 4 could represent no point scored.  She says that she would ignore getting a 5 or 6.  Is Lin's suggestion for a simulation appropriate?  Explain why you would use it *or*, if not, how you would change it.

*Lin correctly suggests that to simulate the result of one throw, a number cube could be used with the 1 representing a hit, 2, 3, 4 representing a missed throw, while ignoring 5 and 6. (As an aside, a tetrahedron could be used by using the side facing down as the result.)*

c.  Suppose a game consists of throwing a dart three times.  A trial consists of three rolls of the number cube.  Based on Lin's suggestion in part (b) and the following simulated rolls, estimate the probability of scoring two points in three darts.

| | | | | |
|---|---|---|---|---|
| 324 | 332 | 411 | 322 | 124 |
| 224 | 221 | 241 | 111 | 223 |
| 321 | 332 | 112 | 433 | 412 |
| 443 | 322 | 424 | 412 | 433 |
| 144 | 322 | 421 | 414 | 111 |
| 242 | 244 | 222 | 331 | 224 |
| 113 | 223 | 333 | 414 | 212 |
| 431 | 233 | 314 | 212 | 241 |
| 421 | 222 | 222 | 112 | 113 |
| 212 | 413 | 341 | 442 | 324 |

*The probability of scoring two points in three darts is $\frac{5}{50} = 0.1$. (Students need to count the number of trials that contain exactly two 1's.)*

d.  The theoretical probability model for winning 0, 1, 2, and 3 points in three throws of the dart as described in this problem is

i.   winning 0 points has a probability of $0.42$;

ii.  winning 1 point has a probability of $0.42$;

iii. winning 2 points has a probability of $0.14$;

iv.  winning 3 points has a probability of $0.02$.

Use the simulated rolls in part (c) to build a model of winning $0, 1, 2,$ and $3$ points, and compare it to the theoretical model.

*To find the estimated probability of 0 points, count the number of trials that have no 1's in them ($\frac{23}{50} = 0.46$).*

*To find the estimated probability of 1 point, count the number of trials that have one 1 in them ($\frac{20}{50} = 0.4$).*

*From part (c), the estimated probability of 2 points is $0.1$.*

*To find the estimated probability of 3 points, count the number of trials that have three 1's in them ($\frac{2}{50} = 0.04$.)*

*The theoretical and simulated probabilities are reasonably close.*

|            | 0    | 1    | 2    | 3    |
|------------|------|------|------|------|
| Theoretical | 0.42 | 0.42 | 0.14 | 0.02 |
| Simulated   | 0.46 | 0.40 | 0.10 | 0.04 |

**Lesson 11:  Using Simulation to Estimate a Probability**

## Student Outcomes

- Students design their own simulations.
- Students learn to use two more devices in simulations:  colored disks and a random number table.

## Classwork

### Example 1 (5 minutes):  Simulation

> **Example 1:  Simulation**
>
> In the last lesson, we used coins, number cubes, and cards to carry out simulations.  Another option is putting identical pieces of paper or colored disks into a container, mixing them thoroughly, and then choosing one.
>
> For example, if a basketball player typically makes five out of eight foul shots, then a colored disk could be used to simulate a foul shot.  A green disk could represent a made shot, and a red disk could represent a miss.  You could put five green and three red disks in a container, mix them, and then choose one to represent a foul shot.  If the color of the disk is green, then the shot is made.  If the color of the disk is red, then the shot is missed.  This procedure simulates one foul shot.

Ask your students what device they would use to simulate problems in which the probability of winning in a single outcome is $\frac{5}{8}$.  A coin or number cube will not work.  A deck of eight cards (with five of the cards designated as winners) would work but shuffling cards between draws can be time-consuming and difficult for many students.  Suggest a new device:  colored disks in which five green disks could represent a win and three red disks could represent a miss.  Put the eight disks in a bag, shake the bag, and choose a disk.  Do this as many times as are needed to comprise a trial, and then do as many trials as needed to carry out the simulation.  Students could also create their own spinner with eight sections, with three sections colored one color, and five sections a different color, to represent the two different outcomes.

### Exercises 1–2 (3–5 minutes)

Let students work on Exercises 1 and 2 independently.  Then discuss and confirm as a class.

> **Exercises 1–2**
>
> 1.  Using colored disks, describe how one at-bat could be simulated for a baseball player who has a batting average of $0.300$.  Note that a batting average of $0.300$ means the player gets a hit (on average) three times out of every ten times at bat.  Be sure to state clearly what a color represents.
>
>     *Put ten disks in a bag, three of which are green (representing a hit), and seven are red (representing a non-hit).*
>
> 2.  Using colored disks, describe how one at-bat could be simulated for a player who has a batting average of $0.273$.  Note that a batting average of $0.273$ means that on average, the player gets $273$ hits out of $1000$ at-bats.
>
>     *Put $1,000$ disks in a bag, $273$ green ones (hit), and $727$ red ones (non-hit).*

### Example 2 (5–7 minutes):  Using Random Number Tables

---

**Example 2:  Using Random Number Tables**

Why is using colored disks not practical for the situation described in Exercise 2?  Another way to carry out a simulation is to use a random-number table, or a random-number generator.  In a random-number table, the digits 0, 1, 2, 3, 4, 5, 6, 7, 8, and 9 occur equally often in the long run.  Pages and pages of random numbers can be found online.

For example, here are three lines of random numbers.  The space after every five digits is only for ease of reading.  Ignore the spaces when using the table.

25256  65205  72597  00562  12683  90674  78923  96568  32177  33855

76635  92290  88864  72794  14333  79019  05943  77510  74051  87238

07895  86481  94036  12749  24005  80718  13144  66934  54730  77140

To use the random-number table to simulate an at-bat for the $0.273$ hitter in Exercise 2, you could use a three-digit number to represent one at bat.  The three-digit numbers from 000–272 could represent a hit, and the three-digit numbers from 273–999 could represent a non-hit.  Using the random numbers above, and starting at the beginning of the first line, the first three-digit random number is 252, which is between 000 and 272, so that simulated at-bat is a hit.  The next three-digit random number is 566, which is a non-hit.

---

Begin the example by posing the question to the class.  Allow for multiple responses.

- Why would using colored disks to simulate a probability of $0.273$ for a win not be practical?

Explain random-number tables and how they work.  Be sure to clarify the following to students:

- Note that in a random-number table, the gaps between every five digits do not mean anything. They are there to help you keep track of where you are.  So, ignore the gaps when generating numbers as though they are not present.

- Continue to the next line as though the lines were contiguous.  Also, in a large page of random digits, start by closing your eyes and placing your finger on the page.  Then, begin generating random numbers from there.

- To help students interpret these numbers, try starting with just single-digit outcomes.  For example, ask students how they could use the digits to simulate a fair coin toss.  Use a set of random numbers to illustrate that an even digit could be "heads," and an odd digit could be "tails."  Another example might be how to simulate the toss of a 6-sided number cube.  Digits 1 to 6 would represent the outcome of a toss, while any other digits are ignored.  If necessary, develop further with students who are still unclear with this process, or create a few sample coin or number-cube questions.

Additional considerations:

- Note that to have 1000 three-digit numbers, 000 needs to be included.  You need to decide what to do with 000.  It's perhaps more natural to use 001–273 to represent 273 numbers for a hit, while putting 000 with 274–999 to represent a non-hit.  The example suggests using the contiguous numbers 000–272 to represent a hit and 273–999 for a non-hit.  Either way is fine; choose whichever way your students feel more comfortable using.

- The zero difficulty is eliminated if you have a calculator that has a random number key on it. This key typically allows you to input the bounds between which you want a random number, e.g., rand(1,1000). It also allows you to specify how many random numbers you want to generate, e.g., if 50 you would use rand(1,1000,50), which is very convenient. Use of the calculator might be considered as an extension of the following exercises.

If you think your students might have difficulty jumping into Exercise 3, consider providing more practice with the random-number table by investigating the situation in which two-digit numbers would be used. For example, if a player has a batting average of 0.300, you could generate two-digit numbers from the random table, specifically 00–99, where a selection of 00–29 would be considered a hit and 30–99 as a non-hit. (Point out to students that 01–100 would not work, because 100 is a three-digit number.) When students are comfortable using two-digit numbers, then you can move to Exercise 3, where they use three-digit numbers to carry out the simulation.

### Exercise 3 (5 minutes)

Give students an opportunity to try using a random-number table on their own. Let students work in pairs to identify the outcomes for the next six at-bats. Then confirm as a class.

---

**Exercise 3**

3.  Continuing on the first line of the random numbers above, what would the hit/non-hit outcomes be for the next six at-bats? Be sure to state the random number and whether it simulates a hit or non-hit.

    *The numbers are 520 (non-hit), 572 (non-hit), 597 (non-hit), 005 (hit), 621 (non-hit), and 268 (hit).*

---

### Example 3 (10–12 minutes): Baseball Player

Work through this example as a class. Students use the random-number table to simulate the probability that a hitter with a 0.273 batting average gets three or four hits in four at-bats. Their simulation should be based on 12 trials.

Pose each question one at a time, and allow for multiple responses.

---

**Example 3: Baseball Player**

A batter typically gets to bat four times in a ballgame. Consider the $0.273$ hitter of the previous example. Use the following steps (and the random numbers shown above) to estimate that player's probability of getting at least three hits (three or four) in four times at-bat.

a.  Describe what one trial is for this problem.

    *A trial consists of four three-digit numbers. For the first trial, 252, 566, 520, 572 constitute one trial.*

b.  Describe when a trial is called a success and when it is called a failure.

    *A success is getting 3 or 4 hits per game; a failure is getting 0, 1, or 2 hits. For the first trial, the hitter got only 1 hit, so it would be a failure.*

c.  Simulate 12 trials. (Continue to work as a class, or let students work with a partner.)

---

MP.5

d.    Use the results of the simulation to estimate the probability that a $0.273$ hitter gets three or four hits in four times at-bat.  Compare your estimate with other groups.

*As a side note, the theoretical probability is calculated by considering the possible outcomes for four at bats that have either 4 hits (HHHH) or 3 hits (HHHM, HHMH, HMHH, and MHHH).  The outcome that consists of 4 hits has probability $(0.273)^4$, and each of the 4 outcomes with 3 hits has probability $(0.273)^3(0.727)$.  The theoretical probability is approximately $0.067$.*

## Example 4 (5 minutes):  Birth Month

**Example 4:  Birth Month**

In a group of more than 12 people, is it likely that at least two people, maybe more, will have the same birth-month?  Why?  Try it in your class.

Now suppose that the same question is asked for a group of only seven people.  Are you likely to find some groups of seven people in which there is a match, but other groups in which all seven people have different birth-months?  In the following exercise, you will estimate the probability that at least two people in a group of seven, were born in the same month.

Begin this example by posing the questions in the text to the class:

- If there are more than 12 people in a group, will at least two people have the same birth month?
    - *Yes*
- Why?
    - *There are 12 months and there are more than 12 people in the group.  So even if the first 12 people in the group have different birth months, the $13^{th}$ member will have to share with someone else.*
- What is the probability this will occur?
    - *The probability is 1.*

Introduce the next scenario where there are only seven people in the group.  Pose the following question to the class:

- Is it clear that there could be some groups of seven people in which there is a match, but other groups in which no one shares the same birth-month?
    - *Yes, it is possible for all seven people to be born in different months.*

Note to teacher:  There is a famous birthday problem that asks:  "What is the probability of finding (at least one) birthday match in a group of $n$ people?"  The surprising result is that there is a 50-50 chance of finding at least one birthday match in as few as 23 people.  Simulating birthdays is a bit time-consuming, so this problem simulates birth-months.

## Exercises 4–7 (10–15 minutes)

This exercise asks students to estimate the probability of finding at least two people with the same birth month in a group of 7 people but does not have students actually carry out the simulation.  If time allows, you could have each student carry out the simulation 20 times, and then either compare or pool their results.

Let students work on the exercises in small groups.  Then, let each group share its design.

**Exercises 4–7**

4.  What might be a good way to generate outcomes for the birth-month problem—using coins, number cubes, cards, spinners, colored disks, or random numbers?

    *Answers will vary; keep in mind that the first thing to do is specify how a birth-month for one person is going to be simulated.  For example, a dodecahedron is a 12-sided solid.  Each of its sides would represent one month.*

    *The following will not work:  coin (only two outcomes), number cube (only six outcomes and need 12.)*

    *The following devices will work:  cards (could label twelve cards January through December), spinners (could make a spinner with twelve equal sectors), colored disks (would need 12 different colors and then remember which color represents which month), 12 disks would work if you could write the name of a month on them, random number table (two-digit numbers 01, 02, …, 12 would work, but 00, 13, through 99 would have to discarded, very laborious).*

5.  How would you simulate one trial of seven birth-months?

    *Answers will vary; suppose students decide to use disks with the names of the months printed on them.  To generate a trial, put the 12 disks in a bag.  Then, shake the bag and choose a disk to represent the first person's birthday.  Then, replace the disk and do the process six more times.  The list of seven birth-months generates a trial.*

6.  How is a success determined for your simulation?

    *A success would be if there is at least one match in the seven.*

7.  How is the simulated estimate determined for the probability that a least two in a group of seven people were born in the same month?

    *Repeat this n times, count the number of successes, and divide it by n to get the estimated probability of having at least one birth-month match in a group of seven people.*

## Closing (5 minutes)

Discuss with students the Lesson Summary.

---

**Lesson Summary**

In the previous lesson, you carried out simulations to estimate a probability.  In this lesson, you had to provide parts of a simulation design.  You also learned how random numbers could be used to carry out a simulation.

To design a simulation:

- Identify the possible outcomes and decide how to simulate them, using coins, number cubes, cards, spinners, colored disks, or random numbers.

- Specify what a trial for the simulation will look like and what a success and a failure would mean.

- Make sure you carry out enough trials to ensure that the estimated probability gets closer to the actual probability as you do more trials.  There is no need for a specific number of trials at this time; however, you want to make sure to carry out enough trials so that the relative frequencies level off.

---

## Exit Ticket (5–7 minutes)

Name _____     Date_____

# Lesson 11: Using a Simulation to Estimate a Probability

Exit Ticket

Liang wants to form a chess club. His principal says that he can do that if Liang can find six players, including him. How would you conduct a simulated model that estimates the probability that Liang will find at least five other players to join the club if he asks eight players who have a 70% chance of agreeing to join the club? Suggest a simulation model for Liang by describing how you would do the following parts.

a. Specify the device you want to use to simulate one person being asked.

b. What outcome(s) of the device would represent the person agreeing to be a member?

c. What constitutes a trial using your device in this problem?

d. What constitutes a success using your device in this problem?

e. Based on 50 trials, using the method you have suggested, how would you calculate the estimate for the probability that Liang will be able to form a chess club?

## Exit Ticket Sample Solutions

> Liang wants to form a chess club. His principal says that he can do that if Liang can find six players, including him. How would you conduct a simulated model that estimates the probability that Liang will find at least five other players willing to join the club if he asks eight players who have a 70% chance of agreeing to join the club? Suggest a simulation model for Liang by describing how you would do the following parts:
>
> a.  Specify the device you want to use to simulate one person being asked.
>
>   *Answers will vary; you may want to discuss what devices can be used to simulate a 70% chance. Using single digits in a random number table would probably be the quickest and most efficient device. 1–7 could represent "yes," and 0, 8, 9 could represent a "no" response.*
>
> b.  What outcome(s) of the device would represent the person agreeing to be a member?
>
>   *Answers will vary based on device from part (a).*
>
> c.  What constitutes a trial using your device in this problem?
>
>   *Using a random-number table, a trial would consist of eight random digits.*
>
> d.  What constitutes a success using your device in this problem?
>
>   *Answers will vary; based on the above, a success is at least five people agreeing to join and would be represented by any set of digits with at least five of the digits being from 1–7. Note that a random string 33047816 would represent 6 of 8 people agreeing to be a member, which is a success. The string 48891437 would represent a failure.*
>
> e.  Based on 50 trials using the method you have suggested, how would you calculate the estimate for the probability that Liang will be able to form a chess club?
>
>   *Based on 50 such trials, the estimated probability that Liang will be able to form a chess club would be the number of successes divided by 50.*

## Problem Set Sample Solutions

Important Note:  You will need to provide your students with a page of random numbers.  For ease of grading, you may want students to generate their outcomes starting from the same place in the table.

> 1.  A model airplane has two engines. It can fly if one engine fails but is in serious trouble if both engines fail. The engines function independently of one another. On any given flight, the probability of a failure is 0.10 for each engine. Design a simulation to estimate the probability that the airplane will be in serious trouble the next time it goes up.
>
> a.  How would you simulate the status of an engine?
>
>   *Answers will vary; it is possible to use a random number table. The failure status of an engine can be represented by the digit 0, while digits 1–9 represent an engine in good status.*

b.  **What constitutes a trial for this simulation?**

*A trial for this problem would be a pair of random digits, one for each engine. The possible equally likely pairings would be* $00, 0x, x0, xx$ *(where* $x$ *stands for any digit from* $1$–$9$*). There are* $100$ *of them.* $00$ *represents both engines failing;* $0x$ *represents the left engine failing, but the right engine is good;* $x0$ *represents the right engine failing but the left engine is good;* $xx$ *represents both engines are in good working order.*

c.  **What constitutes a success for this simulation?**

*A success would be both engines failing, which is represented by* $00$.

d.  **Carry out 50 trials of your simulation, list your results, and calculate an estimate of the probability that the airplane will be in serious trouble the next time it goes up.**

*Answers will vary; divide the number of successes by* $50$.

2.  In an effort to increase sales, a cereal manufacturer created a really neat toy that has six parts to it.  One part is put into each box of cereal.  Which part is in a box is not known until the box is opened.  You can play with the toy without having all six parts, but it is better to have the complete set.  If you are really lucky, you might only need to buy six boxes to get a complete set, but if you are very unlucky, you might need to buy many, many boxes before obtaining all six parts.

a.  **How would you represent the outcome of purchasing a box of cereal, keeping in mind that there are six different parts?  There is one part in each box.**

*Answers will vary; since there are six parts in a complete set, the ideal device to use in this problem is a number cube.  Each number represents a different part.*

b.  **What constitutes a trial in this problem?**

*Students are asked to estimate the probability that it takes* $10$ *or more boxes to get all six parts, so it is necessary to look at the outcomes of the first* $9$ *boxes.  One roll of the number cube represents one box of cereal.  A trial could be a string of* $9$ *digits from* $1$–$6$*, the results of rolling a number cube.*

c.  **What constitutes a success in a trial in this problem?**

*A success would then be looking at the* $9$ *digits and seeing if at least one digit from* $1$–$6$ *is missing.  For example, the string:* $251466645$ *would count as a success since part* $3$ *was not acquired, whereas* $344551262$ *would be considered a failure because it took fewer than* $10$ *boxes to get all six parts.*

d.  **Carry out 15 trials, list your results, and compute an estimate of the probability that it takes the purchase of 10 or more boxes to get all six parts.**

*Students are asked to generate* $15$ *such trials, count the number of successes in the* $15$ *trials, and divide the number by* $15$*.  The result is the estimated probability that it takes* $10$ *or more boxes to acquire all six parts.*

3.  Suppose that a type A blood donor is needed for a certain surgery.  Carry out a simulation to answer the following question:  If 40% of donors have type A blood, what is an estimate of the probability that it will take at least four donors to find one with type A blood?

Note: this problem is taken from the Common Core Standards Grade 7.SP.8c.

a.  How would you simulate a blood donor having or not having type A?

*With 40% taken as the probability that a donor has type A blood, a random digit would be a good device to use.  For example, 1, 2, 3, 4 could represent type A blood, and 0, 5, 6, 7, 8, 9 could represent non-type A blood.*

b.  What constitutes a trial for this simulation?

*The problem asks for the probability that it will take four or more donors to find one with type A blood.  That implies that the first three donors do not have type A blood.  So a trial is three random digits.*

c.  What constitutes a success for this simulation?

*A success is none of the three digits are 1, 2, 3, or 4.  For example, 605 would be a success since none of the donors had type A blood.  An example of a failure would be 662.*

d.  Carry out 15 trials, list your results, and compute an estimate for the probability that it takes at least four donors to find one with type A blood.

*Students are to generate 15 such trials, count the number of successes, and divide by 15 to calculate their estimated probability of needing four or more donors to get one with type A blood.*

# Lesson 12: Using Probability to Make Decisions

## Student Outcomes

- Students will use estimated probabilities to judge whether a given probability model is plausible.
- Students will use estimated probabilities to make informed decisions.

## Lesson Notes

This is an extensive lesson. Note that there are more examples and exercises in this lesson than can be done in one class period. Teachers should work through Exercise 4, and then choose to do either Exercise 5 or 6, depending on time availability.

The examples and problem set in this lesson require students to use several devices in order to perform a simulation. Teachers should plan accordingly before the start of class.

## Classwork

### Example 1 (5 minutes): Number Cube

---

**Example 1: Number Cube**

Your teacher gives you a number cube with numbers 1– 6 on its faces. You have never seen that particular cube before. You are asked to state a theoretical probability model for rolling it once. A probability model consists of the list of possible outcomes (the sample space) and the theoretical probabilities associated with each of the outcomes. You say that the probability model might assign a probability of $\frac{1}{6}$ to each of the possible outcomes, but because you have never seen this particular cube before, you would like to roll it a few times. (Maybe it's a trick cube.) Suppose your teacher allows you to roll it 500 times and you get the following results:

| Outcome | 1 | 2 | 3 | 4 | 5 | 6 |
|---------|----|----|----|----|----|----|
| Frequency | 77 | 92 | 75 | 90 | 76 | 90 |

---

Begin by showing your students a number cube, and ask:

- What is a theoretical probability model for its six outcomes?

   □ *They should respond that the outcomes are equally likely, $\frac{1}{6}$ for each outcome.*

- Is the number cube you are showing them a fair one? Could it be a trick one?

- How could they find out whether it is fair or weighted in a way that makes it not fair?

The example suggests that they rolled the number cube 500 times and observed the following. Clearly, the number cube favors even numbers, and the estimated probabilities cause serious doubt about the conjectured equally likely model.

### Exercises 1–2 (3–5 minutes)

Let students work independently on Exercises 1 and 2. Then confirm as a class.

---

**Exercises 1–2**

1. If the "equally likely" model were correct, about how many of each outcome would you expect to see if the cube is rolled 500 times?

   *If the "equally likely" model were correct, you would expect to see each outcome occur about 83 times.*

2. Based on the data from the 500 rolls, how often were odd numbers observed? How often were even numbers observed?

   *Odd numbers were observed 228 times. Even numbers were observed 272 times.*

---

## Example 2 (5 minutes): Probability Model

---

**Example 2: Probability Model**

Two black balls and two white balls are put in a small cup whose bottom allows the four balls to fit snugly. After shaking the cup well, two patterns of colors are possible as shown. The pattern on the left shows the similar colors are opposite each other, and the pattern on the right shows the similar colors are next to or adjacent to each other.

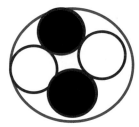

Philippe is asked to specify a probability model for the chance experiment of shaking the cup and observing the pattern. He thinks that because there are two outcomes—like heads and tails on a coin—that the outcomes should be equally likely. Sylvia isn't so sure that the "equally likely" model is correct, so she would like to collect some data before deciding on a model.

---

You will need enough small cups (bathroom disposable-size) for each student to have one. You will also need enough colored balls so that each student has four, two of each color. (If you do not have access to colored balls or blocks, tape a color label on coins.) The two colors do not matter. Students need to be able to identify which of two patterns of colors emerge when the cup containing the four balls is shaken. If your students do not know the word adjacent, use the term side-by-side. When students shake the cup, be sure that they are truly shaking up and down and are not "swirling" the cup around.

Read through the example and poll the class:

- Who agrees with Philippe?
- Who agrees with Sylvia and thinks that data should be collected before making a decision?

 **MP.4** The example suggests a theoretical probability model of opposite and adjacent patterns should be equally likely since there are just two outcomes. Hopefully students will be cautious of that sort of reasoning and ask to collect some data to test the conjectured model.

### Exercise 3 (10–15 minutes)

Direct students to gather empirical evidence by actually shaking their cups and recording the number of times the colors are adjacent, and the number of times the colors are opposite.

Write their totals on the board and calculate the empirical probabilities that the experiment results in an adjacent pattern or that the experiment results in an opposite pattern. If each student generates 20 data points, pooling the data will result in a large number of observations.

**MP.5** When all of the results are recorded, let students estimate the probabilities and answer the question independently. Discuss and confirm as a class.

---

**Exercise 3**

Collect data for Sylvia. Carry out the experiment of shaking a cup that contains four balls, two black and two white, observing, and recording whether the pattern is opposite or adjacent. Repeat this process 20 times. Then, combine the data with that collected by your classmates.

3.  Do your results agree with Philippe's equally likely model, or do they indicate that Sylvia had the right idea? Explain.

*Answers will vary; estimated probabilities of around $\frac{1}{3}$ for the opposite pattern and $\frac{2}{3}$ for the adjacent pattern will emerge, casting serious doubt on the equally likely model.*

---

As an explanation for the observations, you can have your students do a live proof of the $\frac{1}{3}$, $\frac{2}{3}$ model. Choose four students to come to the front of the room, say two boys and two girls. Have them stand as though they were balls in a cup (i.e., in a kind of square array). Fix one of them in position so that rotations are not counted. Have them systematically identify the (six) possible positions of the other three students, counting that there are four patterns for which the boys and girls are adjacent and two for which the boys are opposite each other and the girls are opposite each other.

## Exercises 4–5 (12–15 minutes)

**.2** Students should work together as a class to design a simulation for this example. The simulation procedure they develop is then used on each bag. Have about an equal number of bags of the different percentages to distribute to pairs of students, one bag per pair.

---

**Exercises 4–5**

There are three popular brands of mixed nuts. Your teacher loves cashews, and in his experience of having purchased these brands, he suggests that not all brands have the same percentage of cashews. One has around 20% cashews, one has 25%, and one has 35%.

Your teacher has bags labeled A, B, and C representing the three brands. The bags contain red beads representing cashews and brown beads representing other types of nuts. One bag contains 20% red beads, another 25% red beads, and the third has 35% red beads. You are to determine which bag contains which percentage of cashews. You cannot just open the bags and count the beads.

4. Work as a class to design a simulation. You need to agree on what an outcome is, what a trial is, what a success is, and how to calculate the estimated probability of getting a cashew. Base your estimate on 50 trials.

   - *An outcome is the result of choosing one bead from the given bag.*

   - *A red bead represents a cashew; a brown bead represents a non-cashew.*

   - *In this problem, a trial consists of one outcome.*

   - *A success is observing a red bead. Beads are replaced between trials. 50 trials are to be done.*

   - *The estimated probability of selecting a cashew from the given bag is the number of successes divided by 50.*

5. Your teacher will give your group one of the bags labeled A, B, or C. Using your plan from part (a), collect your data. Do you think you have the 20%, 25%, or 35% cashews bag? Explain.

   *Once estimates have been computed, have the class try to decide which percentage is in which bag. Agreeing on the 35% bag may be the easiest, but there could be disagreement concerning the 20% and 25%. If they can't decide, ask them what they should do. Hopefully they say that they need more data. Typically, about 500 data points in a simulation yields estimated probabilities fairly close to the theoretical ones.*

---

## Exercises 6–8 (12–15 minutes)

You may need to go over this problem carefully with your students as it could be confusing to some of them. There are two bags with the same number of slips of paper in them; on the slips of paper are numbers between 1–75; some numbers occur more than once.

Note that there are 21 primes and 6 powers of 2 between 1 and 75. You may want to have 75 slips of paper in each bag. In the prime bag, number the slips 1–75. The probability of winning would be $\frac{21}{75} = 0.28$. In the power bag, to make the probabilities of winning distinctive, you may want to put each power of 2 in twice, and then put any non-power of two numbers on the other 63 slips. The probability of winning in that bag is $\frac{12}{75} = 0.16$.

**Exercises 6–8**

Suppose you have two bags, A and B, in which there are an equal number of slips of paper. Positive numbers are written on the slips. The numbers are not known, but they are whole numbers between 1 and 75, inclusive. The same number may occur on more than one slip of paper in a bag.

These bags are used to play a game. In this game, you choose one of the bags and then choose one slip from that bag. If you choose bag A, and the number you choose from it is a prime number, then you win. If you choose bag B, and the number you choose from it is a power of 2, you win. Which bag should you choose?

6.  Emma suggests that it doesn't matter which bag you choose because you don't know anything about what numbers are inside the bags. So she thinks that you are equally likely to win with either bag. Do you agree with her? Explain.

    *Without any data, Emma is right. You may as well toss a coin to determine which bag to choose. However, gathering information through empirical evidence will help in making an informed decision.*

7.  Aamir suggests that he would like to collect some data from both bags before making a decision about whether or not the model is equally likely. Help Aamir by drawing 50 slips from each bag, being sure to replace each one before choosing again. Each time you draw a slip, record whether it would have been a winner or not. Using the results, what is your estimate for the probability of drawing a prime number from bag A and drawing a power of 2 from bag B?

    *Answers will vary.*

8.  If you were to play this game, which bag would you choose? Explain why you would pick this bag.

    *Answers will vary.*

## Closing (5 minutes)

Discuss with students the Lesson Summary.

---

**Lesson Summary**

This lesson involved knowing the probabilities for outcomes of a sample space. You used these probabilities to determine whether or not the simulation supported a given theoretical probability. The simulated probabilities, or estimated probabilities, suggested a workable process for understanding the probabilities. Only 50 trials were conducted in some examples; however, as stated in several other lessons, the more trials that are observed from a simulation, the better.

---

## Exit Ticket (5 minutes)

Name _____     Date_____

# Lesson 12: Using Probability to Make Decisions

There are four pieces of bubble gum left in a quarter machine. Two are red and two are yellow. Chandra puts two quarters in the machine. One piece is for her and one is for her friend, Kay. If the two pieces are the same color, she's happy because they won't have to decide who gets what color. Chandra claims that they are equally likely to get the same color because the colors are either the same or they are different. Check her claim by doing a simulation.

a.   Name a device to use to simulate getting a piece of bubble gum. Specify what outcome of the device represents a red piece and what outcome represents yellow.

b.   Define what a trial is for your simulation.

c.   Define what constitutes a success in a trial of your simulation.

d.   Perform and list 50 simulated trials. Based on your results, is Chandra's equally likely model is correct?

## Exit Ticket Sample Solutions

There are four pieces of bubble gum left in a quarter machine. Two are red and two are yellow. Chandra puts two quarters in the machine. One piece is for her and one is for her friend, Kay. If the two pieces are the same color, she's happy because they won't have to decide who gets what color. Chandra claims that they are equally likely to get the same color because the colors are either the same or they are different. Check her claim by doing a simulation.

a.   Name a device to use to simulate getting a piece of bubble gum. Specify what outcome of the device represents a red piece and what outcome represents yellow.

*There are several ways to simulate the bubble gum outcomes. For example, two red and two yellow disks could be put in a bag. A red disk represents a red piece of bubble gum, and a yellow disk represents a yellow piece.*

b.   Define what a trial is for your simulation.

*A trial is choosing two disks (without replacement).*

c.   Define what constitutes a success in a trial of your simulation.

*A success is if two disks are of the same color; a failure is if they differ.*

d.   Perform and list 50 simulated trials. Based on your results, is Chandra's equally likely model is correct?

*50 simulated trials will produce a probability estimate of about $\frac{1}{3}$ for same color and $\frac{2}{3}$ for different colors.*

*Note: Some students may believe that the model is equally likely, but hopefully they realize by now that they should make some observations to make an informed decision. Some may see that this problem is actually similar to Example 3.*

## Problem Set Sample Solutions

1.   Some M&M's are "defective." For example, a defective M&M may have its "m" missing, or it may be cracked, broken, or oddly shaped. Is the probability of getting a defective M&M higher for peanut M&M's than for plain M&M's?

Gloriann suggests the probability of getting a defective plain M&M is the same as the probability of getting a defective peanut M&M. Suzanne doesn't think this is correct because a peanut M&M is bigger than a plain M&M, and, therefore, has a greater opportunity to be damaged.

a.   Simulate inspecting a plain M&M by rolling two number cubes. Let a sum of 7 or 11 represent a defective plain M&M, and the other possible rolls represent a plain M&M that is not defective. Do 50 trials and compute an estimate of the probability that a plain M&M is defective. Record the 50 outcomes you observed. Explain your process.

*A simulated outcome for a plain M&M involves rolling two number cubes. A trial is the same as an outcome in this problem. A success is getting a sum of 7 or 11, either of which represents a defective plain M&M. 50 trials should produce somewhere around 11 successes. A side note is that the theoretical probability of getting a sum of 7 or 11 is $\frac{8}{36} = 0.2222$.*

b.  Simulate inspecting a peanut M&M by selecting a card from a well-shuffled deck of cards.  Let a one-eyed face card and clubs represent a defective peanut M&M, and the other cards represent a peanut M&M that is not defective.  Be sure to replace the chosen card after each trial and to shuffle the deck well before choosing the next card.  Note that the one-eyed face cards are the King of Diamonds, Jack of Hearts, and Jack of Spades.  Do 20 trials and compute an estimate of the probability that a peanut M&M is defective.  Record the list of 20 cards that you observed.  Explain your process.

*A simulated outcome for a peanut M&M involves choosing one card from a deck.  A trial is the same as an outcome in this problem.  A success is getting a one-eyed face card or a club, any of which represents a defective peanut M&M.  20 trials of choosing cards with replacement should produce somewhere around six successes.*

c.  For this problem, suppose that the two simulations provide accurate estimates of the probability of a defective M&M for plain and peanut M&M's.  Compare your two probability estimates and decide whether Gloriann's belief that the defective probability is the same for both types of M&M's is reasonable.  Explain your reasoning.

*Estimates will vary; the probability estimate for finding a defective plain M&M is approximately $0.22$, and the probability estimate for finding a defective peanut M&M is about $0.30$.  Gloriann could possibly be right, as it appears more likely to find a defective peanut M&M than a plain one, based on the higher probability estimate.*

2.  One at a time, mice are placed at the start of the maze shown below.  There are four terminal stations at A, B, C, and D.  At each point where a mouse has to decide in which direction to go, assume that it is equally likely for it to choose any of the possible directions.  A mouse cannot go backwards.

In the following simulated trials, L stands for Left, R for Right, and S for Straight.  Estimate the probability that a mouse finds station C where the food is.  No food is at A, B, or D.  The following data were collected on 50 simulated paths that the mice took.

LR RL RL LL LS LS RL RR RR RL

RL LR LR RR LR LR LL LS RL LR

RR LS RL RR RL LR LR LL LS RR

RL RL RL RR RR RR LR LL LL RR

RR LS RR LR RR RR LL RR LS LS

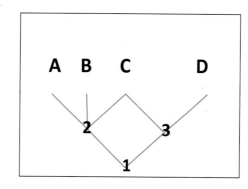

a.  What paths constitute a success and what paths constitute a failure?

*An outcome is the direction chosen by the mouse when it has to make a decision.  A trial consists of a path (two outcomes) that leads to a terminal station.  The paths are LL (leads to A), LS (leads to B), LR (leads to C), RL (leads to C), and RR (leads to D).  A success is a path leading to C, which is LR or RL; a failure is a path leading to A, B, or D, which is LL, LS, or RR.*

b.   Use the data to estimate the probability that a mouse finds food.  Show your calculation.

*Using the given 50 simulated paths:*

|  | A(LL) | B(LS) | C(LR,RL) | D(RR) |
|---|---|---|---|---|
| Simulated | $\frac{6}{50} = 0.12$ | $\frac{8}{50} = 0.16$ | $\frac{10 + 11}{50} = 0.42$ | $\frac{15}{50} = 0.30$ |
| Theoretical | 0.167 | 0.167 | 0.417 | 0.250 |

c.   Paige suggests that it is equally likely that a mouse gets to any of the four terminal stations.  What does your simulation suggest about whether her equally likely model is believable?  If it is not believable, what do your data suggest is a more believable model?

*Paige is incorrect.  The mouse still makes direction decisions that are equally likely, but there is one decision point that has three rather than two emanating paths.  Based on the simulation, the mouse is more likely to end up at terminal C.*

d.   Does your simulation support the following theoretical probability model?  Explain.

The probability a mouse finds terminal point A is $0.167$.

The probability a mouse finds terminal point B is $0.167$.

The probability a mouse finds terminal point C is $0.417$.

The probability a mouse finds terminal point D is $0.250$.

*Yes, the simulation appears to support the given probability model as the estimates are reasonably close.*

|  | A(LL) | B(LS) | C(LR,RL) | D(RR) |
|---|---|---|---|---|
| Simulated | $\frac{6}{50} = 0.12$ | $\frac{8}{50} = 0.16$ | $\frac{10 + 11}{50} = 0.42$ | $\frac{15}{50} = 0.30$ |
| Theoretical | 0.167 | 0.167 | 0.417 | 0.250 |

Name _____  Date _____

Round all decimal answers to the nearest hundredth.

1. Each student in a class of 38 students was asked to report how many siblings (brothers and sisters) he or she has. The data are summarized in the table below.

| Number of Siblings | 0 | 1 | 2 | 3 | 4 | 5 | 6 |
|---|---|---|---|---|---|---|---|
| Count | 8 | 13 | 12 | 3 | 1 | 0 | 1 |

a. Based on these data, estimate the probability that a randomly selected student from this class is an only child.

b. Based on these data, estimate the probability that a randomly selected student from this class has three or more siblings.

c. Consider the following probability distribution for the number of siblings:

| Number of Siblings | 0 | 1 | 2 | 3 | 4 | 5 | 6 |
|---|---|---|---|---|---|---|---|
| Probability | 0.15 | 0.35 | 0.30 | 0.10 | 0.05 | 0.03 | 0.02 |

Explain how you could use simulation to estimate the probability that you will need to ask at least five students if each is an only child before you find the first one that is an only child.

2.  A cell phone company wants to predict the probability of a seventh grader in your city, City A, owning a cell phone. Records from another city, City B, indicate that 201 of 1,000 seventh graders own a cell phone.

    a.  Assuming the probability of a seventh grader owning a cell phone is similar for the two cities, estimate the probability that a randomly selected seventh grader from City A owns a cell phone.

    b.  The company estimates the probability that a randomly selected seventh-grade male owns a cell phone is 0.25. Does this imply that the probability that a randomly selected seventh-grade female owns a cell phone is 0.75? Explain.

    c.  According to these data, which of the following is more likely?

        *   a seventh-grade male owning a cell phone
        *   a seventh grader owning a cell phone

        Explain your choice.

Suppose the cell phone company sells three different plans to its customers:

- Pay-as-you-go: Customer is charged per minute for each call.
- Unlimited minutes: Customer pays a flat fee per month and can make unlimited calls with no additional charges.
- Basic plan: Customer not charged per minute unless the customer exceeds 500 minutes in the month; then, the customer is charged per minute for the extra minutes.

Consider the chance experiment of selecting a customer at random and recording which plan they purchased.

d. What outcomes are in the sample space for this chance experiment?

e. The company wants to assign probabilities to these three plans. Explain what is wrong with each of the following probability assignments.

Case 1: Probability of pay-as-you-go = 0.40, probability of unlimited minutes = 0.40, and probability of basic plan = 0.30.

Case 2: Probability of pay-as-you-go = 0.40, probability of unlimited minutes = 0.70, and probability of basic plan = 0.10.

Now consider the chance experiment of randomly selecting a cell phone customer and recording both the cell phone plan for that customer and whether or not the customer exceeded 500 minutes last month.

f. One possible outcome of this chance experiment is (pay-as-go, over 500). What are the other possible outcomes in this sample space?

g. Assuming the outcomes of this chance experiment are equally likely, what is the probability that the selected cell phone customer had a basic plan and did not exceed 500 minutes last month?

h. Suppose the company random selects 500 of its customers and finds that 140 of these customers purchased the basic plan and did not exceed 500 minutes. Would this cause you to question the claim that the outcomes of the chance experiment described in (g) are equally likely? Explain why or why not.

3. In the game of Darts, players throw darts at a circle divided into 20 wedges. In one variation of the game, the score for a throw is equal to the wedge number that the dart hits. So, if the dart hits anywhere in the 20 wedge, you earn 20 points for that throw.

a. If you are equally likely to land in any wedge, what is the probability you will score 20 points?

b. If you are equally likely to land in any wedge, what is the probability you will land in the "upper right" and score 20, 1, 18, 4, 13, or 6 points?

c. Below are the results of 100 throws for one player. Does this player appear to have a tendency to land in the "upper right" more often than we would expect if the player was equally likely to land in any wedge?

| Points | 1 | 2 | 3 | 4 | 5 | 6 | 7 | 8 | 9 | 10 | 11 | 12 | 13 | 14 | 15 | 16 | 17 | 18 | 19 | 20 |
|--------|---|---|---|---|---|---|---|---|---|----|----|----|----|----|----|----|----|----|----|----|
| Count  | 7 | 9 | 2 | 6 | 6 | 3 | 5 | 2 | 4 | 7  | 2  | 6  | 4  | 6  | 5  | 7  | 4  | 6  | 5  | 4  |

| A Progression Toward Mastery | | | | | |
|---|---|---|---|---|---|
| Assessment Task Item | | STEP 1 Missing or incorrect answer and little evidence of reasoning or application of mathematics to solve the problem. | STEP 2 Missing or incorrect answer but evidence of some reasoning or application of mathematics to solve the problem. | STEP 3 A correct answer with some evidence of reasoning or application of mathematics to solve the problem, or an incorrect answer with substantial evidence of solid reasoning or application of mathematics to solve the problem. | STEP 4 A correct answer supported by substantial evidence of solid reasoning or application of mathematics to solve the problem. |
| 1 | a 7.SP.C.7 | Student does not make use of the table. No evidence of reading the table. | Student incorrectly reads the table. | Student only reports 8. | Student determines the total class size (38) and answers $\frac{8}{38}$. |
| | b 7.SP.C.7 | Student does not make use of the table. No evidence of reading the table. | Student incorrectly reads the table. | Student reports 3/38 or $3 + 1 + 1 = 5$. | Student answers $3 + 1 + 1 = 5$, so $\frac{5}{38}$. |
| | c 7.SP.C.7 | Student does not use simulation in answering this question. | Student discusses simulation, but does not provide a valid way of representing the outcomes of zero siblings or in estimating the final probability. | Student indicates how to represent the probability distribution (or just zero siblings) with simulation but does not address the question of estimating the probability of five or more students. | Student makes a reasonable assignment to represent probabilities or just probability of zero siblings and then describes simulation method, including finding the number of repetitions with five or more students. Student may not be perfectly clear on four trials vs. five trials. |
| 2 | a 7.SP.C.6 | Student provides no understanding of "probability." | Student bases answer solely on intuition (or assumes 50/50) and does not use information from stem. | Student uses the information about 201 out of 1,000 cell phone owners but does not report a probability in decimal or fraction format. | Student uses $\frac{201}{1000} = 0.201$ or approximately $\frac{1}{5}$. |

| | | | | |
|---|---|---|---|---|
| **b**<br><br>7.SP.C.5 | Student does not address probability rules. | Student claims that the probabilities should sum to one. | Student recognizes that the probabilities need not sum to one but does not provide complete justification. For example, student only states that these are not all the possible outcomes. | Student correctly identifies these probabilities as belonging to different sample spaces and/or identifies the correct "complement" events (a male *not* owning a cell phone). |
| **c**<br><br>7.SP.C.5 | Student demonstrates no understanding of probability or likelihood. | Student bases answer solely on intuition or focuses only on the fact that the sizes of the two groups will not be the same. | Student uses the information from the problem but does not provide complete justification. For example, student only states "0.25 is larger" without clearly comparing to 0.201. | Student indicates that the outcome with probability 0.25 is more likely than the outcome with probability 0.201. (Note: Student may correctly use an incorrect value from part (a).) |
| **d**<br><br>7.SP.C.7 | Student demonstrates no understanding of the term "sample space." | Student lists variables related to cell phone plans but does not address the cell phone company definitions of the three plans. | Student addresses the cell phone plan categorizations but does not have three outcomes or focuses on probabilities rather than outcomes. | Student lists three cell phone plans (basic, pay-as-go, and unlimited). |
| **e**<br><br>7.SP.C.5 | Student identifies no errors. | Student claims the three probabilities should be equal. | Student only states that the probability for one of the categories must be larger or smaller, without complete justification based on needing nonnegative probabilities and probabilities summing to one. | Student cites the sum larger than one in Case 1 and cites the negative probability in Case 2. |
| **f**<br><br>7.SP.C.8 | Student provides no understanding of the term "sample space." | Student cites multiple outcomes but not compound outcomes for this chance experiment. | Student cites compound outcomes but not consistently (some are compound but not all) or completely (not all six). | Student cites the six compound outcomes (pay as go, over 500); (pay as go, under 500); (unlimited, under 500); (unlimited, over 500); (basic, under 500); (basic, over 500). |
| **g**<br><br>7.SP.C.7 | Student does not provide a viable probability. | Student assigns an incorrect probability based on faulty assumptions (e.g., 0.5 because equally likely) or focuses on only one variable ($\frac{1}{3}$ because three categories). | Student uses the answer to (f) but incorrectly or inconsistently (e.g., only focuses on pay as go). | Student correctly considers the six outcomes and which one corresponds to this compound event ($\frac{1}{6}$). Answer may be incorrect but may follow from answer in (f). |

| | | | | | |
|---|---|---|---|---|---|
| | **h**<br><br>**7.SP.C.7** | Student does not make use of 140/500. | Student computes relative frequency and comments on its value but does not address equally likelihood of outcomes in sample space. | Student computes relative frequency, $140/500 = 0.280$, and makes a judgment about equally likely but does not give a clear comparison value (e.g., compares to 0.250 but not how 0.250 was obtained). | Student compares observed relative frequency, $140/500 = 0.280$, to value corresponding to his or her answer to (f) $\left(\frac{1}{6}\right)$ or compares 140 to $\frac{500}{6} = 83.33$. |
| **3** | **a**<br><br>**7.SP.C.7** | Student is not able to calculate a probability. | Student attempts to estimate a probability but does not see connection to number of wedges and equal probability. | Student claims the outcomes are equally likely with no justification or value provided. | Student specifies $\frac{1}{20}$. |
| | **b**<br><br>**7.SP.C.8** | Student is not able to calculate a probability. | Student attempts to estimate a probability but does not see connection to number of wedges in upper right region. | Student answer uses either equal probability condition or number of regions in upper right but does not connect the two (e.g., $\frac{1}{6}$). | Student specifies $\frac{6}{20}$ or $\frac{3}{10}$ or 0.30. |
| | **c**<br><br>**7.SP.C.6** | Student does not answer the question. | Student makes errors in total counts and fails to compare the result to the answer from (b). | Student makes errors in totaling the counts but is focused on the counts in those six regions or fails to compare the result to the answer in (b). | Student determines the number of hits in the upper right and divides by the number of throws: $\frac{4+7+6+6+4+3}{100} = 0.30$. This answer is then compared to $\frac{6}{20} = 0.30$, and student concludes there is no evidence that this player is landing in that region more often than expected. |

Name _____     Date _____

1. Each student in a class of 38 students was asked to report how many siblings (brothers and sisters) he or she has. The data are summarized in the table below.

| Number of Siblings | 0 | 1 | 2 | 3 | 4 | 5 | 6 |
|---|---|---|---|---|---|---|---|
| Count | 8 | 13 | 12 | 3 | 1 | 0 | 1 |

(a) Based on these data, estimate the probability that a randomly selected student from this class is an only child.

$8/38 = .21$

(b) Based on these data, estimate the probability that a randomly selected student from this class has 3 or more siblings.

$$\frac{3+1+0+1}{38} = \frac{5}{38} = .13$$

(c) Consider the following probability distribution for the number of siblings:

| Number of Siblings | 0 | 1 | 2 | 3 | 4 | 5 | 6 |
|---|---|---|---|---|---|---|---|
| Probability | .15 | .35 | .30 | .10 | .05 | .03 | .02 |

Explain how you could use simulation to estimate the probability that you will need to ask at least 5 students before you find the first one that is an only child.

Put 100 chips in a bag with 15 red & 85 blue. Pull out chips until you get a red one (putting chip back each time) & record the # of chips. Repeat 1000 times & see how often need 5 or more chips

2. A cell phone company wants to predict the probability of a seventh grader in your city owning a cell phone. Records from another city indicated that 201 of 1000 seventh graders owned a cell phone.

a. Assuming the probability of a seventh grader owning a cell phone is similar for the two cities, estimate the probability that a randomly selected seventh grader from City A owns a cell phone.

$$\frac{201}{1000} = .201$$

b. The company estimates the probability that a randomly selected seventh grade *male* owns a cell phone is 0.25. Does this imply that the probability that a randomly selected seventh grade *female* owns a cell phone equals 0.75? Explain.

no, .75 would be the probability
a male does not own a cell phone.
The probabilities for females could be
very different.

c. According to these data, which of the following is more likely?
   • a seventh grade male owning a cell phone
   • a seventh grader owning a cell phone
   Explain your choice.

male = 25 > overall = 201

This implies the probability is a
little higher for males so males
are more likely.

Suppose the cell phone company sells three different plans to their customers:
- Pay-as-you-go: Customer is charged per minute for each call
- Unlimited minutes: Customer pays a flat fee per month and can make unlimited calls with no additional charges
- Basic plan: Customer not charged per minute unless the customer exceeds 500 minutes in the month, then the customer is charged per minute for the extra minutes

Consider the chance experiment of selecting a customer at random and recording which plan they purchased.

d.  What outcomes are in the sample space for this chance experiment?

① pay-as-you-go
② unlimited minutes
③ basic plan

e.  The company wants to assign probabilities to these three plans. Explain what is wrong with each of the following probability assignments.

Case 1: Probability of pay-as-you-go = 0.40, probability of unlimited minutes = 0.40, and probability of basic plan = 0.30.

the sum of the three possibilities in the sample space is larger than one which can't happen.

Case 2: Probability of pay-as-you-go = 0.40, probability of unlimited minutes = 0.70, and probability of basic plan = -0.10.

can't have negative probabilities.

Now consider the chance experiment of randomly selecting a cell phone customer and recording for that person both the cell phone plan for that customer and whether or not the customer exceeded 500 minutes last month.

f.  One possible outcome of this chance experiment is (pay-as-go, over 500). What are the other possible outcomes in this sample space?

(pay as go, under 500)          (put exactly 500
(basic plan, over 500)              in "under" 500)
(basic plan, under 500)
(unlimited, under 500)
(unlimited, over 500)

g.  Assuming the outcomes of this chance experiment are equally likely, what is the probability that the selected cell phone customer had a basic plan and didn't exceed 500 minutes last month?

1/6 ≈ .167

h.  Suppose the company random selects 500 of its customers and finds that 140 of these customers purchased the basic plan and did not exceed 500 minutes. Would this cause you to question the claim that the outcomes of the chance experiment described in (g) are equally likely? Explain why or why not.

$\frac{140}{500}$ = .280

Yes, .280 is a lot higher than .167,
Seems to indicate this outcome is
more likely than some of the others

3. In the game of Darts, players throw darts at a circle divided into 20 wedges. In one variation of the game, the score for a throw is equal to the wedge number the dart hits. So if you land anywhere in the 20 wedge, you earn 20 points for that throw.

(a) If you are equally likely to land in any wedge, what is the probability you will score 20 points?

$$^4/20 = .05$$

(b) If you are equally likely to land in any wedge, what is the probability you will land in the "upper right" and score 20, 1, 18, 4, 13, or 6 points?

$$6/20 = .30$$

(c) Below are the results of 100 throws for one player. Does this player appear to have a tendency to land in the "upper right" more often than we would expect if the player was equally likely to land in any wedge?

| Pts | 1 | 2 | 3 | 4 | 5 | 6 | 7 | 8 | 9 | 10 | 11 | 12 | 13 | 14 | 15 | 16 | 17 | 18 | 19 | 20 |
|-------|---|----|---|---|---|---|---|---|---|----|----|----|----|----|----|----|----|----|----|----|
| count | 7 | 10 | 2 | 6 | 6 | 3 | 5 | 2 | 4 | 8  | 2  | 7  | 4  | 6  | 5  | 7  | 4  | 6  | 5  | 4  |

$$\frac{7 + 6 + 3 + 4 + 6 + 4}{100} = \frac{30}{100}$$

This is exactly how often we would predict this to happen.

GRADE

# Mathematics Curriculum

## Topic C:

# Random Sampling and Estimating Population Characteristics

## 7.SP.A.1, 7.SP.A.2

| Focus Standard: | 7.SP.A.1 | Understand that statistics can be used to gain information about a population by examining a sample of the population; generalizations about a population from a sample are valid only if the sample is representative of that population. Understand that random sampling tends to produce representative samples and support valid inferences. |
| --- | --- | --- |
| | 7.SP.A.2 | Use data from a random sample to draw inferences about a population with an unknown characteristic of interest.  Generate multiple samples (or simulated sample) of the same size to gauge the variation in estimates or predictions.  *For example, estimate the mean word length in a book by randomly sampling words from the book; predict the winner of a school election based on randomly sampled survey data. Gauge how far off the estimate or prediction might be.* |
| Instructional Days: | 8 | |
| Lesson 13: | Populations, Samples, and Generalizing from a Sample to a Population (P)[1] | |
| Lesson 14: | Selecting a Sample (P) | |
| Lesson 15: | Random Sampling (P) | |
| Lesson 16: | Methods for Selecting a Random Sample (P) | |
| Lesson 17: | Sampling Variability (P) | |
| Lesson 18: | Estimating a Population Mean (P) | |
| Lesson 19: | Understanding Variability when Estimating a Population Proportion (P) | |
| Lesson 20: | Estimating a Population Proportion (P) | |

Topic C begins developing the concept of generalizing from a sample to a larger population.  In Lesson 13, students are introduced to the following terminology: *population, sample, population characteristic*, and *sample statistic*.  Students distinguish between the population and a sample and between a population characteristic and a sample statistic as they investigate statistical questions.  In Lesson 14, students learn the

---

[1] Lesson Structure Key:  **P**-Problem Set Lesson, **M**-Modeling Cycle Lesson, **E**-Exploration Lesson, **S**-Socratic Lesson

importance of random sampling and of using a random mechanism in the sample selection process. In Lesson 15, students select a random sample from a population and begin to develop an understanding of sampling variability. In Lesson 16, students develop a plan for selecting a random sample from a specified population. Students see a more formal introduction to "sampling variability" in Lesson 17, where several samples are randomly selected from the sample population. They compute sample means and use collected data to develop a sense of variation in the values of a sample mean and an understanding of how this variability is related to the size of a sample. In Lesson 18, students use data from a sample to estimate a population mean. In Lesson 19, they develop an understanding of the term *sampling variability* in the context of estimating a population mean. In Lesson 20, they estimate a population proportion using categorical data from a random sample.

**Lesson 13:  Populations, Samples, and Generalizing from a Sample to a Population**

### Student Outcomes

- Students differentiate between a population and a sample.
- Students differentiate between a population characteristic and a sample statistic.
- Students investigate statistical questions that involve generalizing from a sample to a larger population.

### Lesson Notes

This lesson continues focusing on using data to answer a statistical question. Remind students that a statistical question is one that can be answered by collecting data, and where there will be variability in the data. In this lesson, students plan to collect sample data from a population to answer a statistical question. Thus, this lesson begins the development of students' thinking about how to select the sample. What does the sample tell us about the population? Is this sample representative of the population? How likely is it that the summary values calculated from the sample also reflect the population? These are big ideas that are first encountered in this grade level and developed further in later grades.

### Classwork

> In this lesson, you will learn about collecting data from a sample that is selected from a population. You will also learn about summary values for both a population and a sample and think about what can be learned about the population by looking at a sample from that population.

### Exercises 1–4 (14 minutes):  Collecting Data

Data are used to find answers to questions or to make decisions. People collect data in many different ways for many different reasons.

 Be sure students answer the questions as if they had no other source available; they could not go to the internet and ask for the average home cost, for example. They would have to figure out how to get enough information to estimate an average cost.

Pose questions from this exercise one at a time, and allow for multiple responses. As you discuss the answers, point out the difference between a population and a sample and how that might be related to each part of this exercise. A population is the entire set of objects (people, animals, plants, etc.) from which data might be collected. A sample is a subset of the population. Consider organizing a table similar to the following for selected parts of this exercise as students discuss their answers.

| Population | Sample |
|---|---|
| pot of soup | teaspoon of soup |
| all batteries of a certain brand | The group of batteries put in flashlights and timed to determine how long they last |

**Exercises 1–4: Collecting Data**

1. Describe what you would do if you had to collect data to investigate the following statistical questions using either a sample statistic or a population characteristic. Explain your reasoning in each case.

   a. How might you collect data to answer the question, "Does the soup taste good?"

   *Take a teaspoon of soup used to check the seasoning.*

   b. How might you collect data to answer the question, "How many movies do students in your class see in a month?"

   *Ask each student to write down how many movies they went to and collect their responses.*

   c. How might you collect data to answer the question, "What is the median price of a home in our town?"

   *Find the price of homes listed in the newspaper and use that data to estimate the median price. Another option is to go to a realty office and get prices for the homes they have listed for sale.*

   d. How might you collect data to answer the question, "How many pets do people own in my neighborhood?"

   *Answers might vary depending on where students live. Students living in urban areas with high rise apartment buildings might ask some people on each floor of the building, or people as they go to work in the morning; those living in suburban or rural areas might go door-to-door and ask their neighbors.*

   e. How might you collect data to answer the question, "What is the typical number of absences in math classes at your school on a given day?"

   *Ask each math teacher how many students were absent in each of his or her math classes for a given day.*

   f. How might you collect data to answer the question, "What is the typical lifetime of a particular brand flashlight battery?"

   *Put some batteries in flashlights and time how long they last.*

   g. How might you collect data to answer the question, "What percentage of girls and of boys in your school have a curfew?"

   *Ask all students if they have a curfew. Note: Students may find it challenging to ask everyone in the school the question (especially students at large schools). Let students describe how they could use a sample (for example, asking a group of students selected at random from the school directory) to answer the question.*

   h. How might you collect data to answer the question, "What is the most common blood type of students in my class?"

   *Prick the fingers of everyone in class to get some blood and have it tested.*

Remind students that numerical summary values calculated using data from an entire population are called *population characteristics*. Numerical summary values calculated using data from a sample are called *statistics*. Let students work with partners on Exercises 2–3. Discuss as a class.

A <u>population</u> is the entire set of objects (people, animals, plants, etc.) from which data might be collected. A <u>sample</u> is a subset of the population. Numerical summary values calculated using data from an entire population are called <u>population characteristics</u>. Numerical summary values calculated using data from a sample are called <u>statistics</u>.

2.  For which of the scenarios in Exercise 1 did you describe collecting data from a population, and which from a sample?

    *Answers will vary depending on the responses. Students should indicate that data on the soup seasoning, battery life, median home cost, and number of pets would be collected from a sample. Data on the number absent from math classes, the average number of movies, and most common blood type might be collected from the population.*

3.  Think about collecting data in the scenarios above. Give at least two reasons you might want to collect data from a sample rather than from the entire population.

    *If you used the whole population, you might use it all up like in the soup and batteries examples. In some cases, a sample can give you all of the information you need. For instance, you only need a sample of soup to see that the seasoning of the soup in the pot is good because it is the same all the way through. Sometimes it is too hard to collect the data for an entire population. It might cost too much or take too long to ask everyone in a population.*

4.  Make up a result you might get in response to the situations in Exercise 1, and identify whether the result would be based on a population characteristic or a sample statistic.

    a.  Does the soup taste good?

        *"Yes, but it needs more salt." The spoonful seasoning would be similar to a statistic. (Although it is not really a "statistic," it is based on a sample, so in that way it is like a statistic.)*

    b.  How many movies do your classmates see in a month?

        *The mean number of movies was 5; population characteristic.*

    c.  What is the median price of a home in our town?

        *$150,000; statistic.*

    d.  How many pets do people in my neighborhood own?

        *1 pet, either a population characteristic or a statistic (depending on method used to collect the data).*

    e.  What is the typical number of absences in math classes at your school on a given day?"

        *4 absences (mean or median representing typical); population characteristic.*

    f.  What is the typical lifetime of a particular brand of flashlight batteries?

        *54 hours; sample statistics.*

    g.  What percentage of girls and of boys in your school that have a curfew?

        *65% of the girls, and 58% of the boys have a curfew; either a population characteristic or statistic depending on method of collecting data.*

> **h.**  What is the most common blood type of my classmates?
>
> *Type O + is the most common, with 42% of the class having O +; the class is the population, so the 42% would be a population characteristic.  (It could be possible to think of the class as a sample of all seventh graders, and then it would be a sample statistic.)*

## Exercise 5 (9 minutes):  Population or Sample?

Let students continue to work with their partners on Exercise 5.  Confirm answers as a class.

---

**Exercise 5:  Population or Sample?**

5.  Indicate whether the following statements are summarizing data collected to answer a statistical question from a population or from a sample.  Identify references in the statement as population characteristics or sample statistics.

   **a.**  54% of the responders to a poll at a university indicated that wealth needed to be distributed more evenly among people.

     *The population would be all students attending the university; poll respondents would be a sample, not the population.  54% would be a sample statistic.*

   **b.**  Are students in the Bay Shore school district proficient on the state assessments in mathematics?  After all the tests taken by the students in the Bay Shore schools were evaluated, over 52% of those students were at or above proficient on the state assessment.

     *The population would be all of the students in the Bay Shore school district; 52% would be a population characteristic.*

   **c.**  Does talking on mobile phones while driving distract people?  Researchers measured the reaction times of 38 study participants as they talked on mobile phones and found that the average level of distraction from their driving was rated 2.25 out of 5.

     *The study participants would be a sample.  All drivers would be the population; the 2.25 out of 5 would be a sample statistic.*

   **d.**  Did most people living in New York in 2010 have at least a high school education?  Based on the data collected from all New York residents in 2010 by the United States Census Bureau, 84.6% of people living in New York had at least a high school education.

     *The population is all of the people living in New York in 2010; the 84.6% would be a population characteristic.*

   **e.**  Were there more deaths than births in the United States between July 2011 and July 2012?  Data from a health service agency indicated that there were 2% more deaths than births in the U.S. during that timeframe.

     *This is a good question to discuss with students.  The population would be all people in the U.S., but the data probably came from a sample of the population.  If necessary, point out to students that a nearly complete census of the United State did not occur in 2011 or 2012; the 2% would be a sample statistic.  (Although obtaining number of births and deaths out of everyone in the U.S. would be possible for 2011 or 2012, it would be very difficult and is generally done when a national census is conducted.)*

   **f.**  What is the fifth best-selling book in the United States?  Based on the sales of books in the United States, the fifth best-selling book was *Oh, the Places You'll Go!* by Dr. Seuss.

     *The population would be a list of all best-selling books in the U.S. (using some subjective benchmark for "best"); the number of copies sold for each book would need to be known to determine the fifth best-selling book, so this is a population characteristic.*

---

## Exercises 6–8 (14 minutes): A Census

This exercise would be most effective if students can access the census web site to look at the questionnaires and data from the 2010 census: www.census.gov/2010census/data/ . They can access the form and data from the entire United States related to the census questions. Students should understand that the census data is used to allocate the number of representatives each state has in the House of Representatives, and to allocate federal funds for various services and programs.

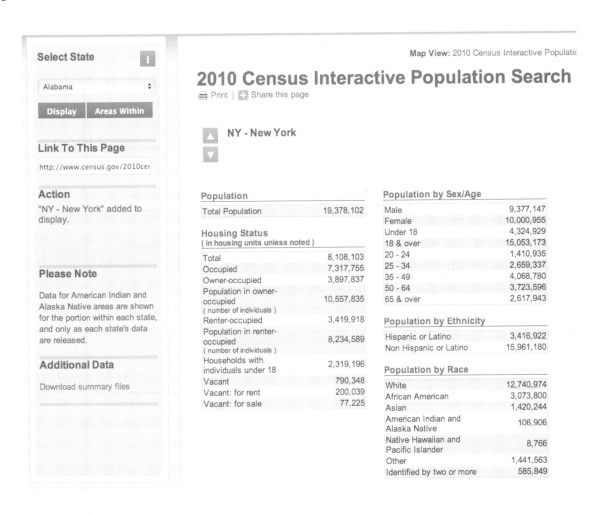

If it is not possible for students to access the site, make a copy of the history of the United States Census for students to use in responding to the questions. If needed, a copy of the pages can be found at the end of the lesson. Let students work with a partner. If class time is running short, you may choose to work on these exercises as a class.

Exercises 6–8: A Census

6.  When data are collected from an entire population, it is called a census.  The United States takes a census of its population every ten years, with the last one in 2010.  Go to http://ri.essortment.com/unitedstatesce_rlta.htm to find the history of the U.S. census.

    a.  Identify three things that you found to be interesting.

        *Students might suggest (1) the idea of a census dates back to ancient Egyptian times; (2) the U.S. constitution mandates a census every 10 years; (3) the first censuses only counted the number of men; (4) until 1950, all of the counting was done manually.*

    b.  Why is the census important in the United States?

        *According to the constitution, it is important for taxation purposes and in determining the number of representatives in Congress.  Other reasons might include planning for things such as roads and schools.*

7.  Go to the site: www.census.gov/2010census/popmap/ipmtext.php?fl=36
    Select the state of New York.

    a.  How many people were living in New York for the 2010 census?

        $19,378,102$ *people.*

    b.  Estimate the ratio of those 65 and older to those under 18 years old.  Why is this important to think about?

        *The ratio is $2,627,943$ to $4,324,929$ or about $2.6$ to $4.3$, a bit less than $\frac{1}{2}$.  It is important because when there are a greater number of older people than younger people, there will be fewer workers than people who have to be supported.*

    c.  Is the ratio a population characteristic or a statistic?  Explain your thinking.

        *The ratio is a population characteristic because it is based on data from the entire population of New York state.*

8.  The American Community Survey (ACS) takes samples from a small percentage of the US population in years between the Censuses.  (www.census.gov/acs/www/about_the_survey/american_community_survey/)

    a.  What is the difference between the way the ACS collects information about the U.S. population and the way the U.S. Census Bureau collects information?

        *The ACS obtains its results from a small percentage of the U.S. population while the Census Bureau attempts to obtain its results from the entire population.*

    b.  In 2011, the ACS sampled workers living in New York about commuting to work each day.  Why do you think these data are important for the state to know?

        *In order to plan for the best ways to travel, communities need to know how many people are using the roads, which roads, whether they travel by public transportation, and so on.*

    c.  Suppose that from a sample of $200,000$ New York workers, $32,400$ reported traveling more than an hour to work each day.  From this information, statisticians determined that between $16\%$ and $16.4\%$ of the workers in the state traveled more than an hour to work every day in 2011.  If there were $8,437,512$ workers in the entire population, about how many traveled more than an hour to work each day?

        *Between about $1,350,002$ and $1,383,752$ people.*

d. Reasoning from a sample to the population is called making an inference about a population characteristic. Identify the statistic involved in making the inference in part (c).

*The sample statistic is $\frac{32,400}{200,000}$ or $16.2\%$.*

e. The data about traveling time to work suggest that across the United States typically between $79.8$ and $80\%$ of commuters travel alone, $10$ to $10.2\%$ carpool, and $4.9$ to $5.1\%$ use public transportation. Survey your classmates to find out how a worker in their family gets to work. How do the results compare to the national data? What might explain any differences?

*Answers will vary. Reasons for the differences will largely depend on the type of community in which students live. Those living near a large metropolitan area, such as Washington, D.C. or New York City, may have lots of commuters using public transportation; while other areas, such as Milwaukee, WI, do not have many public transportation options.*

## Closing (4 minutes)

Consider posing the following questions; allow a few student responses for each.

- Describe two examples you have read about or experienced where data were collected from a population and two examples where data were collected from a sample drawn from a population.

  □ *Data collected from a population could be grades on a test in class or the amount of taxes paid on homes in a given area. Data collected from a sample could be polls about an election or studies of certain kinds of new medicines or treatments.*

- What does it mean to make an inference in statistics?

  □ *You get information from a sample and draw conclusions from that information as to what that same information might be for the whole population.*

---

### Lesson Summary

The focus of this lesson was on collecting information either from a <u>population</u>, which is the entire set of elements in the group of interest, or a subset of the population, called a <u>sample</u>. One example of data being collected from a population is the US Census, which collects data from every person in the United States every ten years. The Census Bureau also samples the population to study things that affect the economy and living conditions of people in the U.S. in more detail. When data from a population are used to calculate a numerical summary, the value is called a <u>population characteristic</u>. When data from a sample are used to calculate a numerical summary, the value is called a <u>sample statistic</u>. Sample statistics can be used to learn about population characteristics.

---

## Exit Ticket (4 minutes)

Name _____    Date_____

# Lesson 13:  Populations, Samples, and Generalizing from a Sample to a Population

Exit Ticket

What is the difference between a population characteristic and a sample statistic?  Give an example to support your answer.  Clearly identify the population and sample in your example.

## Exit Ticket Sample Solutions

> What is the difference between a population characteristic and a sample statistic? Give an example to support your answer. Clearly identify the population and sample in your example.
>
> *A population characteristic is a summary measure that describes some feature of population, the entire set of things or objects from which data might be collected. A sample statistic is a summary measure that describes a feature of some subset of the population. For example, the population could be all of the students in school, and a population characteristic could be the month in which most of the students were born. A sample of students could be those that had fifth-hour mathematics, and a sample statistic could be their grade-point average.*

## Problem Set Sample Solutions

The problem set is intended to support students' emerging understanding of the difference between a population and a sample, and summary measures for each. Students should do at least problems 1, 3, and 4 of the set below.

1. The lunch program at Blake Middle School is being revised to align with the new nutritional standards that reduce calories and increase servings of fruit and vegetables. The administration decided to do a census of all students at Blake Middle School by giving a survey to all students about the school lunches.

   http://frac.org/federal-foodnutrition-programs/school-breakfast-program/school-meal-nutrition-standards

   a. Name some questions that you would include in the survey. Explain why you think those questions would be important to ask.

      *Answers will vary. Possibilities include: How often do you eat the school lunch? Do you ever bring your lunch from home? What is your favorite food? Would you eat salads if they were served? What do you drink with your lunch? What do you like about our lunches now? What would you change? Explanations would vary but might include the need to find out how many students actually eat lunch, and if the lunches were different, would more students eat? What types of food should be served so more people will eat it?*

   b. Read through the paragraph below that describes some of the survey results. Then, identify the population characteristics and the sample statistics.

      About $\frac{3}{4}$ of the students surveyed eat the school lunch regularly. The median number of days per month that students at Blake Middle School ate a school lunch was 18 days. 36% of students responded that their favorite fruit is bananas. The survey results for Tanya's 7th grade homeroom showed that the median number of days per month that her classmates ate lunch at school was 22 and only 20% liked bananas. The fiesta salad was approved by 78% of the group of students who tried it, but when it was put on the lunch menu, only 40% of the students liked it. Of the seventh graders as a whole, 73% liked spicy jicama strips, but only 2 out 5 of all the middle school students liked them.

      *Population characteristics: $\frac{3}{4}$ eat school lunch; median number of days is 18; 36% like bananas; 40% liked fiesta salad; 2 out of 5 liked jicama strips.*

      *Sample statistics: homeroom median number of days 22; 20% liked bananas; 78% liked fiesta salad in trial; 73% liked spicy jicama strips.*

2.  For each of the following questions: (1) describe how you would collect data to answer the question, and (2) describe whether it would result in a sample statistic or a population characteristic.

   a.  Where should the eighth grade class go for their class trip?

      *Sample response: All 8th grade students would be surveyed. The result would be a population characteristic.*

      *Possibly only students in a certain classroom or students of a particular teacher would be surveyed. The students surveyed would be a sample, and the result would be a sample statistic.*

   b.  What is the average number of pets per family for families that live in your town?

      *Sample response: Data collected from families responding to a survey at a local food store. Data would be a sample and the result a sample statistic.*

      *Possibly a town is small enough that each family owning a pet could be surveyed. Then, the people surveyed would be the population, and the result would be a population characteristic.*

   c.  If people tried a new diet, what percentage would have an improvement in cholesterol reading?

      *Sample response: Data collected from people at a local health center. The people surveyed using the new diet would be a sample, and the result would be a sample statistic.*

      *Possibly all people involved with this new diet were identified and agreed to complete the survey. The people surveyed would then be the population, and the result would be a population characteristic.*

   d.  What is the average grade point of students who got accepted to a particular state university?

      *Sample response: The data would typically come from the grade point averages of all entering freshman. The result would be a population characteristic.*

      *It may have been possible to survey only a limited number of students who registered or applied. The students responding to the survey would be a sample, and the result would be a sample statistic.*

   e.  What is a typical number of home runs hit in a particular season for major league baseball players?

      *Sample response: This answer would come from examining the population of all major league hitters for that season; it would be a population characteristic.*

3.  Identify a question that would lead to collecting data from the given set as a population, and one where the data could be a sample from a larger population.

   a.  All students in your school

      *The school might be the population when considering what to serve for school lunch, or what kind of speaker to bring for an all-school assembly.*

      *The school might be a sample in considering how students in the state did on the algebra portion of the state assessment, or what percent of students engage in extracurricular activities.*

   b.  Your state

      *The percent of students who drop out of school would be calculated from data for the population of all students in schools; how people were likely to vote in the coming election could use the state as a sample of an area of the country.*

4.  Suppose that researchers sampled attendees of a certain movie and found that the mean age was 17-years old. Based on this observation, which of the following would be most likely.

    a.  The mean ages of all of the people who went to see the movie was 17-years old.

    b.  About a fourth of the people who went to see the movie were older than 51.

    c.  The mean age of all people who went to see the movie would probably be in an interval around 17, say maybe between 15 and 19.

    d.  The median age of those who attended the movie was 17-years old as well.

    *Answer 'c' would be most likely because the sample would not give an exact value for the whole population.*

5.  The headlines proclaimed: "Education Impacts Work-Life Earnings Five Times More Than Other Demographic Factors, Census Bureau Reports."  According to a U.S. Census Bureau study, education levels had more effect on earnings over a 40-year span in the workforce than any other demographic factor. www.census.gov/newsroom/releases/archives/education/cb11-153.html

    a.  The article stated that the estimated impact on annual earnings between a professional degree and an 8th grade education was roughly five times the impact of gender, which was  $13,000.  What would the difference in annual earnings be with a professional degree and with an eighth grade education?

    *About* $65,000 *a year.*

    b.  Explain whether you think the data are from a population or a sample, and identify either the population characteristic or the sample statistic.

    *The data probably come from a sample because the report was a study and not just about the population, so the numbers are probably sample statistics.*

### History of the U.S. Census

Do you realize that our census came from the Constitution of the United States of America? It was nowhere near being the very first census ever done, though. The word "census" is Latin, and it means to "tax." Archeologists have found ancient records from the Egyptians dating as far back as 3000 B.C.

In the year of 1787, the United States became the first nation to make a census mandatory in its constitution. Article One, Section Two of this historic document directs that "Representatives and direct taxes shall be apportioned among the several states ... according to their respective numbers ..." It then goes on to describe how the "numbers" or "people" of the United States would be counted and when.

Therefore, the first census was started in the year of 1790. The members of Congress gave the responsibility of visiting every house and every establishment and filling out the paperwork to the federal marshals. It took a total of eighteen months, but the tally was finally in on March of 1792. The results were given to President Washington. The first census consisted of only a few simple questions about the number of people in the household and their ages. (The census was primarily used to determine how many young men were available for wars.)

Every ten years thereafter, a census has been performed. In 1810, however, Congress decided that, while the population needed to be counted, other information needed to be gathered, also. Thus, the first census in the field of manufacturing began. Other censuses [are also conducted] on: agriculture, construction, mining, housing, local governments, commerce, transportation, and business.

Over the years, the census changed and evolved many times over. Finally, in 1950, the UNIVAC (which stood for Universal Automatic Computer) was used to tabulate the results of that year's census. It would no longer be done manually. Then, in 1960, censuses were sent via the United States Postal Service. The population was asked to complete the censuses and then wait for a visit from a census official.

Nowadays, as with the 2000 census, questionnaires were sent out to the population. The filled-out censuses are supposed to be returned via the mail to the government Census Bureau. There are two versions of the census, a short form and then a longer, more detailed form. If the census is not filled out and sent back by the required date, a census worker pays a visit to each and every household who failed to do so. The worker then verbally asks the head of the household the questions and fills in the appropriate answers on a printed form.

Being an important part of our Constitution, the census is a useful way for the Government to find out, not only the number of the population, but also other useful information.

eSsortment  http://ri.essortment.com/unitedstatesce_rlta.htm

**Lesson 14:  Selecting a Sample**

## Student Outcomes

- Students understand that how a sample is selected is important if the goal is to generalize from the sample to a larger population.
- Students understand that random selection from a population tends to produce samples that are representative of the population.

## Lesson Notes

It is important for students to understand how samples are obtained from a population.  Students may think that getting a sample is relatively easy.  For example, ask students to consider the statistical question, "How many hours of network television does the typical student in our school watch each week?"  Also, ask students how they would go about getting responses to that question that would be representative of the school population.  Often students will indicate that they will ask their friends or the students in their math class.  This type of sample is referred to as a "convenience sample."  Ask the class to think about why convenience samples generally do not result in responses that represent the population.  Students should begin to understand that there is something unfair about a convenience sample, and therefore, might not be the best sample to represent the entire population.

This lesson sets the stage for selecting and summarizing random samples.  Explain that in the above example, a sample is obtained where not all students are equally likely to be selected into the sample.  A sampling method that avoids this is called a random sampling.  In this lesson, students will observe how a random sample is representative of the population.

Read through the directions for Exercises 3–11.  Exercises 3–11 involve analyzing the words of the poem *Casey at the Bat*.  Prepare the materials needed for students to complete the exercises as indicated in the directions.  The samples obtained in the exercises are an important part in understanding why random samples are used to learn about population characteristics.

## Classwork

> As you learned in Lesson 13, sampling is a central concept in statistics.  Examining every element in a population is usually impossible.  So research and articles in the media typically refer to a "sample" from a population.  In this lesson, you will begin to think about how to choose a sample.

### Exercises 1–2 (10 minutes):  What is Random?

Students often do not understand what the word "random" means in statistics.  Before random samples are developed, explore the ideas of random behavior that are developed in the first two exercises.  These exercises indicate that random behavior may not necessarily look like what students expect.  Similarly, random samples are generally representative of the population despite not always looking as expected.

Students should write down a sequence of 20 H's and T's without tossing a coin but based on what they think the tosses might be. After they have written down and talked about their sequences, convey the idea that personal decisions on what is random typically produce behavior that is not really random.

---

**Exercises 1–2: What is Random?**

1.  Write down a sequence of heads/tails you think would typically occur if you tossed a coin 20 times. Compare your sequence to the ones written by some of your classmates. How are they alike? How are they different?

    *Students might notice a lot of variability in the sequences. Most students will have at the most two heads or two tails in a row and few will have more than one streak of three or more. Responses might look like the following:*

    H T H T H T H T H T H T H T H T H T        H T H H T H H T H H T H H T H H T H H T

    T T H T H T T H TTTH H T H T H H T T        H H T H T T H T H T T H T H T T H T H H

2.  Working with a partner, toss a coin 20 times and write down the sequence of heads and tails you get.

    a.  Compare your results with your classmates.

        *Students should notice that streaks of three or more heads and tails typically appear twice in the 20 tosses.*

        T T H T HHHHTTTT H T HHH T H T        T H H TTTT H H T H T H TTT H T T H

        H T HHHHH T T H T T H H TTT H T T        H TTT H H T H T H T TTTT H T H T H H

    b.  How are your results from actually tossing the coin different from the sequences you and your classmates wrote down?

        *The results are different because when we generated the sequence, we did not have a lot of heads or tails in a row.*

    c.  Toni claimed she could make up a set of numbers that would be random. What would you say to her?

        *She could try, but probably she would not have some of the characteristics that a real set of random numbers would have, such as three consecutive numbers or four even numbers in a row.*

---

### Exercises 3–11 (25 minutes): Letter Distribution in a Random Sample of Words

Preparation for the following exercises: Students should work with a partner. Each pair should have a copy of the poem "Casey at the Bat" (located at the end of this lesson file) with the words partitioned into 29 groups with 20 words in each group, a bag containing the numbers 1 to 29 for the group numbers and a bag containing the numbers 1 to 20 for the words within a group. Students draw a number from the first bag that identifies a group number and a number from the second bag that identifies a word in the group. For example, the combination 21 and 7 will identify group 21 and the seventh word counting from left to right in that group. Record the number of letters in that word. Put the numbers back in the bags and mix the bags. Students will now draw another number from the group bag and another number from the bag containing the numbers 1 to 20. They identify the word based on the second selection of numbers and record the number of letters in their second word. If, by chance, they pick the same word within a group, ignore the word, place the numbers back in the bag, and repeat the process to get a different word. Students continue the process until they have eight numbers representing the number of letters in eight randomly selected words. The eight numbers represent a student's random sample.

Exercises 4, 5, and 6 are intended to make students aware that selecting a sample they think will be random is usually not. Summarize what the students have done by organizing the means in the chart that is suggested in Exercise 7.

(Note: Teachers may also consider organizing the class responses to Exercises 4 and 6 in a back-to-back stem and leaf plot or on two parallel number lines with the same scale to make the difference between a self-selected sample and a random sample more visible. These visuals may also show that the sample distribution of the mean number of letters in the words in the random samples will tend to center around the population mean, 4.2 letters. If you collect the set of all of the means from the different random samples in the class, the distribution of the sample means will usually center around 4.2 letters.) The set of numbers selected by students just picking "representative" words will typically be larger than the population mean because people usually do not recognize how many two and three letter words are in the population. The graph below shows the population distribution of the length of the words in the poem.

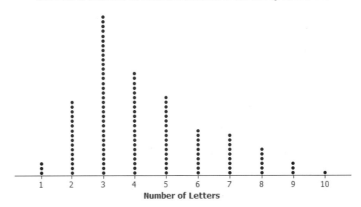

**Dot Plot of Number of Letters in Words from Casey at the Bat**

Number of Letters

Each symbol represents up to 5 observations.

---

**Exercises 3–11: Length of Words in the Poem *Casey at the Bat***

3. Suppose you wanted to learn about the lengths of the words in the poem *Casey at the Bat*. You plan to select a sample of eight words from the poem and use these words to answer the following statistical question: On average, how long is a word in the poem? What is the population of interest here?

   *The population of interest is all of the words in the poem.*

4. Look at the poem, *Casey at the Bat* by Ernest Thayer, and select eight words you think are representative of words in the poem. Record the number of letters in each word you selected. Find the mean number of letters in the words you chose.

   *Answers will vary. Sample response: The words "their, while, thousand, ball, bat, strike, muscles, grow" would have a mean of 5.25 letters.*

5. A random sample is a sample in which every possible sample of the same size has an equal chance of being chosen. Do you think the set of words you wrote down was random? Why or why not?

   *Answers will vary. Sample response: I thought it was random because I tried to use some little words and some long ones.*

---

6.  Working with a partner, follow your teacher's instruction for randomly choosing eight words. Begin with the title of the poem and count a hyphenated word as one word.

    a.  Record the eight words you randomly selected and find the mean number of letters in those words.

        *Sample response: We drew group 1, 12 (nine); group 1, 18 (four); group 5, 7 (seemed); group 3, 17 (in); group 23,6 (fraud); group 27, 11 (is); group 27,10 (air); group 16, 16 (close). The mean length of the words was 3.875.*

    b.  Compare the mean of your random sample to the mean you found in Exercise 4. Explain how you found the mean for each sample.

        *Sample response: The mean of the sample from Exercise 4 is based on the length of eight words I selected. The mean of the sample in this exercise is the mean of eight words randomly selected from the poem. Anticipate that for most students, the mean from the random sample will be greater than the mean for the self-selected sample.*

7.  As a class, compare the means from Exercise 4 and the means from Exercise 6. Your teacher will provide a chart to compare the means. Record your mean from Exercise 4 and your mean for Exercise 6 on this chart.

    *Organize the responses in a table posted in the front of class. Have students add their means to the poster. Consider the following example:*

    | Mean from the sample in Exercise 4 | Mean from the random sample in Exercise 6 |
    | --- | --- |
    | 6.3 | 4.5 |
    | 4.9 | 4.1 |
    | 6.3 | 4.4 |

8.  Do you think means from Exercise 4 or the means from Exercise 6 are more representative of the mean of all of the words in the poem? Explain your choice.

    *Sample response: The means in the random sample seem to be similar. As a result, I think the means from the random sample are more representative of the words in the poem.*

9.  The actual mean of the words in the poem *Casey at the Bat* is 4.2 letters. Based on the fact that the population mean is 4.2 letters, are the means from Exercise 4 or means from Exercise 6 a better representation the mean of the population. Explain your answer.

    *Sample response: The means from the random samples are similar and are closer to the mean of 4.2. Also, the means from Exercise 4 are generally larger than the mean of the population.*

10. How did population mean of 4.2 letters compare to the mean of your random sample from Exercise 6 and to the mean you found in Exercise 4?

    *Sample response: The mean number of letters in all of the words in the poem is 4.2, which is about four letters per word and the mean of my random sample, 3.875, was also about four letters per word. The mean of my sample in Exercise 6 was about five letters per word.*

11. Summarize how you would estimate the mean number of letters in the words of another poem based on what you learned in the above exercises.

    *Sample response: Students should summarize a process similar to what they did in this lesson. They may simply indicate that they would number each word in the poem. They would make slips of paper from 1 to the number of words in the poem, place the slips of paper in a bag or jar, and select a sample of eight or more slips of paper. Students would then record the number of letters in the words identified by the slips of paper. As in the exercises, the mean of the sample would be used to estimate the mean of all of the words in the poem.*

## Closing (5 minutes)

Discuss the following items with students. Allow a few student responses for each item.

- Ask students whether or not the sample of words based on picking numbers from the bag for the poem *Casey at the Bat* was a random sample.

    □ *Yes, the sample of words was a random sample because each group had $\frac{1}{29}$ chance of being selected and each word in each group had $\frac{1}{20}$ chance of being selected, so every word had the same chance of being selected.*

- Ask students to identify the population, the sample, and the sample statistic involved in the exercises.

    □ *Sample response: The population was all of the words in the poem. The sample was the set of eight words we randomly chose. The sample statistic was the mean number of letters in the eight words.*

- Ask students to consider the following situation. You are interested in whether students *at your school* enjoyed the poem *Casey at the Bat*? A survey asking students if they liked the poem was given to a sample consisting of the players on the school's baseball team. Is this sample a random sample? Do you think this sample is representative of the population? Explain.

    □ *Sample response: The data from the baseball team would represent a sample. The proportion of students who indicted they like the poem would be a sample statistic. However, selecting the baseball team to complete the survey did not give an equal chance for all students in our school to complete the survey. It is not a random sample of the school. I would not trust the results as there is a chance baseball players would tend to have a different answer to this question than students who do not play baseball.*

---

### Lesson Summary

When choosing a sample, you want the sample to be representative of a population. When you try to select a sample just by yourself, you do not usually do very well, like the words you chose from the poem to find the mean number of letters. One way to help ensure that a sample is representative of the population is to take a random sample, a sample in which every element of the population has an equal chance of being selected. You can take a random sample from a population by numbering the elements in the population, putting the numbers in a bag and shaking the bag to mix the numbers. Then draw numbers out of a bag and use the elements that correspond to the numbers you draw in your sample, as you did to get a sample of the words in the poem.

---

## Exit Ticket (5 minutes)

Name _____    Date_____

# Lesson 14: Selecting a Sample

Exit Ticket

Write down three things you learned about taking a sample from the work we have done today.

## Exit Ticket Sample Solutions

> Write down three things you learned about taking a sample from the work we have done today.
>
> *A random sample is one where every element in the set has an equal chance of being selected.*
>
> *When people just choose a sample they think will be random, it will usually be different from a real random sample.*
>
> *Random samples are usually similar to the population.*

## Problem Set Sample Solutions

The Problem Set is intended to reinforce material from the prior lesson and have students think about examples of samples that are random and those that are not.

1. Would any of the following provide a random sample of letters used in text of the book *Harry Potter and the Sorcerer's Stone* by J.K. Rowling? Explain your reasoning.

    a. Use the first letter of every word of a randomly chosen paragraph

    *This is not a random sample. Some common letters, like 'u', don't appear very often as the first letter of a word and may tend to be underrepresented in the sample.*

    b. Number all of the letters in the words in a paragraph of the book, cut out the numbers, and put them in a bag. Then, choose a random set of numbers from the bag to identify which letters you will use.

    *This would give you a random sample of the letters.*

    c. Have a family member or friend write down a list of their favorite words and count the number of times each of the letters occurs.

    *This would not be a random sample. They might like words that rhyme or that all start with the same letter. The list might also include words not in the book.*

2. Indicate whether the following are random samples from the given population and explain why or why not.

    a. Population: All students in school; sample includes every fifth student in the hall outside of class.

    *Sample response: No, because not everyone in school would be in our hall before class—our hall only has sixth graders in It, so the seventh and eighth graders would not have a chance to be chosen.*

    b. Population: Students in your class; sample consists of students that have the letter "s" in their last name.

    *Sample response: No, because students that do not have the letter "s" in their last name would not have a chance to be chosen*

    c. Population: Students in your class; sample selected by putting their names in a hat and drawing the sample from the hat.

    *Sample response: Yes, everyone would have the same chance to be chosen.*

    d. Population: People in your neighborhood; sample includes those outside in the neighborhood at 6:00 p.m.

    *Sample response: No, because people who are not in the neighborhood at that time have no chance of being selected.*

e.  Population: Everyone in a room; sample selected by having everyone tosses a coin and those with heads are the sample.

*Sample response: Yes, everyone would have the same chance to be chosen.*

3.  Consider the two sample distributions of the number of letters in randomly selected words shown below:

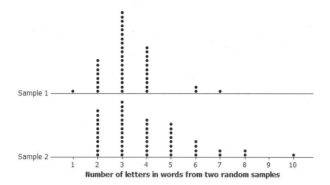

Number of letters in words from two random samples

a.  Describe each distribution using statistical terms as much as possible.

*Answers will vary; the top distribution seems to have both a median and balance point, or mean, at 3, with a minimum of 1 letter in a word and a maximum of 7 letters. Most of the words in the sample were 2 to 4 letters long. The bottom distribution seems more skewed with the median of about 4 letters. The smallest number of letters was 2 and the largest was 10 letters. Most of the letters in this sample had between 2 and 5 letters.*

b.  Do you think the two samples came from the same poem? Why or why not?

*Sample response: The samples could have come from the same poem, but the distributions seem different both with respect to shape and to a measure of center, so it seems more likely that they were from two different populations.*

4.  What questions about samples and populations might you want to ask if you saw the following headlines in a newspaper?

a.  "Peach Pop is the top flavor according to 8 out of 10 people."

*Sample response: How the people were selected? How many people were surveyed? What were the choices and how many did not like Peach Pop?*

b.  "Candidate X looks like a winner! 10 out of 12 people indicate they will vote for Candidate X."

*Sample response: How was the sample chosen? Were the people selected at random or were they friends of candidate X?*

c.  "Students Overworked. Over half of 400 people surveyed think students spend too many hours on homework."

*Sample response: Who was surveyed and how were they selected? Was the survey given to students in a school?*

d.  "Action/adventure was selected as the favorite movie type by an overwhelming 75% of those surveyed."

*Sample response: Was the survey given at a movie theater showing an action/adventure movie where people were there because they like that kind of movie?*

Handout 1

\1\ Casey at the Bat

The Outlook wasn't brilliant for the Mudville nine that day:  The score stood four to two, \2\ with but one inning more to play.  And then when Cooney died at first, and Barrows did the same,  A \3\ sickly silence fell upon the patrons of the game.

A straggling few got up to go in deep despair. The \4\ rest  Clung to that hope which springs eternal in the human breast;  They thought, if only Casey could get but \5\ a whack at that -  We'd put up even money, now, with Casey at the bat.

But Flynn preceded Casey, as \6\ did also Jimmy Blake,  And the former was a lulu and the latter was a cake;  So upon that stricken \7\ multitude grim melancholy sat, For there seemed but little chance of Casey's getting to the bat.

But Flynn let drive \8\ a single, to the wonderment of all,  And Blake, the much despised, tore the cover off the ball;  And when \9\ the dust had lifted, and the men saw what had occurred,  There was Jimmy safe at second and Flynn a \10\ hugging third.

Then from five thousand throats and more there rose a lusty yell;  It rumbled through the valley, it \11\ rattled in the dell;  It knocked upon the mountain and recoiled upon the flat,  For Casey, mighty Casey, was advancing \12\ to the bat.

There was ease in Casey's manner as he stepped into his place;  There was pride in Casey's \13\ bearing and a smile on Casey's face.  And when, responding to the cheers, he lightly doffed his hat,  No stranger \14\ in the crowd could doubt 'twas Casey at the bat.

Ten thousand eyes were on him as he rubbed his \15\ hands with dirt;   Five thousand tongues applauded when he wiped them on his shirt.  Then while the writhing pitcher ground \16\ the ball into his hip,  Defiance gleamed in Casey's eye, a sneer curled Casey's lip.

And now the leather covered \17\ sphere came hurtling through the air,  And Casey stood a-watching it in haughty grandeur there.  Close by the sturdy batsman \18\ the ball unheeded sped-  "That ain't my style," said Casey. "Strike one," the umpire said.

From the benches, black with \19\ people, there went up a muffled roar,  Like the beating of the storm waves on a stern and distant shore. \20\  "Kill him! Kill the umpire!" shouted someone on the stand;  And its likely they'd a-killed him had not Casey raised \21\ his hand.

With a smile of Christian charity great Casey's visage shone;  He stilled the rising tumult; he bade the **\22\** game go on;  He signaled to the pitcher, and once more the spheroid flew;  But Casey still ignored it, and **\23\** the umpire  said, "Strike two."

"Fraud!" cried the maddened thousands, and echo answered fraud;  But one scornful look from Casey **\24\** and the audience was awed.  They saw his face grow stern and cold, they saw his muscles strain,  And they **\25\** knew that Casey wouldn't let that ball go by again.

The sneer is gone from Casey's lip, his teeth are **\26\** clenched in hate;  He pounds with cruel violence his bat upon the plate.  And now the pitcher holds the ball, **\27\**  and now he lets it go,  And now the air is shattered by the force of Casey's blow.

Oh, somewhere **\28\** in this favored land the sun is shining bright;  The band is playing somewhere, and somewhere hearts are light,   And **\29\** somewhere men are laughing, and somewhere children shout;  But there is no joy in Mudville - mighty Casey has struck out.

**by Ernest Lawrence Thayer**

**Lesson 15:  Random Sampling**

### Student Outcomes

- Students select a random sample from a population.
- Students begin to develop an understanding of sampling variability.

### Lesson Notes

This lesson continues students' work with random sampling. The lesson engages students in two activities that investigate random samples and how random samples vary. The first investigation (Sampling Pennies) looks at the ages of pennies based on the dates in which they were made or minted. Students select random samples of pennies and examine the age distributions of several samples. Based on the data distributions, students think about what the age distribution of all of the pennies in the population might look like.

In the second investigation, students examine the statistical question, "Do store owners price the groceries with cents that are closer to a higher value, or to a lower value?" Students select random samples of the prices from a population of grocery items, and indicate how the samples might represent the population of all grocery items. In this lesson, students begin to see sampling variability and how it must be considered when using sample data to learn about a population.

### Classwork

> In this lesson, you will investigate taking random samples and how random samples from the same population vary.

### Exercises 1–5 (14 minutes):  Sampling Pennies

**Preparation**: If possible, collect 150 pennies from your students and record each penny's "age," where age represents the number of years since the coin was minted. (If students are not familiar with the word minted, explain the values used in this exercise as the age of the coin.) Put the pennies in a jar and shake it well. If it is not possible to collect your own pennies, make up a jar filled with small folded pieces of paper (so that the ages are not visible) with the ages of the pennies used in this example. The data are below (note that 0 represents a coin that was minted in the current year):

Penny Age (in years)

12, 0 , 0, 32, 10, 13, 48, 23, 0, 38, 2, 30, 21, 0, 37, 10, 16, 32, 43, 0, 25, 0, 17, 22, 7, 23, 10, 38, 0, 33, 22, 43, 24, 9, 2, 22, 4, 0, 8, 17, 14, 0, 1, 4, 1, 13, 10, 10, 26, 16, 27, 37, 41, 17, 34, 0, 29, 10, 16, 7, 25, 37, 6, 12, 31, 30, 30, 8, 7, 42, 2, 19, 16, 0, 0, 24, 38, 0, 32, 0, 38, 38, 1, 28, 18, 19, 1, 29, 48, 14, 20, 16, 6, 21, 23, 3, 29, 24, 8, 53, 24, 2, 34, 32, 46, 19, 0, 34, 16, 4, 32, 1, 30, 23, 11, 9, 17, 15, 28, 7, 22, 5, 33, 4, 31, 5, 5, 1, 5, 10, 38, 39, 23, 21, 26, 16, 1, 0, 54, 39, 5, 9, 0, 30, 19, 10, 37, 17, 20, 24

In Exercise 4, students select samples of pennies. Pennies are drawn one at a time until each student has ten pennies (without replacement). Students record the ages of the pennies, and then return them to the jar. To facilitate the sampling process, you might have students work in pairs, with one student selecting the pennies and the other recording the ages. They could begin to draw samples when they enter the room before class begins, or they could draw samples as the whole class works on a warm-up or review task. Put five number lines with the scale similar to the one in the plot below, on the board, document camera, or poster paper. (Note: The scale should go from 0 to 55 by fives.) If necessary, you may need to review with students that the collection of data represents a *data distribution*. Data distributions were a key part of students' work in Grade 6 as they learned to think about shape, center, and spread.

Before students begin the exercises, ask the following: "You are going to draw a sample of 10 pennies to learn about the population of all the pennies in the jar. How can you make sure it will be a random sample?" Make sure students understand that as they select the pennies, they want to make sure each penny in the jar had the same chance in being picked. Let students work with a partner on Exercises 1–4.

---

**Exercises 1–5: Sampling Pennies**

1.  Do you think different random samples from the same population will be fairly similar? Explain your reasoning.

    *Most of the samples will probably be about the same because they come from the same distribution of pennies in the jar. They are random samples, so we expect them to be representative of the population.*

2.  The plot below shows the number of years since being minted (the penny age) for 150 pennies that JJ had collected over the past year. Describe the shape, center, and spread of the distribution.

    **Dot Plot of Population of Penny Ages**

    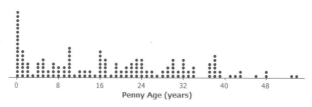

    Penny Age (years)

    *The distribution is skewed with many of the pennies minted fairly recently. The minimum is 0, and the maximum is about 54 years since the penny was minted. Thinking about the mean as a balance point, the mean number of years since a penny in this population was minted seems like it would be about 18 years.*

3.  Place ten dots on the number line that you think might be the distribution of a sample of 10 pennies from the jar.

    *Answers will vary. Most of the ages in the sample will be between 0 and 25 years. The maximum might be somewhere between 40 and 54.*

    Penny Age (years)

4.  Select a random sample of 10 pennies and make a dot plot of the ages. Describe the distribution of the penny ages in your sample. How does it compare to the population distribution?

    *Answers will vary. Sample response: The median is about 21 years. Two of the pennies were brand new, and one was about 54 years old. The distribution was not as skewed as I thought it would be based on the population distribution.*

---

After students have thought about the relationship between their samples and the population, choose five groups to put their dot plots on the axes you have provided. See example below.

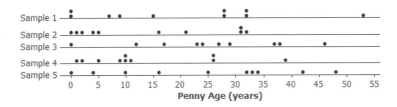

Discuss the following question as a class:

> 5. Compare your sample distribution to the sample distributions on the board.
>
>    a. What do you observe?
>
>    *Answers will vary. Sample response: Most of them seem to have the same minimum, 0 years, but the maximums vary from about 35 to 54 years. Overall, the samples look fairly different: one median is at 25, but several are less than 10. All but two of the distributions seem to be skewed like the population (i.e., with more of the years closer to 0 than to the larger number of years).*
>
>    b. How does your sample distribution compare to those on the board?
>
>    *Answers will vary. Sample response: It is pretty much the same as sample 3, with some values at 0 and the maximum is 45 years. The median is around 25 years.*

If time permits, you might want to have students consider sample size. After students draw a sample of ten pennies, direct them to create a dot plot of the sample data distribution. Ask them to draw another sample of 20 pennies and draw the data distribution of this sample. Discuss the differences. In most cases, the data distribution of the 20 pennies will more closely resemble the population distribution.

## Exercises 6–9 (23 minutes): Grocery Prices and Rounding

In the following exercises, students, again, select random samples from a population and use the samples to gain information about the population to further their understanding of sampling variability.

**Preparation:** The following exercises investigate several facets of the cents (i.e., the 45 in $0.45) in a set of prices advertised by a grocery store. Obtain enough flyers from a local grocery store, or from several different stores, so that students can work in pairs. (Flyers can be found either through the local paper, the store itself, or online.) Avoid big superstores that sell everything. We suggest using items that would typically be found in a grocery store. Have students cut out at least 100 items advertised in the flyer with their prices, and then put the slip of paper with the item and its price in a bag. Omit items advertised as "buy two, get one free." Students may complete this step as homework, and come to class with the items in a bag. If it is difficult to obtain a list of grocery items or to get students to prepare a list, 100 items are provided at the end of this lesson with prices based on flyers from several grocery stores in 2013. (Students might be more interested in this activity if they are able to prepare their own grocery lists.)

> **Exercises 6–9: Grocery Prices and Rounding**
>
> 6.   Look over some of the grocery prices for this activity.  Consider the following statistical question, "Do the store the store owners price the merchandise with cents that are closer to a higher dollar value or a lower dollar value?"  Describe a plan that might answer that question that does not involve working with all 100 items.
>
> *Sample response:  I would place all of the items in a bag.  The prices in the bag represent the population.  I would begin by selecting items from the bag, and record the prices of each item I select.  I would get a sample of at least 10 items.*

Discuss various plans suggested by students.  Discuss concerns such as how the prices from the list are selected, how many prices are selected (the sample size), and how the results are recorded.  Based on the previous lesson, it is anticipated that students will suggest obtaining random samples of the grocery items.  Remind students that random samples select items in which each item in the population has an equal chance in being selected.  Also, point out to students that the larger the sample size, the more likely it is that it will be representative of the population.  The sample size, however, should also be reasonable enough to answer the questions posed in the exercises (e.g., selecting samples of sizes 10–25).  If students select sample sizes of 50 or greater, remind them that selecting samples of that size may take a long time.  (Investigations regarding varying sample sizes will be addressed in other lessons of this module.)

To proceed with the exercises, develop a plan that all students will follow.  If students are unable to suggest a workable plan, try using the following plan.  After students cut out the prices of grocery items from ads or from the suggested list, they should shake the bag containing the slips thoroughly, draw one of the slips of paper from the bag, record the item and its price, and set the item aside.  They should then re-shake the bag and continue the process until they have a sample (without replacement) of 25 items.

If pressed for time, select a sample from one bag as a whole-class activity.  To enable students to see how different samples from the same population differ, generate a few samples of 25 items and their prices ahead of time.  Display, or hand out summarizes of these samples for the class to see.  Be sure to emphasize how knowing what population your sample came from is important in summarizing the results.

The exercises present questions that highlight how different random samples produce different results.  This illustrates the concept of *sampling variability*.  The exercises are designed to help students understand sampling variability and why it must be taken into consideration when using sample data to draw conclusions about a population.

> 7.   Do the store the store owners price the merchandise with cents that are closer to a higher dollar value or a lower dollar value?  To investigate this question in one situation, you will look at some grocery prices in weekly flyers and advertising for local grocery stores.
>
>   a.   How would you round $3.49 and $4.99 to the nearest dollar?
>
>   $3.49 *would round to* $3 *and* $4.99 *would round to* $5.
>
>   b.   If the advertised price was three for $4.35, how much would you expect to pay for one item?
>
>   $1.45
>
>   c.   Do you think more grocery prices will round up or round down?  Explain your thinking.
>
>   *Sample response:  Prices are probably set up to round up so the store owners can get more money.*

8. Follow your teacher's instructions to cut out the items and their prices from the weekly flyers, and put them in a bag. Select a random sample of 25 items without replacement, and record the items and their prices in the table below.

   *(Possible responses shown in table.)*

| Item | Price | rounded |
|---|---|---|
| grapes | $1.28/lb. | $1 down |
| peaches | $1.28/lb. | $1 down |
| melon | $1.69/lb. | $2 up |
| tomatoes | $1.49/lb. | $1 down |
| shredded cheese | $1.88 | $2 up |
| bacon | $2.99 | $3 up |
| asparagus | $1.99 | $2 up |
| soda | $0.88 | $1 up |
| Ice cream | $2.50 | $3 up |
| roast beef | $6.49 | $6 down |
| feta cheese | $4.99 | $5 up |
| mixed nuts | $6.99 | $7 up |
| coffee | $6.49 | $6 down |

| item | price | rounded |
|---|---|---|
| paper towels | $2.00 | $2 |
| laundry soap | $1.97 | $2 up |
| paper plates | $1.50 | $2 up |
| caramel rolls | $2.99 | $3 up |
| ground beef | $4.49/lb. | $4 down |
| sausage links | $2.69/lb. | $3 up |
| pork chops | $2.99/lb. | $3 up |
| cheese | $5.39 | $5 down |
| apple juice | $2.19 | $2 down |
| crackers | $2.39 | $2 down |
| soda | $0.69 | $1 up |
| pickles | $1.69 | $2 up |
|  |  |  |

Give students time to develop their answers for Exercise 9 (a) individually. After a few minutes, ask students to share their answers with the entire class. Provide a chart that displays the number of times prices were rounded up, and the number of times the prices were rounded down for each student. Also, ask students to calculate the percent of the prices that they rounded to the higher value and display that value in the chart. As students record their results, discuss how the results differ, and what that might indicate about the entire population of prices in the bag or jar. Develop the following conversation with students: "We know that these samples are all random samples, so we expect them to be representative of the population. Yet, the results are not all the same. Why not?" Highlight that the differences they are seeing is an example of *sampling variability*.

**Example of chart suggested:**

| Student | Number of times prices were rounded to the higher value. | Number of times the prices were rounded to the lower value. | Percent of prices rounded up. |
|---|---|---|---|
| Bettina | 20 | 5 | 80% |

9. Round each of the prices in your sample to the nearest dollar and count the number of times you rounded up and the number of times you rounded down.

   a. Given the results of your sample, how would you answer the question: Are grocery prices in the weekly ads at the local grocery closer to a higher dollar value or a lower dollar value?

      *Answers will vary. Sample response: In our sample, we found 16 out of 25, or 64%, of the prices rounded to the higher value, so the evidence seems to suggest that more prices are set to round to a higher dollar amount than to a lower dollar amount.*

   b. Share your results with classmates who used the same flyer or ads. Looking at the results of several different samples, how would you answer the question in part (a)?

      *Answers will vary. Sample response: Different samples had between 54% and 70% of the prices rounded to a higher value, so they all seem to support the notion that the prices typically are not set to round to a lower dollar amount.*

c.   Identify the population, sample and sample statistic used to answer the statistical question.

*Answers will vary.  Sample response:  The population was the set of all items in the grocery store flyer, or ads that we cut up and put in the bag; the sample was the set of items we drew out of the bag, and the sample statistic was the percent of prices that would be rounded up.*

d.   Bettina says that over half of all the prices in the grocery store will round up.  What would you say to her?

*Answers will vary.  Sample response:  While she might be right, we cannot tell from our work.  The population we used was the prices in the ad or flyer.  These may be typical of all of the store prices, but we do not know because we never looked at those prices.*

## Closing (4 minutes)

Consider posing the following questions and allow a few student responses for each:

- Suppose everyone in class drew a different random sample from the same population.  How do you think the samples will compare?  Explain your reasoning.

  - *Answers will vary.  The samples will vary with different medians, maximums, and minimums, but should have some resemblance to the population.*

- Explain how you know the sample of the prices of grocery store items was a random sample.

  - *Answers will vary.  To be a random sample, each different sample must have the same chance of being selected, which means that every item has to have the same chance of being chosen.  Each item in the paper bag had the same chance of being chosen.  We even shook the bag between each draw so we wouldn't get two that were stuck together, or all the ones on the top.*

---

### Lesson Summary

In this lesson, you took random samples in two different scenarios.  In the first scenario, you physically reached into a jar and drew a random sample from a population of pennies.  In the second scenario, you drew items from a bag and recorded the prices.  In both activities, you investigated how random samples of the same size from the same population varied.  Even with sample sizes of 10, the sample distributions of pennies were somewhat similar to the population (the distribution of penny ages was skewed right, and the samples all had 0 as an element).  In both cases, the samples tended to have similar characteristics.  For example, the samples of prices from the same store had about the same percent of prices that rounded to the higher dollar value.

---

## Exit Ticket (4 minutes)

 **EUREKA MATH**™  | Lesson 15:   Random Sampling

Name _____    Date_____

# Lesson 15:  Random Sampling

Exit Ticket

Identify each as true or false.  Explain your reasoning in each case.

1.  The values of a sample statistic for different random samples of the same size from the same population will be the same.

2.  Random samples from the same population will vary from sample to sample.

3.  If a random sample is chosen from a population that has a large cluster of points at the maximum, the sample is likely to have at least one element near the maximum.

## Exit Ticket Sample Solutions

Identify each as true or false. Explain your reasoning in each case.

1.  The values of a sample statistics for different random samples of the same size from the same population will be the same.

    *False, because by chance the samples will have different elements, so the values of summary statistics may be different.*

2.  Random samples from the same population will vary from sample to sample.

    *True, because each element has the same chance of being selected, and you cannot tell which ones will be chosen; it could be any combination.*

3.  If a random sample is chosen from a population that has a large cluster of points at the maximum, the sample is likely to have at least one element near the maximum.

    *True, because if many of the elements are near the same value, it seems the chance of getting one of those elements in a random sample would be high.*

## Problem Set Sample Solutions

The Problem Set is intended to reinforce material from the lesson and have students think about sample variation and how random samples from the same population might differ.

1.  Look at the distribution of years since the pennies were minted from Example 1. Which of the following box plots seem like they might not have come from a random sample from that distribution? Explain your thinking.

    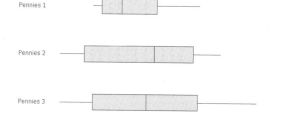

    Box Plots of Three Random Samples of Penny Ages

    *Sample response: Given that the original distribution had a lot of ages that were very small, the "Pennies 1" sample seems like it might not come from that population. The middle half of the ages are close together with a small IQR (about 12 years). The other two samples both have small values and a much larger IQR than Pennies 1, which both seem more likely to happen in a random sample given the spread of the original data.*

2. Given the following sample of scores on a physical fitness test, from which of the following populations might the sample have been chosen? Explain your reasoning.

**Dot Plots of Four Populations and One Sample**

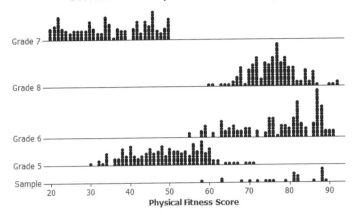

*Sample response: These sample values were not in Grades 5 or 7, so the sample could not have come from those grades. It could have come from either of the other two grades (Grades 6 or 8). The sample distribution looks skewed like Grade 6, but the sample size is too small to be sure.*

3. Consider the distribution below:

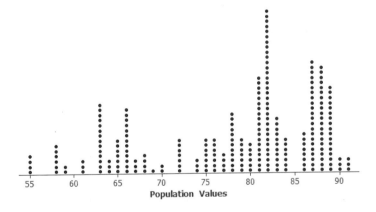

a. What would you expect the distribution of a random sample of size 10 from this population to look like?

*Sample response: The sample will probably have at least one or two elements between 80 and 90, and might go as low as 60. The samples will vary a lot, so it is hard to tell.*

b.    Random samples of different sizes that were selected from the population in part (a) are displayed below. How did your answer to part (a) compare to these samples of size 10?

**Dot Plots of Five Samples of Different Sizes**

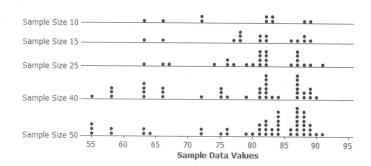

*Sample response: My description was pretty close.*

c.    Why is it reasonable to think that these samples could have come from the above population?

*Sample response: Each of the samples is centered about where the population is centered, although this is easier to see with a larger sample size. The spread of each sample also looks like the spread of the population.*

d.    What do you observe about the sample distributions as the sample size increases?

*Sample response: As the sample size increases, the sample distribution more closely resembles the population distribution.*

4.    Based on your random sample of prices from Exercise 2, answer the following questions:

a.    It looks like a lot of the prices end in 9. Do your sample results support that claim? Why or why not?

*Sample response: Using the prices in the random sample, about $84\%$ of them end in a 9. The results seem to support the claim.*

b.    What is the typical price of the items in your sample? Explain how you found the price and why you chose that method.

*Sample response: The mean price is $2.50, and the median price is $2. The distribution of prices seems slightly skewed to the right, so I would probably prefer the median as a measure of the typical price for the items advertised.*

5. The sample distributions of prices for three different random samples of 25 items from a grocery store are shown below.

   a. How do the distributions compare?

   **Dot Plots of Three Samples**

   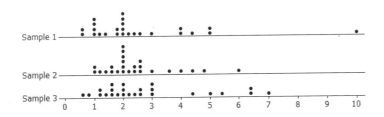

   *Sample response: The samples are slightly skewed right. They all seem to have a mean around $2.50$ and a median around $2.00$. Sample 1 has one item that costs a lot more than the others. Most of the prices vary from a bit less than $1$ to around $5$.*

   b. Thomas says that if he counts the items in his cart at that grocery store and multiplies by $2$, he will have a pretty good estimate of how much he will have to pay. What do you think of his strategy?

   *Answers will vary. Sample response: Looking at the three distributions, $2$ is about the median, so half of the items will cost less than $2$ and half will cost more, but that will not tell you how much they cost. The mean would be a better estimate of the total cost because that is how a mean is calculated. In this case, the mean (or balance point) of the distributions looks like it is about $2.50$, so he would have a better estimate of the total cost if he multiplied the number of items by $2.50$.*

## 100 Grocery Items (2013 prices)

| | | | |
|---|---|---|---|
| T-Bone steaks $6.99 (1 lb.) | Porterhouse steaks $7.29 (1 lb.) | Pasta sauce $2.19 (16 oz.) | Ice cream cups $7.29 (6 cups) |
| Hot dog buns $0.88 (6 buns) | Baking chips $2.99 (12 oz.) | Cheese chips $2.09 (12 oz.) | Cookies $1.77 (15 oz.) |
| Kidney beans $0.77 (15 oz.) | Box of oatmeal $1.77 (18 oz.) | Soup $0.77 (14 oz.) | Chicken breasts $7.77 (1.5 lb.) |
| Pancake syrup $2.99 (28 oz.) | Cranberry juice $2.77 (64 oz.) | Asparagus $3.29 | Seedless cucumbers $1.29 |
| Avocado $1.30 | Sliced pineapple $2.99 | Box of tea $4.29 (16 tea bags) | Cream cheese $2.77 (16 oz.) |
| Italian roll $1.39 (1 roll) | Turkey breast $4.99 (1 lb.) | Meatballs $5.79 (26 oz.) | Chili $1.35 (15 oz.) |
| Peanut butter $1.63 (12 oz.) | Green beans $0.99 (1 lb.) | Apples $1.99 (1 lb.) | Mushrooms $0.69 (8 oz.) |
| Brown sugar $1.29 (32 oz.) | Confectioners' sugar $1.39 (32 oz.) | Zucchini $0.79 (1 lb.) | Yellow onions $0.99 (1 lb.) |
| Green peppers $0.99 (1 count) | Mozzarella cheese $2.69 (8 oz.) | Frozen chicken $6.49 (48 oz.) | Olive oil $2.99 (17 oz.) |
| Dark chocolate $2.99 (9 oz.) | Cocoa mix $3.33 (1 package) | Margarine $1.48 (16 oz.) | Mac and Cheese $0.66 (6 oz. box) |
| Birthday cake $9.49 (7 in.) | Crab legs $19.99 (1 lb.) | Cooked shrimp $12.99 (32 oz.) | Crab legs $19.99 (1 lb.) |
| Cooked shrimp $12.99 (32 oz.) | Ice cream $4.49 (1 qt.) | Pork chops $1.79 (1 lb.) | Bananas $0.44 (1 lb.) |
| Chocolate milk $2.99 (1 gal.) | Beef franks $3.35 (1 lb.) | Sliced bacon $5.49 (1 lb.) | Fish fillets $6.29 (1 lb.) |

Lesson 15:  Random Sampling

| | | | |
|---|---|---|---|
| Pears<br>$ 1.29 (1 lb.) | Tangerine<br>$ 3.99 (3 lb.) | Orange juice<br>$ 2.98 (59 oz.) | Cherry pie<br>$ 4.44 (8 in.) |
| Grapes<br>$ 1.28 (1 lb.) | Peaches<br>$ 1.28 (1 lb.) | Melon<br>$ 1.69 (1 melon) | Tomatoes<br>$ 1.49 (1 lb.) |
| Shredded cheese<br>$ 1.88 (12 oz.) | Soda<br>$ 0.88 (1 can) | Roast beef<br>$ 6.49 (1 lb.) | Coffee<br>$ 6.49 (1 lb.) |
| Feta cheese<br>$ 4.99 (1 lb.) | Pickles<br>1.69 (12 oz. jar) | Load of rye bread<br>$ 2.19 | Crackers<br>$ 2.69 (7.9 oz.) |
| Purified water<br>$ 3.47 (35 pk.) | BBQ sauce<br>$ 2.19 (24 oz.) | Ketchup<br>$ 2.29 (34 oz.) | Chili sauce<br>$ 1.77 (12 oz.) |
| Sugar<br>$ 1.77 (5 lb.) | Flour<br>$ 2.11 (4 lb.) | Breakfast cereal<br>$ 2.79 (9 oz.) | Cane sugar<br>$ 2.39 (4 lb.) |
| Cheese sticks<br>$ 1.25 (10 oz.) | Cheese spread<br>$ 2.49 (45 oz.) | Coffee creamer<br>$ 2.99 (12 oz.) | Candy bars<br>$ 7.77 (40 oz.) |
| Pudding mix<br>$ 0.98 (6 oz.) | Fruit drink<br>$ 1.11 (24 oz.) | Biscuit mix<br>$ 0.89 (4 oz.) | Sausages<br>$ 2.38 (13 oz.) |
| Meat balls<br>$ 1.98 (10 oz.) | Apple juice<br>$ 1.48 (64 oz.) | Ice cream sandwich<br>$1.98 (12 ct.) | Cottage cheese<br>$ 1.98 (24 oz.) |
| Frozen vegetables<br>$ 0.88 (10 oz.) | English muffins<br>$ 1.68 (6 ct.) | String cheese<br>$ 6.09 (24 oz.) | Baby greens<br>$ 2.98 (10 oz.) |
| Caramel apples<br>$ 3.11 (1 apple) | Pumpkin mix<br>$ 3.50 (1 lb.) | Chicken salad<br>$ 0.98 (2 oz.) | Whole wheat bread<br>$ 1.55 (1 loaf) |
| Tuna<br>$ 0.98 (2.5 oz.) | Nutrition bar<br>$ 2.19 (1 bar) | Potato chips<br>$ 2.39 (12 oz.) | 2% milk<br>$3.13 (gal.) |

**Lesson 16:  Methods for Selecting a Random Sample**

## Student Outcomes

- Students select a random sample from a population.
- Given a description of a population, students design a plan for selecting a random sample from that population.

## Lesson Notes

In this lesson, students will use random numbers to select a random sample.  Unlike the previous lessons where students selected at random from a physical population, in this lesson the random selection will be based on using random numbers to identify the specific individuals in the population that should be included in the sample.  Students will also design a plan for selecting a random sample to answer a statistical question about a population.

A variety of methods can be used to generate random numbers.  The TI graphing calculators have a random number generator that can be used if calculators are available.  For this lesson, students will generate random integers.  Random number generators can also be found on a number of websites.  (A specific website is referenced below that could be used to create a list of random numbers.  Directions needed to use the random number generator are provided at the website.)  If calculators or access to websites are not possible, simply create a random number bag.  Write the numbers from 1 to 150 on small slips of paper.  Students will select slips of paper from the bag to form their list of random numbers.  Having access to technology will make it easier for students to concentrate on the concepts rather than counting and sorting numbers.  If you are not able to use technology, each student or pair of students should do only one or two examples, and the class data should be collected and displayed.

## Classwork

> In this lesson, you will obtain random numbers to select a random sample.  You will also design a plan for selecting a random sample to answer a statistical question about a population.

### Example 1 (2 minutes):  Sampling Children's Books

Introduce the data in the table and examine the histogram.

> **Example 1:  Sampling Children's Books**
>
> What is the longest book you have ever read?  *The Hobbit* has $95,022$ words and *The Cat in the Hat* has $830$ words.  Popular books vary in the number of words they have—not just the number of *different* words, but the total number of words.  The table below shows the total number of words in some of those books.  The histogram displays the total number of words in 150 best-selling children's books with fewer than $100,000$ words.

| Book | Words | Book | Words | Book | Words |
|------|-------|------|-------|------|-------|
| Black Beauty | 59,635 | Charlie and the Chocolate Factory | 30,644 | The Hobbit | 95,022 |
| Catcher in the Rye | 73,404 | Old Yeller | 35,968 | Judy Moody was in a Mood | 11,049 |
| The Adventures of Tom Sawyer | 69,066 | Cat in the Hat | 830 | Treasure Island | 66,950 |
| The Secret Garden | 80,398 | Green Eggs and Ham | 702 | Magic Tree House Lions at Lunchtime | 5,313 |
| Mouse and the Motorcycle | 22,416 | Little Bear | 1,630 | Philosopher's Stone | 77,325 |
| The Wind in the Willows | 58,424 | Red Badge of Courage | 47,180 | Chamber of Secrets | 84,799 |
| My Father's Dragon | 7,682 | Anne Frank: The Diary of a Young Girl | 82,762 | Junie B. Jones and the Stupid Smelly Bus | 6,570 |
| Frog and Toad All Year | 1,727 | Midnight for Charlie Bone | 65,006 | White Mountains | 44,763 |
| Book of Three | 46,926 | The Lion, The Witch and the Wardrobe | 36,363 | Double Fudge | 38,860 |

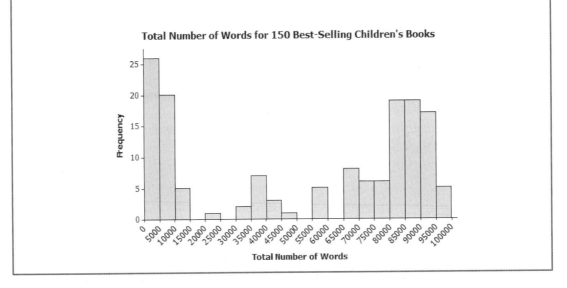

Total Number of Words for 150 Best-Selling Children's Books

- In which interval is *Black Beauty*? *Cat in the Hat*?
  - *Black Beauty is in the interval* 55,000 *to* 60,000 *words, while* Cat in the Hat *is in the first interval of* 0 *to* 5,000 *words.*
- What is the meaning of the first bar in the histogram?
  - *The first bar indicates the number of books with total number of words between* 0 *and* 4,999.

## Exercises 1–2 (3 minutes)

Let students work independently on Exercises 1 and 2.  Have students compare their dot plots with a neighbor.

---

**Exercises 1–2**

1.  From the table, choose two books with which you are familiar and describe their locations in the data distribution shown in the histogram.

    *Answers will vary.  Sample response:  I read The Mouse and the Motorcycle, and that has 22,416 words. It is below the median number of words and maybe below the lower quartile. The Chamber of Secrets has 84,799 words.  It is one of the books with lots of words, but maybe not in the top quarter for the total number of words.*

2.  Put dots on the number line below that you think would represent a random sample of size 10 from the number of words distribution above.

    ```
    ├──────┬──────┬──────┬──────┬──────┬──────
    0     2000   4000   6000   8000   10000
    ```

    *Answers will vary.  The sample distribution might have more values near the maximum and minimum than in the center.*

---

### Example 2 (3–5 minutes):  Using a Random Number Generator to Select a Sample

---

**Example 2:  Using Random Numbers to Select a Sample**

The histogram indicates the differences in the number of words in the collection of 150 books.  How many words are typical for a best-selling children's book?  Answering this question would involve collecting data, and there would be variability in that data.  This makes the question a statistical question.  Think about the 150 books used to create the histogram above as a population.  How would you go about collecting data to determine the typical number of words for the books in this population?

*Sample response:  I would add up all of the words in the 150 books and divide by 150.  This would be the mean number of words for the 150 books.  As the data distribution is not symmetrical, I could also find the median of the 150 books, as it would be a good description of the typical number of words.  (Note:  Discuss with students that using data for all 150 books is very tedious.  As a result, students may indicate that selecting a random sample of the 150 books might be a good way to learn about the number of words in these children's books.)*

---

Discuss students' suggestions for choosing a random sample.  Be sure to bring out the following points:  to choose a random sample, you could number all of the books, put the numbers in a bag, and then draw your sample from the bag.  Another way is to use a random number generator, where instead of pulling numbers out of a bag, the generator selects the numbers to use in the sample.  (In this discussion, observe how students make sense of random selection and generating a random sample.)

---

How would you choose a random sample from the collection of 150 books discussed in this lesson?

*Sample response:  I would make 150 slips of paper that contained the names of the books.  I would then put the slips of paper in a bag and selection 10 or 15 books.  The number of pages of the books selected would be my sample.*

The data for the number of words in the 150 best-selling children's books are listed below.  Select a random sample of the number of words for 10 books.

---

If necessary, explain how to use ten numbers selected from a bag that contains the numbers from 1 to 150 to select the books for the sample.

If students need more direction in finding a random sample, develop the following example: Consider the following random numbers obtained by drawing slips from a bag that contained the numbers 1 to 150: $\{114, 65, 77, 38, 86, 105, 50, 1, 56, 85\}$. These numbers represent the randomly selected books. To find the number of words in those books, order the random numbers $\{1, 38, 50, 56, 75, 77, 85, 86, 105, 114\}$. Count from left to right across the first row of the list of the number of words, then down to the second row, and so on. The sample will consist of the 1st element in the list, the 38th, the 50th, and so on.

Use the above example of random numbers to help students connect the random numbers to the books selected, and to the number of words in those books.

- What number of words corresponds to the book identified by the random number 1?
  - 59,635; *the first children's book listed has* 59,635 *words.*
- What number of words corresponds to the book identified by the random number 38?
  - 3,252; *the 38th children's book listed has* 3,252 *words.*

| | | | | | | | | | | |
|---|---|---|---|---|---|---|---|---|---|---|
| Books 1–10 | 59,635 | 82,762 | 92,410 | 75,340 | 8,234 | 59,705 | 92,409 | 75,338 | 8,230 | 82,768 |
| Books 11–20 | 73,404 | 65,006 | 88,250 | 2,100 | 81,450 | 72,404 | 88,252 | 2,099 | 81,451 | 65,011 |
| Books 21–30 | 69,066 | 36,363 | 75,000 | 3,000 | 80,798 | 69,165 | 75,012 | 3,010 | 80,790 | 36,361 |
| Books 31–40 | 80,398 | 95,022 | 71,200 | 3,250 | 81,450 | 80,402 | 71,198 | 3,252 | 81,455 | 95,032 |
| Books 41–50 | 22,416 | 11,049 | 81,400 | 3,100 | 83,475 | 22,476 | 81,388 | 3,101 | 83,472 | 11,047 |
| Books 51–60 | 58,424 | 66,950 | 92,400 | 2,750 | 9,000 | 58,481 | 92,405 | 2,748 | 9,002 | 66,954 |
| Books 61–70 | 7,682 | 5,313 | 83,000 | 87,000 | 89,170 | 7,675 | 83,021 | 87,008 | 89,167 | 5,311 |
| Books 71–80 | 1,727 | 77,325 | 89,010 | 862 | 88,365 | 1,702 | 89,015 | 860 | 88,368 | 77,328 |
| Books 81–90 | 46,926 | 84,799 | 88,045 | 927 | 89,790 | 46,986 | 88,042 | 926 | 89,766 | 84,796 |
| Books 91–100 | 30,644 | 6,570 | 90,000 | 8,410 | 91,010 | 30,692 | 90,009 | 8,408 | 91,015 | 6,574 |
| Books 101–110 | 35,968 | 44,763 | 89,210 | 510 | 9,247 | 35,940 | 89,213 | 512 | 9,249 | 44,766 |
| Books 111–120 | 830 | 8,700 | 92,040 | 7,891 | 83,150 | 838 | 92,037 | 7,889 | 83,149 | 8,705 |
| Books 121–130 | 702 | 92,410 | 94,505 | 38,860 | 81,110 | 712 | 94,503 | 87,797 | 81,111 | 92,412 |
| Books 131–140 | 1,630 | 88,250 | 97,000 | 7,549 | 8,245 | 1,632 | 97,002 | 7,547 | 8,243 | 88,254 |
| Books 141–150 | 47,180 | 75,000 | 89,241 | 81,234 | 8,735 | 47,192 | 89,239 | 81,238 | 8,739 | 75,010 |

## Exercises 3–6 (10–12 minutes)

In this set of exercises, students select a random sample by using a random-number generator or slips of paper in a bag. To generate a set of random numbers, consider directing students to a site, such as www.rossmanchance.com/applets/RandomGen/GenRandom01.htm, or to a random-number generator on a graphing calculator. Be sure that the numbers generated are *unique,* or without replacement, so that no number is used twice. If students' access to technology is problematic, demonstrate a random-number generator for them and have them copy down the numbers you generate. If you use a graphing calculator similar to the TI-84, be sure to seed the calculator by completing the command: RandSeed #, where # is any number unique to the student, such as the last four digits of a phone number. If this is not done, all of the "random" numbers may begin at the same place and the samples will all be the same. However, once they have seeded their random number generator one time, they do not have to do this again unless you are in a situation where you want the whole class to use the same seed so that the entire class will produce the same set of numbers.

Students should work in pairs with one student counting and the other recording the sample values.

---

Exercises 3–6

3. Follow your teacher's instructions to generate a set of 10 random numbers. Find the total number of words corresponding to each book identified by your random numbers.

   *Answers will vary. Sample response: I generated random numbers $123, 25, 117, 119, 93, 135, 147, 69, 48, 46,$ which produces the sample $94505, 80798, 92037, 83149, 90000, 8245, 89239, 89167, 3101, 22476.$*

4. Choose two more different random samples of size 10 from the data and make a dot plot of each of the three samples.

   *Answers will vary. One possible response is displayed below.*

   **Dot Plot of Total Number of Words for Three Random Samples**

   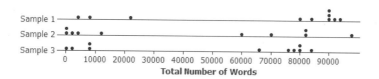

5. If your teacher randomly chooses 10 books from your summer vacation reading list, would you be likely to get many books with a lot of words? Explain your thinking using statistical terms.

   *Answers will vary. Sample response: From my samples, it looks like I probably would get at least one book that had over $90,000$ words because the maximum in each of the samples approached or exceeded $90,000$ words. The three samples vary a lot, probably because the sample size is only $10$. The median numbers of words for the three samples were about $86,000, 35,000,$ and $70,000,$ respectively, so it seems like at least half of the books would contain about $50,000$ or more words.*

---

6.  If you were to compare your samples with your classmates' samples, do you think your answer to Exercise 5 would change?  Why or why not?

*Answers will vary.  Sample response:  The sample size is pretty small so different samples might be different.  I still think that I would have to read some books with a lot of words because of the shape of the population distribution.*

## Exercises 7–9 (19 minutes):  A Statistical Study of Balance

The following exercises ask students to develop a plan to investigate a statistical question regarding balance.  Actually carrying out the activity is optional given the challenges summarized in this explanation.  The goal of the activity is for students to collect real data from a random sample to help them learn about a population.

This activity may take different forms in different schools, depending on the size of the school.  It is important to select a population that is large enough that taking a random sample would be a reasonable way to investigate the population.  The statistical question is framed to investigate whether seventh graders have better balance than sixth graders.  A sample of 10 sixth graders and a sample of 10 seventh graders would work, which would mean starting with populations of at least approximately 100 sixth graders and 100 seventh graders ideal.  This may not be possible in some schools.  In very small schools, teachers might find another school with which to partner for the activity.  If pooling together students is not possible, and the number of sixth and seventh graders is too small, you can use different populations, but be explicit as to which populations students are using.  Clearly state that the populations of interest are all of the sixth and seventh graders in the school, and use a reasonable sample size for the selection of the random samples.  You may also modify the statistical question so that the groups compared represent a larger number of the students in your school (for example, students in Grades 7 and 8 compared to students in Grades 5 and 6 or do students in Grades 7 and 8 spend more time on homework than students in Grades 5 and 6).  Students can complete the exercises that involve planning this activity, even if it is not possible to actually carry out the data collection.

In completing the exercises, students have to think about how to use random numbers to select a random sample.  The need for a random sample is based on the premise that it would take too much time and it would be too difficult to carry out a study using all students in the two populations.  If possible, you many want students to actually carry out the data collection, but that will take some time, probably two days to plan, collect, and analyze the data.  If you can find the time to do this, it will be time well spent as the activity engages students in important aspects of the entire statistical process: beginning with a statistical question, designing a study, collecting data, analyzing the data, and using the results to answer the question.

Cell phones or stopwatches can be used for timing balance.  (Some science departments have stopwatches they might lend to the students.)

 Students should put their complete plans for Exercise 9 on chart paper or some other public display, and each group should share their thinking.  The other students might anonymously write on a $3 \times 5$ notecard one thing they like about the plan being presented and one thing that concerns them.  If the class actually carries out the activity, Exercise 9 could be done as a whole class or the class could vote to determine which of the groups' plans they would like to use.

Exercises 7–9:  A Statistical Study of Balance and Grade

7.  Is the following question a statistical question, "Do sixth graders or seventh graders tend to have better balance?"  Why or why not?

*Yes, this is a statistical question because it would be answered by collecting data on balance and there would be variability in the data collected.  It would be important to think about how balance will be measured.*

8.  Berthio's class decided to measure balance by finding out how long people can stand on one foot.

    a.  How would you rephrase the question above to create a statistical question using this definition of balance? Explain your reasoning.

        *Answers will vary.  Sample response:  Can sixth graders balance on one foot longer than seventh graders? The data collected to answer this question will have some variability – 1 min., 2 min., 2 min. 10 sec., and so on.  So, it is also a statistical question.*

    b.  What should the class think about to be consistent in how they collect the data if they actually have people stand on one foot and measure the time?

        *Sample response:  Would it make a difference if students stood on their right foot or on their left?  How high do they have to hold the foot off the ground?  Can they do it barefoot or with shoes on?  Would tennis shoes be better than shoes with higher heels?  What can we use to measure the time?*

9.  Work with your class to devise a plan to select a random sample of sixth graders and a random sample of seventh graders to measure their balance using Berthio's method.  Then, write a paragraph describing how you will collect data to determine whether there is a difference in how long sixth graders and seventh graders can stand on one leg. Your plan should answer the following questions:

    ▪  What is the population? How will samples be selected from the population? And, why is it important that they be random samples?

       *Sample response:  The populations will be all of the sixth graders and all of the seventh graders in our school. To get a random sample, we will find the number of sixth graders, say 62, and generate a list of 15 random numbers from the set 1 to 62, i.e., $\{4, 17, 19, 25, \dots\}$.  Then we will go into one classroom and count off the students beginning with 1 and use student 4, 17 and 19; go into the next classroom and count off the students beginning where we left off in the first room and so on.  We will do the same for the seventh graders.  This will give random samples because it offers every sixth and seventh grader the same chance of being selected (if using this plan with both grades).*

    ▪  How would you conduct the activity?

       *Sample response:  Students will stand for as long as they can using whichever leg they choose in their stocking or bare feet with their eyes open.  We will time them to the nearest second using stop watches from our science class.  We will have students do the activity one at a time out in the hall so they cannot see each other.*

    ▪  What sample statistics will you calculate, and how you would display and analyze the data?

       *Sample response:  The sample statistics will be the mean time (in seconds) standing on one leg for the sixth graders, seventh graders, and the corresponding MADs.  We will make a dot plot of the times for the sixth graders, and for the seventh graders using parallel number lines with the same scale.*

    ▪  What you would accept as evidence that there actually is a difference in how long sixth graders can stand on one leg compared to seventh graders.

       *We will compare the shape, center and spread of the sample distributions of times for the sixth graders, and do the same for the seventh graders.  If the mean times are fairly close together, and the spreads not that different, there is not really evidence to say one has better balance.*

## Closing (4 minutes)

Consider posing the following questions. Allow a few student responses for each:

- Sallee argued that the set {20, 24, 27, 32, 35, 40, 45, 50, 120, 500} could not possibly be a random sample of ten numbers from 1 to 500 because the set had too many small numbers. Do you agree or disagree with Sallee? Explain your thinking.
  - *Every possible set of ten numbers from* 1 *to* 500 *would be a possible random sample so Sallee is not correct.*
- Why is it important to choose a random sample when you are collecting data?
  - *If you do not have a random sample, your sample may not reflect the population, and therefore, would not offer accurate information about the population.*

> **Lesson Summary**
>
> In this lesson, you collected data on total number of words by selecting a random sample of children's books. You also observed that several different samples of the same size had some characteristics in common with each other and with the population. In the second activity, you designed a statistical study. First, you considered a statistical question. Then, you went through the statistical process beginning with the question, and then thinking about how to choose a random sample, how students would take part, what data you would collect to answer the question, and how you would display and analyze the data.

## Exit Ticket (4 minutes)

Name _____      Date_____

# Lesson 16: Methods for Selecting a Random Sample

Exit Ticket

1. Name two things to consider when you are planning how to select a random sample.

2. Consider a population consisting of the 200 seventh graders at a particular middle school. Describe how you might select a random sample of 20 students from a list of the students in this population.

## Exit Ticket Sample Solutions

1. Name two things to consider when you are planning how to select a random sample.

   *Answers will vary.*

   - *What it means to be a random sample—that everyone in the population will have the same chance to be selected.*

   - *Is there a way to use a random number generator to make it easier to select the sample?*

2. Consider a population consisting of the 200 seventh graders at a particular middle school. Describe how you might select a random sample of 20 students from a list of the students in this population.

   *Answers will vary. Number the students on the list from 001 to 200. Using a random-number generator, get 20 different random numbers between 001 and 200, and then select the students corresponding to those numbers on the list. It would also be correct for a student to say that they would write the 200 names on slips of paper, put them in a bag, mix them well, and then select 20 names from the bag*

## Problem Set Sample Solutions

Students should do problems 1 and 3 to be sure they understand the concepts in the lesson. Problem 2 (b) can be used to engage students in generating and analyzing random samples using technology. You might have them work in pairs to generate and record the numbers in the random samples to see whether the teacher's method of collecting homework will turn out to be relatively "fair" for the students.

1. The suggestions below for how to choose a random sample of students at your school were made and vetoed. Explain why you think each was vetoed.

   a. Use every fifth person you see in the hallway before class starts.

   *Students who are not in the hallway because they have a class in another part of the building would not have a chance to be selected, so the sample would not be a random sample.*

   b. Use all of the students taking math the same time as your class meets.

   *The students not taking math at that time would not have a chance of being selected so the sample would not be a random sample.*

   c. Have students who come to school early do the activity before school starts.

   *The sample would be not be a random sample because some students would not be able to get to school early, so they could not be selected.*

   d. Have everyone in the class find two friends to be in the sample.

   *Choosing people that members of the class know would not be a random sample because people that members of the class do not know have no chance to be chosen.*

2. A teacher decided to collect homework from a random sample of her students, rather than grading every paper every day.

   a. Describe how she might choose a random sample of five students from her class of 35 students.

   *Sample response: You could assign each student a number from 1 to 35, generate five random numbers from 1 to 35 and choose the corresponding students.*

   b. Suppose every day for 75 days throughout an entire semester she chooses a random sample of five students. Do you think some students will never get selected? Why or why not?

   *Sample response: Over that many days, it should almost even out. If you think about 375 numbers generated all together with each number from 1 to 35 having an equal chance of showing up each time, then each number should be in the overall set about 10 or 11 times. I generated 75 random samples of numbers from 1 to 35 and looked at how the numbers showed up. Every number from 1 to 35 showed up at least three times and most of the numbers showed up about 10 or 11 times.*

3. Think back to earlier lessons in which you chose a random sample. Describe how you could have used a random-number generator to select a random sample in each case.

   a. A random sample of the words in the poem "Casey at the Bat"

   *Sample response: You could have generated the random numbers from 1 to 26 for the block of words and the random numbers 1 to 20 to choose a word in the block. Or, you could number all of the words from 1 to 520 and then generate random numbers between 1 and 520 to choose the words.*

   b. A random sample of the grocery prices on a weekly flyer.

   *Sample response: Instead of cutting out all of the prices and putting them in a bag, you could just number them on the flyer and use the random-number generator to select numbers to identify the items in the sample and use the price of those items.*

4. Sofia decided to use a different plan for selecting a random sample of books from the population of 150 top-selling children's books from Example 2. She generated ten random numbers between 1 and 100,000 to stand for the possible number of pages in any of the books. Then she found the books that had the number of pages specified in the sample. What would you say to Sofia?

*Sample response: She would have to reject the numbers in the sample that referred to pages that were not in her list of 150 books. For example, if she gets the random numbers 4 or 720, she would have to generate new numbers because no books on the list had either 4 or 720 pages. She would have to throw out a lot of random numbers that did not match the number of pages in the books in the list. It would take her a long time. But if there were no two books that had the same total number of words in the population, it would be a random sample if she wanted to do it that way. However, because there are quite a few books that have the same number of words as other books in the population, this method would not work for selecting a random sample of the books.*

5. Find an example from a newspaper, magazine or another source that used a sample. Describe the population, the sample, the sample statistic, how you think the sample might have been chosen, and whether or not you think the sample was random.

*Responses will vary depending on the articles students find. For example, "an estimated 60% of the eligible children in Wisconsin did not attend preschool in 2009." The population would be all of the children in Wisconsin eligible for preschool in 2009, and the sample would be the ones selected by the study. The sample statistic would be 60%. The article did not tell how the sample was chosen but said the source was from the Census Bureau, so it was probably a random sample.*

**Lesson 17:  Sampling Variability**

## Student Outcomes

- Students use data from a random sample to estimate a population mean.
- Students understand the term "sampling variability" in the context of estimating a population mean.

## Lesson Notes

Consider starting this lesson with a review of the activities from the last lesson.  The number of total number of words for books in a random sample of children's books was used to estimate the mean total number of words for the population of 150 children's books.  A random sample of prices from a population of grocery prices was used to estimate the proportion of grocery prices that would round up to the nearest dollar.  This lesson and the lessons that follow continue to develop the idea of using data from a random sample to estimate a population mean or a population proportion.  Sampling variability in sample means and sample proportions is explored in order to further student understanding of this important concept.  In later grades, students learn how to take sampling variability into account when generalizing from a random sample to the population from which it was selected.  In this lesson, students build on their developing understanding of sampling variability from the previous lessons by constructing a dot plot of the means of different random samples selected from the same population.

In previous lessons, students have selected random samples by selecting items from a bag or by using a random-number generator from a website or graphing calculator.  This lesson adds another tool that can be used in selecting random samples, namely a random-number table.  A random-number table is just a table of digits from 0 to 9 that are arranged in a random order.  A random-number table and directions for using it to select a random sample are provided as part of the lesson.  Students may ask how the numbers in a random-number table are generated.  Today, computer programs are used to generate random number tables, but the idea behind these tables is relatively simple.  The digits 0, 1, 2, 3, 4, 5, 6, 7, 8, 9 are placed in a bag or jar (with each digit written on a slip of paper or a chip).  The 10 digits are thoroughly mixed.  A slip or chip is picked.  The number on the slip or chip is recorded, and then returned to the bag or jar.  This process is repeated, generating a random list of 0's to 9's.

## Classwork

### Example 1 (5 minutes):  What is a Sample Mean?

---

**Example 1:  Estimating a Population Mean**

The owners of a gym have been keeping track of how long each person spends at the gym.  Eight hundred of these times (in minutes) are shown in the population tables located at the end of the lesson.  These 800 times will form the <u>population</u> that you will investigate in this lesson.

Look at the values in the population.  Can you find the longest time spent in the gym in the population?  Can you find the shortest?

On average, roughly how long do you think people spend at the gym?  In other words, by just looking at the numbers in the two tables, make an estimate of the <u>population mean</u>.

You could find the population mean by typing all 800 numbers into a calculator or a computer, adding them up, and dividing by 800.  This would be extremely time-consuming, and usually it is not possible to measure every value in a population.

Instead of doing a calculation using every value in the population, we will use a <u>random sample</u> to find the mean of the sample.  The sample mean will then be used as an estimate of the population mean.

---

In this example, students see a population of 800 times spent by people at a gym.  The times are displayed in the population tables at the end of the lesson.  Convey to students that it is not practical to use all 800 times to calculate the mean.

It is important to make students aware how unusual it would be to know all of the population values.  Indeed, if you do have all the population values, then there is little point in taking a sample because you could calculate the population mean (using a computer if necessary).

If there is time, ask the students to:

- Think of a population where you do not have all the population values.
  - *Virtually any population will suffice in answer to this question; for example, the heights of all the adults in the town where the school is located.*
- Think of another population where you *do* have all the population values.
  - *This is less common, but one example might be the salaries of all the employees of a company.  All of the employee salaries will be stored in the company's computer.*

Discuss the questions in the example.  Provide students time to think about and respond to the questions.  The following responses provide a summary of the important points students should see as they work through the exercise.

- Look at the values in the population.  Can you find the longest time spent in the gym in the population?  Can you find the shortest?
  - *The longest time is 92 min., and the shortest time is 9 min.  Students will provide other estimates, but direct them to these times.*
- On average, roughly how long do you think people spend at the gym?  In other words, by just looking at the numbers in the two tables, make an estimate of the *population mean.*
  - *It seems from the numbers in the tables that the mean time spent at the gym is around 60 min.*

---

## Example 2 (5–7 minutes):  Selecting a Sample Using a Table of Random Digits

Carefully discuss the sample directions outlined in this section.  The goal is to recognize that each of the population values has an equal chance in being selected into the sample.  Introduce the random-number table and as students work through this example, make sure that they understand how the table is used to identify individuals to be included in the sample.  Once a sample is selected, the sample mean is calculated for this sample, and students are encouraged to think about how different samples would result in different values of the sample mean (this being an example of sampling variability).

---

**Example 2:  Selecting a Sample Using a Table of Random Digits**

The table of random digits provided with this lesson will be used to select items from a population to produce a random sample from the population.  The list of digits are determined by a computer program that simulates a random selection of the digits 0, 1, 2, 3, 4, 5, 6, 7, 8, or 9.  Imagine that each of these digits is written on a slip of paper and placed in a bag.  After thoroughly mixing the bag, one slip is drawn and its digit is recorded in this list of random digits.  The slip is then returned to the bag and another slip is selected.  The digit on this slip is recorded, and then returned to the bag.  The process is repeated over and over.  The resulting list of digits is called a random-number table.

---

Encourage students to think about the following question and to possibly share their suggestions prior to discussing the subsequent steps with all students.

---

**How could you use a table of random digits to take a random sample?**

**Step 1:  Place the table of random digits in front of you.  Without looking at the page, place the eraser end of your pencil somewhere on the table.  Start using the table of random digits at the number closest to where your eraser touched the paper.  This digit and the following two specify which observation from the population tables will be the first observation in your sample.**

For example, suppose the eraser end of your pencil lands on the twelfth number in row 3 of the random digit table.  This number is 5 and the two following numbers are 1 and 4.  This means that the first observation in your sample is observation number 514 from the population.  Find observation number 514 in the population table.  Do this by going to Row 51 and moving across to the column heading "4."  This observation is 53, so the first observation in your sample is 53.

If the number from the random-number table is any number 800 or greater, you will ignore this number and use the next three digits in the table.

---

Point out that four more selections are to be made to complete the random sample of five.  Emphasize that the next four observations are determined by continuing on from the starting position in the random number table.  The following directions and observations are important for students to understand:

---

**Step 2:  Continue using the table of random digits from the point you reached, and select the other four observations in your sample like you did above.**

For example, continuing on from the position in the example given in Step 1:

- The next number from the random-digit table is 716, and observation 716 is 63
- The next number from the random-digit table is 565, and observation 565 is 31.
- The next number from the random-digit table is 911, and there is no observation 911.  So we ignore these three digits.
- The next number from the random-digit table is 928, and there is no observation 928.  So we ignore these three digits.
- The next number from the random-digit table is 303, and observation 303 is 70.
- The next number from the random-digit table is 677, and observation 677 is 42.

---

## Exercises 1–4 (8–10 minutes)

Students should work independently on Exercises 1–4. Then discuss and confirm as a class.

---

**Exercises 1–4**

Initially, you will select just five values from the population to form your sample. This is a very small sample size, but it is a good place to start to understand the ideas of this lesson.

The values found in Example 2 are used to illustrate the answers to the following questions, but each student should select their own random sample of five.

1. Use the table of random number to select five values from the population of times. What are the five observations in your sample?

   *Make sure students write down these values and understand how they were selected. In the example given, the following observations were selected: 53 min., 63 min., 31 min., 70 min., and 42 min.*

2. For the sample that you selected, calculate the sample mean.

   *For the given example, the sample mean is* $\dfrac{53+63+31+70+42}{5} = 51.8\ min.$

3. You selected a random sample and calculated the sample mean in order to estimate the population mean. Do you think that the mean of these five observations is exactly correct for the population mean? Could the population mean be greater than the number you calculated? Could the population mean be less than the number you calculated?

   *Make sure that students see that the value of their sample mean ($51.8$ minutes in the given example, but students will have different sample means) is not likely to be exactly correct for the population mean. The population mean could be greater or less than the value of the sample mean.*

4. In practice, you only take one sample in order to estimate a population characteristic. But, for the purposes of this lesson, suppose you were to take another random sample from the same population of times at the gym. Could the new sample mean be closer to the population mean than the mean of these five observations? Could it be further from the population mean?

   *Make sure students understand that if they were to take a new random sample, the new sample mean is unlikely to be equal to the value of the sample mean of their first sample. It could be closer to or further from the population mean.*

---

## Exercises 5–7 (7 minutes)

Before students begin the following exercises, put all of their sample means on the board or poster paper. Then, ask the following questions: What do you notice about the sample means? Why do you think this is true?

Anticipate that students will indicate that the sample means posted are different and that several of them are very close to each other. They may also point out how different some of them are from each other. Students should recognize that the samples come from the same population, but that the actual selections can often be different. The different selections result in different sample means. Point out that the sample mean is an example of a sample statistic. The variation of sample statistics from sample to sample is called sampling variability.

Here, each student selects a random sample from the population and calculates the sample mean. Students should work individually, helping their neighbors as needed. The use of a calculator to calculate the sample mean should be encouraged to save time and increase accuracy. Point out that for this exercise, students should determine the mean to the nearest tenth of a minute.

---

**Exercises 5–7**

As a class, you will now investigate sampling variability by taking several samples from the same population. Each sample will have a different sample mean. This variation provides an example of sampling variability.

5.  Place the table of random digits in front of you and, without looking at the page, place the eraser end of your pencil somewhere on the table of random numbers. Start using the table of random digits at the number closest to where your eraser touches the paper. This digit and the following two specify which observation from the population tables will be the first observation in your sample. Write this three-digit number and the corresponding data value from the population in the space below.

    *Response: Answers will vary based on the numbers selected.*

6.  Continue moving to the right in the table of random digits from the place ended in Exercise 5. Use three digits at a time. Each set of three digits specifies which observation in the population is the next number in your sample. Continue until you have four more observations, and write these four values in the space below.

    *Response: Answers will vary based on the numbers selected.*

7.  Calculate the mean of the five values that form your sample. **Round your answer to the nearest tenth. Show your work and your sample mean in the space below.**

    *Response: Answers will vary based on the numbers selected.*

---

## Exercises 8–11 (11 minutes)

---

**Exercises 8–11**

You will now use the sample means from Exercise 7 from the entire class to make a dot plot.

8.  Write the sample means for everyone in the class in the space below.

    *Response: Answers will vary based on collected sample means.*

9.  Use all the sample means to make a dot plot using the axis given below. (Remember, if you have repeated or close values, stack the dots one above the other.)

10. What do you see in the dot plot that demonstrates sampling variability?

    *Sample response: The dots are spread out indicating that the sample means are not the same. The results indicate what we discussed as sampling variability. (See explanation provided at the end of this lesson.)*

---

11.  Remember that in practice you only take one sample.  (In this lesson, many samples were taken in order to
      demonstrate the concept of sampling variability.)  Suppose that a statistician plans to take a random sample of size
      5 from the population of times spent at the gym and that he or she will use the sample mean as an estimate of the
      population mean.  Approximately how far can the statistician expect the sample mean to be from the population
      mean?

      *Answers will vary.  Allow students to speculate as to what the value of the population mean might be and how far a
      sample mean would be from that value.  The distance students indicate will be based on the dot plot.  Students may
      indicate the sample mean to be exactly equal to the population and that the distance would be $0$.  They may also
      indicate the sample mean to be one of the dots that is a minimum or a maximum of the distribution, and suggest a
      distance from the minimum or maximum to the center of the distribution.  Students should begin to see that the
      distribution has a center that they suspect is close to the population's mean.*

Here the sample means for the class are combined in order to form a visual illustration of sampling variability (in the
form of a dot plot of the class's sample means.  Students are asked to show how the pattern in the dot plot supports the
notion of sampling variability.  Students will need to refer to the dot plot in their work in the next lesson, so every
student will need to complete the dot plot.

Exercise 10 summarizes an important part of this lesson.  The fact that the sample mean varies from sample to sample is
an example of *sampling variability*.  Clearly emphasize this summary.

Exercise 11 covers an important concept that is worth thinking about in advance of the class.  When you take a sample
and calculate the sample mean, you know that the value of the sample mean is very unlikely to be *exactly* equal to the
population mean.  But how close can you expect it to be?  The dot plot constructed in this lesson provides a way to
answering this question (at least for a random sample of size 5 from the population presented in this class.)  The
population mean is approximately in the *center* of the dots in the dot plot constructed in Exercise 9.  Indicate that the
distances of the dots from the population mean provides information about how far the statistician can expect the
sample mean to be from the population mean.

An important point is also made at the beginning of Exercise 11, which is worth emphasizing whenever the opportunity
arises in the remaining lessons.  In practice, you only take *one* sample from a population.  If you took two samples, then
the sensible thing would be to combine them into one larger sample.

We only take multiple samples in this lesson in order to demonstrate the concept of sampling variability.

- Suppose that you are going to take *one* sample from a population.  Once you have your sample, you will
  calculate the sample mean.  Explain what is meant by sampling variability in this context.
  - *Before taking your sample, you know that there are many different sets of individuals who could form
    your sample.  Every different possible sample could give a different value for the sample mean.  This is
    known as sampling variability.*

It is important that students be aware that the idea of sampling variability does not apply only to the sample mean.
Suppose, for example, an advertising agency plans to take a random sample of people and ask those people whether
they have seen a particular commercial.  The population is all the people who could be asked the question.  The
*proportion* of people in the sample who have seen the commercial is a sample statistic, and the value of this sample
statistic will vary from sample to sample.  This statistic also has sampling variability.  Simulations will be developed in
later lessons to illustrate sampling variability of proportions calculated using data from random samples of a population.
Lesson 20 begins students' investigations with the proportion as the statistic of interest.

**Population**

|    | 0  | 1  | 2  | 3  | 4  | 5  | 6  | 7  | 8  | 9  |
|----|----|----|----|----|----|----|----|----|----|----|
| 00 | 45 | 58 | 49 | 78 | 59 | 36 | 52 | 39 | 70 | 51 |
| 01 | 50 | 45 | 45 | 66 | 71 | 55 | 65 | 33 | 60 | 51 |
| 02 | 53 | 83 | 40 | 51 | 83 | 57 | 75 | 38 | 43 | 77 |
| 03 | 49 | 49 | 81 | 57 | 42 | 36 | 22 | 66 | 68 | 52 |
| 04 | 60 | 67 | 43 | 60 | 55 | 63 | 56 | 44 | 50 | 58 |
| 05 | 64 | 41 | 67 | 73 | 55 | 69 | 63 | 46 | 50 | 65 |
| 06 | 54 | 58 | 53 | 55 | 51 | 74 | 53 | 55 | 64 | 16 |
| 07 | 28 | 48 | 62 | 24 | 82 | 51 | 64 | 45 | 41 | 47 |
| 08 | 70 | 50 | 38 | 16 | 39 | 83 | 62 | 50 | 37 | 58 |
| 09 | 79 | 62 | 45 | 48 | 42 | 51 | 67 | 68 | 56 | 78 |
| 10 | 61 | 56 | 71 | 55 | 57 | 77 | 48 | 65 | 61 | 62 |
| 11 | 65 | 40 | 56 | 47 | 44 | 51 | 38 | 68 | 64 | 40 |
| 12 | 53 | 22 | 73 | 62 | 82 | 78 | 84 | 50 | 43 | 43 |
| 13 | 81 | 42 | 72 | 49 | 55 | 65 | 41 | 92 | 50 | 60 |
| 14 | 56 | 44 | 40 | 70 | 52 | 47 | 30 | 9  | 58 | 53 |
| 15 | 84 | 64 | 64 | 34 | 37 | 69 | 57 | 75 | 62 | 67 |
| 16 | 45 | 58 | 49 | 78 | 59 | 36 | 52 | 39 | 70 | 51 |
| 17 | 50 | 45 | 45 | 66 | 71 | 55 | 65 | 33 | 60 | 51 |
| 18 | 53 | 83 | 40 | 51 | 83 | 57 | 75 | 38 | 43 | 77 |
| 19 | 49 | 49 | 81 | 57 | 42 | 36 | 22 | 66 | 68 | 52 |
| 20 | 60 | 67 | 43 | 60 | 55 | 63 | 56 | 44 | 50 | 58 |
| 21 | 64 | 41 | 67 | 73 | 55 | 69 | 63 | 46 | 50 | 65 |
| 22 | 54 | 58 | 53 | 55 | 51 | 74 | 53 | 55 | 64 | 16 |
| 23 | 28 | 48 | 62 | 24 | 82 | 51 | 64 | 45 | 41 | 47 |
| 24 | 70 | 50 | 38 | 16 | 39 | 83 | 62 | 50 | 37 | 58 |
| 25 | 79 | 62 | 45 | 48 | 42 | 51 | 67 | 68 | 56 | 78 |
| 26 | 61 | 56 | 71 | 55 | 57 | 77 | 48 | 65 | 61 | 62 |
| 27 | 65 | 40 | 56 | 47 | 44 | 51 | 38 | 68 | 64 | 40 |
| 28 | 53 | 22 | 73 | 62 | 82 | 78 | 84 | 50 | 43 | 43 |
| 29 | 81 | 42 | 72 | 49 | 55 | 65 | 41 | 92 | 50 | 60 |
| 30 | 56 | 44 | 40 | 70 | 52 | 47 | 30 | 9  | 58 | 53 |
| 31 | 84 | 64 | 64 | 34 | 37 | 69 | 57 | 75 | 62 | 67 |
| 32 | 45 | 58 | 49 | 78 | 59 | 36 | 52 | 39 | 70 | 51 |
| 33 | 50 | 45 | 45 | 66 | 71 | 55 | 65 | 33 | 60 | 51 |
| 34 | 53 | 83 | 40 | 51 | 83 | 57 | 75 | 38 | 43 | 77 |
| 35 | 49 | 49 | 81 | 57 | 42 | 36 | 22 | 66 | 68 | 52 |
| 36 | 60 | 67 | 43 | 60 | 55 | 63 | 56 | 44 | 50 | 58 |
| 37 | 64 | 41 | 67 | 73 | 55 | 69 | 63 | 46 | 50 | 65 |
| 38 | 54 | 58 | 53 | 55 | 51 | 74 | 53 | 55 | 64 | 16 |
| 39 | 28 | 48 | 62 | 24 | 82 | 51 | 64 | 45 | 41 | 47 |

**Population, continued**

|  | 0 | 1 | 2 | 3 | 4 | 5 | 6 | 7 | 8 | 9 |
|---|---|---|---|---|---|---|---|---|---|---|
| 40 | 53 | 70 | 59 | 62 | 33 | 31 | 74 | 44 | 46 | 68 |
| 41 | 37 | 51 | 84 | 47 | 46 | 33 | 53 | 54 | 70 | 74 |
| 42 | 35 | 45 | 48 | 45 | 56 | 60 | 66 | 60 | 65 | 57 |
| 43 | 42 | 81 | 67 | 64 | 60 | 79 | 46 | 48 | 67 | 56 |
| 44 | 41 | 21 | 41 | 58 | 48 | 38 | 50 | 53 | 73 | 38 |
| 45 | 35 | 28 | 43 | 43 | 55 | 39 | 75 | 45 | 68 | 36 |
| 46 | 64 | 31 | 31 | 40 | 84 | 79 | 47 | 63 | 48 | 46 |
| 47 | 34 | 36 | 54 | 61 | 33 | 16 | 50 | 60 | 52 | 55 |
| 48 | 53 | 52 | 48 | 47 | 77 | 37 | 66 | 51 | 61 | 64 |
| 49 | 40 | 44 | 45 | 22 | 36 | 64 | 50 | 49 | 64 | 39 |
| 50 | 45 | 69 | 67 | 33 | 55 | 61 | 62 | 38 | 51 | 43 |
| 51 | 55 | 39 | 46 | 56 | 53 | 50 | 44 | 42 | 40 | 60 |
| 52 | 11 | 36 | 56 | 69 | 72 | 73 | 71 | 48 | 58 | 52 |
| 53 | 81 | 47 | 36 | 54 | 81 | 59 | 50 | 42 | 80 | 69 |
| 54 | 40 | 43 | 30 | 54 | 61 | 13 | 73 | 65 | 52 | 40 |
| 55 | 71 | 78 | 71 | 61 | 54 | 79 | 63 | 47 | 49 | 73 |
| 56 | 53 | 70 | 59 | 62 | 33 | 31 | 74 | 44 | 46 | 68 |
| 57 | 37 | 51 | 84 | 47 | 46 | 33 | 53 | 54 | 70 | 74 |
| 58 | 35 | 45 | 48 | 45 | 56 | 60 | 66 | 60 | 65 | 57 |
| 59 | 42 | 81 | 67 | 64 | 60 | 79 | 46 | 48 | 67 | 56 |
| 60 | 41 | 21 | 41 | 58 | 48 | 38 | 50 | 53 | 73 | 38 |
| 61 | 35 | 28 | 43 | 43 | 55 | 39 | 75 | 45 | 68 | 36 |
| 62 | 64 | 31 | 31 | 40 | 84 | 79 | 47 | 63 | 48 | 46 |
| 63 | 34 | 36 | 54 | 61 | 33 | 16 | 50 | 60 | 52 | 55 |
| 64 | 53 | 52 | 48 | 47 | 77 | 37 | 66 | 51 | 61 | 64 |
| 65 | 40 | 44 | 45 | 22 | 36 | 64 | 50 | 49 | 64 | 39 |
| 66 | 45 | 69 | 67 | 33 | 55 | 61 | 62 | 38 | 51 | 43 |
| 67 | 55 | 39 | 46 | 56 | 53 | 50 | 44 | 42 | 40 | 60 |
| 68 | 11 | 36 | 56 | 69 | 72 | 73 | 71 | 48 | 58 | 52 |
| 69 | 81 | 47 | 36 | 54 | 81 | 59 | 50 | 42 | 80 | 69 |
| 70 | 40 | 43 | 30 | 54 | 61 | 13 | 73 | 65 | 52 | 40 |
| 71 | 71 | 78 | 71 | 61 | 54 | 79 | 63 | 47 | 49 | 73 |
| 72 | 53 | 70 | 59 | 62 | 33 | 31 | 74 | 44 | 46 | 68 |
| 73 | 37 | 51 | 84 | 47 | 46 | 33 | 53 | 54 | 70 | 74 |
| 74 | 35 | 45 | 48 | 45 | 56 | 60 | 66 | 60 | 65 | 57 |
| 75 | 42 | 81 | 67 | 64 | 60 | 79 | 46 | 48 | 67 | 56 |
| 76 | 41 | 21 | 41 | 58 | 48 | 38 | 50 | 53 | 73 | 38 |
| 77 | 35 | 28 | 43 | 43 | 55 | 39 | 75 | 45 | 68 | 36 |
| 78 | 64 | 31 | 31 | 40 | 84 | 79 | 47 | 63 | 48 | 46 |
| 79 | 34 | 36 | 54 | 61 | 33 | 16 | 50 | 60 | 52 | 55 |

**Table of Random Digits**

| Row | | | | | | | | | | | | | | | | | | | | |
|-----|---|---|---|---|---|---|---|---|---|---|---|---|---|---|---|---|---|---|---|---|
| **1** | 6 | 6 | 7 | 2 | 8 | 0 | 0 | 8 | 4 | 0 | 0 | 4 | 6 | 0 | 3 | 2 | 2 | 4 | 6 | 8 |
| **2** | 8 | 0 | 3 | 1 | 1 | 1 | 1 | 2 | 7 | 0 | 1 | 9 | 1 | 2 | 7 | 1 | 3 | 3 | 5 | 3 |
| **3** | 5 | 3 | 5 | 7 | 3 | 6 | 3 | 1 | 7 | 2 | 5 | 5 | 1 | 4 | 7 | 1 | 6 | 5 | 6 | 5 |
| **4** | 9 | 1 | 1 | 9 | 2 | 8 | 3 | 0 | 3 | 6 | 7 | 7 | 4 | 7 | 5 | 9 | 8 | 1 | 8 | 3 |
| **5** | 9 | 0 | 2 | 9 | 9 | 7 | 4 | 6 | 3 | 6 | 6 | 3 | 7 | 4 | 2 | 7 | 0 | 0 | 1 | 9 |
| **6** | 8 | 1 | 4 | 6 | 4 | 6 | 8 | 2 | 8 | 9 | 5 | 5 | 2 | 9 | 6 | 2 | 5 | 3 | 0 | 3 |
| **7** | 4 | 1 | 1 | 9 | 7 | 0 | 7 | 2 | 9 | 0 | 9 | 7 | 0 | 4 | 6 | 2 | 3 | 1 | 0 | 9 |
| **8** | 9 | 9 | 2 | 7 | 1 | 3 | 2 | 9 | 0 | 3 | 9 | 0 | 7 | 5 | 6 | 7 | 1 | 7 | 8 | 7 |
| **9** | 3 | 4 | 2 | 2 | 9 | 1 | 9 | 0 | 7 | 8 | 1 | 6 | 2 | 5 | 3 | 9 | 0 | 9 | 1 | 0 |
| **10** | 2 | 7 | 3 | 9 | 5 | 9 | 9 | 3 | 2 | 9 | 3 | 9 | 1 | 9 | 0 | 5 | 5 | 1 | 4 | 2 |
| **11** | 0 | 2 | 5 | 4 | 0 | 8 | 1 | 7 | 0 | 7 | 1 | 3 | 0 | 4 | 3 | 0 | 6 | 4 | 4 | 4 |
| **12** | 8 | 6 | 0 | 5 | 4 | 8 | 8 | 2 | 7 | 7 | 0 | 1 | 0 | 1 | 7 | 1 | 3 | 5 | 3 | 4 |
| **13** | 4 | 2 | 6 | 4 | 5 | 2 | 4 | 2 | 6 | 1 | 7 | 5 | 6 | 6 | 4 | 0 | 8 | 4 | 1 | 2 |
| **14** | 4 | 4 | 9 | 8 | 7 | 3 | 4 | 3 | 8 | 2 | 9 | 1 | 5 | 3 | 5 | 9 | 8 | 9 | 2 | 9 |
| **15** | 6 | 4 | 8 | 0 | 0 | 0 | 4 | 2 | 3 | 8 | 1 | 8 | 4 | 0 | 9 | 5 | 0 | 9 | 0 | 4 |
| **16** | 3 | 2 | 3 | 8 | 4 | 8 | 8 | 6 | 2 | 9 | 1 | 0 | 1 | 9 | 9 | 3 | 0 | 7 | 3 | 5 |
| **17** | 6 | 6 | 7 | 2 | 8 | 0 | 0 | 8 | 4 | 0 | 0 | 4 | 6 | 0 | 3 | 2 | 2 | 4 | 6 | 8 |
| **18** | 8 | 0 | 3 | 1 | 1 | 1 | 1 | 2 | 7 | 0 | 1 | 9 | 1 | 2 | 7 | 1 | 3 | 3 | 5 | 3 |
| **19** | 5 | 3 | 5 | 7 | 3 | 6 | 3 | 1 | 7 | 2 | 5 | 5 | 1 | 4 | 7 | 1 | 6 | 5 | 6 | 5 |
| **20** | 9 | 1 | 1 | 9 | 2 | 8 | 3 | 0 | 3 | 6 | 7 | 7 | 4 | 7 | 5 | 9 | 8 | 1 | 8 | 3 |
| **21** | 9 | 0 | 2 | 9 | 9 | 7 | 4 | 6 | 3 | 6 | 6 | 3 | 7 | 4 | 2 | 7 | 0 | 0 | 1 | 9 |
| **22** | 8 | 1 | 4 | 6 | 4 | 6 | 8 | 2 | 8 | 9 | 5 | 5 | 2 | 9 | 6 | 2 | 5 | 3 | 0 | 3 |
| **23** | 4 | 1 | 1 | 9 | 7 | 0 | 7 | 2 | 9 | 0 | 9 | 7 | 0 | 4 | 6 | 2 | 3 | 1 | 0 | 9 |
| **24** | 9 | 9 | 2 | 7 | 1 | 3 | 2 | 9 | 0 | 3 | 9 | 0 | 7 | 5 | 6 | 7 | 1 | 7 | 8 | 7 |
| **25** | 3 | 4 | 2 | 2 | 9 | 1 | 9 | 0 | 7 | 8 | 1 | 6 | 2 | 5 | 3 | 9 | 0 | 9 | 1 | 0 |
| **26** | 2 | 7 | 3 | 9 | 5 | 9 | 9 | 3 | 2 | 9 | 3 | 9 | 1 | 9 | 0 | 5 | 5 | 1 | 4 | 2 |
| **27** | 0 | 2 | 5 | 4 | 0 | 8 | 1 | 7 | 0 | 7 | 1 | 3 | 0 | 4 | 3 | 0 | 6 | 4 | 4 | 4 |
| **28** | 8 | 6 | 0 | 5 | 4 | 8 | 8 | 2 | 7 | 7 | 0 | 1 | 0 | 1 | 7 | 1 | 3 | 5 | 3 | 4 |
| **29** | 4 | 2 | 6 | 4 | 5 | 2 | 4 | 2 | 6 | 1 | 7 | 5 | 6 | 6 | 4 | 0 | 8 | 4 | 1 | 2 |
| **30** | 4 | 4 | 9 | 8 | 7 | 3 | 4 | 3 | 8 | 2 | 9 | 1 | 5 | 3 | 5 | 9 | 8 | 9 | 2 | 9 |
| **31** | 6 | 4 | 8 | 0 | 0 | 0 | 4 | 2 | 3 | 8 | 1 | 8 | 4 | 0 | 9 | 5 | 0 | 9 | 0 | 4 |
| **32** | 3 | 2 | 3 | 8 | 4 | 8 | 8 | 6 | 2 | 9 | 1 | 0 | 1 | 9 | 9 | 3 | 0 | 7 | 3 | 5 |
| **33** | 6 | 6 | 7 | 2 | 8 | 0 | 0 | 8 | 4 | 0 | 0 | 4 | 6 | 0 | 3 | 2 | 2 | 4 | 6 | 8 |
| **34** | 8 | 0 | 3 | 1 | 1 | 1 | 1 | 2 | 7 | 0 | 1 | 9 | 1 | 2 | 7 | 1 | 3 | 3 | 5 | 3 |
| **35** | 5 | 3 | 5 | 7 | 3 | 6 | 3 | 1 | 7 | 2 | 5 | 5 | 1 | 4 | 7 | 1 | 6 | 5 | 6 | 5 |
| **36** | 9 | 1 | 1 | 9 | 2 | 8 | 3 | 0 | 3 | 6 | 7 | 7 | 4 | 7 | 5 | 9 | 8 | 1 | 8 | 3 |
| **37** | 9 | 0 | 2 | 9 | 9 | 7 | 4 | 6 | 3 | 6 | 6 | 3 | 7 | 4 | 2 | 7 | 0 | 0 | 1 | 9 |
| **38** | 8 | 1 | 4 | 6 | 4 | 6 | 8 | 2 | 8 | 9 | 5 | 5 | 2 | 9 | 6 | 2 | 5 | 3 | 0 | 3 |
| **39** | 4 | 1 | 1 | 9 | 7 | 0 | 7 | 2 | 9 | 0 | 9 | 7 | 0 | 4 | 6 | 2 | 3 | 1 | 0 | 9 |
| **40** | 9 | 9 | 2 | 7 | 1 | 3 | 2 | 9 | 0 | 3 | 9 | 0 | 7 | 5 | 6 | 7 | 1 | 7 | 8 | 7 |

## Closing (4 minutes)

If there is time, ask students these questions:

- You have learned today about sampling variability. Do you want sampling variability to be large or small?
    - *We want sampling variability to be small.*
- Why do you want sampling variability to be small?
    - *Suppose you are using the sample mean to estimate a population mean. You want the sampling variability of the sample mean to be small, so that you can expect your value of the sample mean to be close to the population mean.*

---

### Lesson Summary

A population characteristic is estimated by taking a random sample from the population and calculating the value of a statistic for the sample. For example, a population mean is estimated by selecting a random sample from the population and calculating the sample mean.

The value of the sample statistic (e.g., the sample mean) will vary based on the random sample that is selected. This variation from sample to sample in the values of the sample statistic is called <u>sampling variability</u>.

---

## Exit Ticket (5–8 minutes)

Name _____   Date_____

# Lesson 17:  Sampling Variability

Suppose that you want to estimate the mean time per evening students at your school spend doing homework.  You will do this using a random sample of 30 students.

1.  Suppose that you have a list of all the students at your school.  The students are numbered $1, 2, 3, \ldots$.  One way to select the random sample of students is to use the random-digit table from today's class, taking three digits at a time.  If you start at the third digit of row 9, what is the number of the first student you would include in your sample?

2.  Suppose that you have now selected your random sample and you have asked each student how long he or she spends doing homework each evening.  How will you use these results to estimate the mean time spent doing homework for *all* students?

3.  Explain what is meant by *sampling variability* in this context.

## Exit Ticket Sample Solutions

Suppose that you want to estimate the mean time per evening students at your school spend doing homework. You will do this using a random sample of 30 students.

1.  Suppose that you have a list of all the students at your school. The students are numbered 1, 2, 3, …. One way to select the random sample of students is to use the random-digit table from today's class, taking three digits at a time. If you start at the third digit of row 9, what is the number of the first student you would include in your sample?

    *The first student in the sample would be student number* 229.

2.  Suppose that you have now selected your random sample, and you have asked each student how long he or she spends doing homework each evening. How will you use these results to estimate the mean time spent doing homework for *all* students?

    *I would calculate the mean time spent doing homework for the students in my sample.*

3.  Explain what is meant by *sampling variability* in this context.

    *Different samples of students would result in different values of the sample mean. This is sampling variability of the sample mean.*

## Problem Set Sample Solutions

1.  Yousef intends to buy a car. He wishes to estimate the mean fuel efficiency (in miles per gallon) of all cars available at this time. Yousef selects a random sample of 10 cars and looks up their fuel efficiencies on the Internet. The results are shown below.

    22  25  29  23  31  29  28  22  23  27

    a.  Yousef will estimate the mean fuel efficiency of all cars by calculating the mean for his sample. Calculate the sample mean, and record your answer below. (Be sure to show your work.)

    $$\frac{22 + 25 + 29 + 23 + 31 + 29 + 28 + 22 + 23 + 27}{10} = 25.9$$

    b.  In practice, you only take one sample to estimate a population characteristic. However, if Yousef were to take another random sample of 10 cars from the same population, would he likely get the same value for the sample mean?

    *No, it is not likely that Yousef would get the same value for the sample mean.*

    c.  What if Yousef were to take *many* random samples of 10 cars? Would the entire sample means be the same?

    *No, he could get many different values of the sample mean.*

    d.  Using this example, explain what sampling variability means.

    *The fact that the sample mean will vary from sample to sample is an example of sampling variability.*

2.  Think about the mean number of siblings (brothers and sisters) for all students at your school.

    a.  Approximately what do you think is the value of the mean number of siblings for the population of all students at your school?

    *Answers will vary.*

    b.  How could you find a better estimate of this population mean?

    *I could take a random sample of students, ask each student in my sample how many siblings he or she has and find the mean for my sample.*

    c.  Suppose that you have now selected a random sample of students from your school. You have asked all of the students in your sample how many siblings they have. How will you calculate the sample mean?

    *I will add up all of the values in the sample and divide by the number of students in the sample.*

    d.  If you had taken a different sample, would the sample mean have taken the same value?

    *No, a different sample would generally produce a different value of the sample mean. It is possible, but unlikely, that the sample mean for a different sample would have the same mean.*

    e.  There are many different samples of students that you could have selected. These samples produce many different possible sample means. What is the phrase used for this concept?

    *Sampling variability.*

    f.  Does the phrase you gave in part (e) apply only to sample means?

    *No, the concept of sampling variability applies to any sample statistic.*

**Lesson 18: Sampling Variability and the Effect of Sample Size**

## Student Outcomes

- Students use data from a random sample to estimate a population mean.
- Students know that increasing the sample size decreases the sampling variability of the sample mean.

## Lesson Notes

In this lesson, students will need their work from the previous lesson. In particular, they will need the population tables, the random-digit table, and the dot plot of student sample mean values. Remind students that the goal of the investigation in the previous lesson was to answer the statistical question, "What is the typical time spent at the gym for people in a population of gym members?" Remind students that a statistical question is a question that can be answered by collecting data *and* anticipating variability in the data collected. (Students were introduced to statistical questions in Grade 6.) The question is revisited in this lesson, where students will see the connection between sampling variability and the size of the sample.

## Classwork

### Example 1 (6 minutes): Sampling Variability

This example reminds students of the concept of sampling variability, and introduces the idea that you want sampling variability to be small.

Let the students think about the questions posed Example 1. In this lesson, students will begin to understand the connection between sampling variability and sample size. For this example, however, have students simply focus on the variability of the two dot plots and the statistical question they are answering.

Discuss the questions posed as a class. Pay careful attention to terminology.

---

**Example 1: Sampling Variability**

The previous lesson investigated the statistical question, "What is the typical time spent at the gym?" by selecting random samples from the population of 800 gym members. Two different dot plots of sample means calculated from random samples from the population are displayed below. The first dot plot represents the means of 20 samples with each sample having 5 data points. The second dot plot represents the means of 20 samples with each sample having 15 data points.

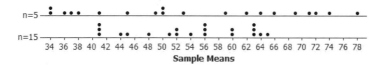

---

Based on the first dot plot, Jill answered the statistical question by indicating the mean time people spent at the gym was between 34 and 78 minutes. She decided that a time approximately in the middle of that interval would be her estimate of the mean time the 800 people spent at the gym. She estimated 52 minutes. Scott answered the question using the second dot plot. He indicated that the mean time people spent at the gym was between 41 and 65 minutes. He also selected a time of 52 minutes to answer the question.

- Describe the differences in the two dot plots.

   *The first dot plot has a more variability in the sample means than the second dot plot. The sampling variability is greater for the first dot plot.*

- Which dot plot do you feel more confident in using to answer the statistical question? Explain your answer.

   *Possible response: The second dot plot gives me more confidence because the sample means do not differ as much from one another. They are more tightly clustered, so I think I have a better idea of where the population mean is located.*

- In general, do you want sampling variability to be large or small? Explain.

   *The larger the sampling variability, the more that the value of a sample statistic will vary from one sample to another and the further you can expect a sample statistic value to be from the population characteristic. You want the value of the sample statistic to be close to the population characteristic. So, you want sampling variability to be small.*

## Exercises 1–3 (8 minutes)

Before students begin the exercises, ask them what would minimize sampling variability. Write down some of their suggestions and discuss them. A possible response might be that they would take lots of random samples. Indicate that taking more random samples would not necessarily reduce the sampling variability. If students suggest that increasing the number of observations in a sample would reduce the sampling variability, indicate that this will be covered in the next set of exercises.

**Exercises 1–3**

In the previous lesson, you saw a population of 800 times spent at the gym. You will now select a random sample of size 15 from that population. You will then calculate the sample mean.

1. Start by selecting a three-digit number from the table of random digits. Place the random-digit table in front of you. Without looking at the page, place the eraser end of your pencil somewhere on the table of random digits. Start using the table of random digits at the digit closest to your eraser. This digit and the following two specify which observation from the population will be the first observation in your sample. Write the value of this observation in the space below. (Discard any three-digit number that is 800 or larger, and use the next three digits from the random-digit table.)

   *Answers will vary.*

2. Continue moving to the right in the table of random digits from the point that you reached in part (a). Each three-digit number specifies a value to be selected from the population. Continue in this way until you have selected 14 more values from the population. This will make 15 values altogether. Write the values of all 15 observations in the space below.

   *Answers will vary.*

3. Calculate the mean of your 15 sample values. Write the value of your sample mean below. Round your answer to the nearest tenth. (Be sure to show your work.)

   *Answers will vary.*

In the previous lesson, students selected a random sample of size 5 from the population. Here they use the same process to select a random sample of size 15. As in the previous lesson, students may want to use calculators in this exercise. Students should work on this exercise individually, but should help their neighbor if that person is having difficulty remembering the process for selecting a random sample from the given population.

Either students or the teacher should write the sample means on the board.

### Exercises 4–6 (12 minutes)

Here, all the students' sample means are shown on a dot plot, so that a comparison can be made with the dot plot of sample means from samples of size 5 made in the previous lesson. The fact that the new dot plot (of sample means for samples of size 15) shows a smaller spread of values than the dot plot from the previous lesson (of sample means for samples of size 5) is a demonstration of the main point of this lesson: that increasing the sample size decreases the sampling variability of the sample mean. Students are creating a coherent representation of what is happening as the sample size increases. This representation provides students a way to attend to the meaning of what smaller spread of values indicate as the sample size is increased.

---

**Exercises 4–6**

You will now use the sample means from Exercise 3 for the entire class to make a dot plot.

4.  Write the sample means for everyone in the class in the space below.

5.  Use all the sample means to make a dot plot using the axis given below. (Remember, if you have repeated values or values close to each other, stack the dots one above the other.)

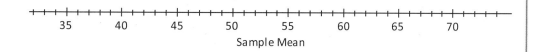

Sample Mean

6.  In the previous lesson, you drew a dot plot of sample means for samples of size 5. How does the dot plot above (of sample means for samples of size 15) compare to the dot plot of sample means for samples of size 5? For which sample size (5 or 15) does the sample mean have the greater sampling variability?

    *The dot plots will vary depending on the results of the random sampling. Dot plots for one set of sample means for 20 random samples of size 5 and for 20 random samples of size 15 are shown below. The main thing for students to notice is that there is less variability from sample to sample for the larger sample size.*

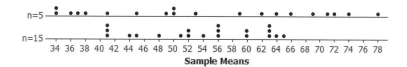

This exercise illustrates the notion that the greater the sample size, the smaller the sampling variability of the sample mean.

---

If you feel that there is time, and if you think the students will be able to understand the concepts involved, consider asking the question below. (This is a potentially complicated question, so avoid it unless you think students are ready for it.)

- Why do you think the sample mean has less variability when the sample size is large?

  - *When the sample size is small, you could get a sample that contains many large values (making the sample mean large), or many small values (making the sample mean small). But for a larger sample size, it is more likely that the values are equally spread between large and small, and any extreme values are averaged-out with the other values to produce a more central value for the sample mean.*

## Exercises 7–8 (5 minutes)

The following exercises ask students to estimate, using the dot plot, a typical degree of accuracy of the sample mean when used as an estimate of the population mean. Use a sample of size 15 from the population used in this lesson as an example. This is a tricky concept, so you could ask them to work in pairs, or if you think students will have difficulty, you could present this exercise as an example.

The exercise starts with the reminder that in practice you only take *one* sample. Multiple samples are taken in these lessons in order to demonstrate (and analyze) the sampling variability of the sample mean. This is a point that is worth repeating when the opportunity arises.

---

**Exercises 7–8**

7.  Remember that in practice you only take one sample. Suppose that a statistician plans to take a random sample of size 15 from the population of times spent at the gym and will use the sample mean as an estimate of the population mean. Based on the dot plot of sample means that your class collected from the population, approximately how far can the statistician expect the sample mean to be from the population mean? (The actual population mean is 53.9 minutes.)

    *Answers will vary according to the degree of variability that appears in the dot plot and a student's estimate of an average distance from the population mean. Allow students to use an approximation of 54 minutes for the population mean. In the example above, the 20 samples could be used to estimate the mean distance of the sample means to the population mean of 54 minutes.*

| Sample Mean | Distance from Population Mean | | Sample Mean | Distance from Population Mean |
|---|---|---|---|---|
| 41 | 13 | | 56 | 2 |
| 41 | 13 | | 56 | 2 |
| 41 | 13 | | 56 | 2 |
| 44 | 10 | | 60 | 6 |
| 45 | 9 | | 60 | 6 |
| 48 | 6 | | 63 | 9 |
| 51 | 3 | | 63 | 9 |
| 52 | 2 | | 63 | 9 |
| 52 | 2 | | 64 | 10 |
| 54 | 0 | | 65 | 11 |

*The sum of the distances from the mean in the above example is 137. The mean of these distances, or the expected distance of a sample mean from the population mean, is 6.85 minutes.*

8.  How would your answer in Exercise 7 compare to the equivalent mean of the distances for a sample of size 5?

    *Response: My answer for Exercise 7 is smaller than the expected distance for the samples of size 5. For samples of size 5, several dots are farther from the mean of 54 minutes. The mean of the distances for samples of size 5 would be a larger.*

---

## Exercises 9–11 (8 minutes)

This exercise may be omitted if you are short of time. Here, students are asked to imagine what the dot plot of sample means would look like for samples of size 25.

---

**Exercises 9–11**

Suppose everyone in your class selected a random sample of size 25 from the population of times spent at the gym.

9. What do you think the dot plot of the class's sample means would look like? Make a sketch using the axis below.

   *The student's sketch should show dots that have less spread than those in the dot plot for samples of size 15. For example, the student's dot plot might look like this:*

   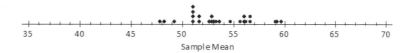

   Sample Mean

10. Suppose that a statistician plans to estimate the population mean using a sample of size 25. According to your sketch, approximately how far can the statistician expect the sample mean to be from the population mean?

    *We were told in Exercise 7 that the population mean is 53.9. If you calculate the mean of the distances from the population mean (in the same way you did in Exercise 7), the average or expected distance of a sample mean from 53.9 is approximately 3 minutes for a dot plot similar to the one above. This estimate is made by approximating the average distance of each dot from 53.9 or 54. Note: If necessary, you can make a chart similar to what was suggested in Exercise 7. Using the dot plot, direct students to estimate the distance of each dot from 54 (a rounding to the nearest whole number is adequate for this question), add up the distances and divide by the number of dots.*

11. Suppose you have a choice of using a sample of size 5, 15, or 25. Which of the three makes the sampling variability of the sample mean the smallest? Why would you choose the sample size that makes the sampling variability of the sample mean as small as possible?

    *Choosing a sample size of 25 will make the sampling variability of the sample mean the smallest, which is preferable because the sample mean is then more likely to be close to the population mean than it would be for the smaller sample sizes.*

---

## Closing (2 minutes)

This question would be particularly beneficial if Exercises 9–11 have been omitted.

- When using a sample mean to estimate a population mean, why do we prefer a larger sample?
  - *A larger sample is preferable because then the sample mean has smaller sampling variability, which means that the sample mean is likely to be closer to the true value of the population mean than it would be for a smaller sample.*

---

Lesson Summary

Suppose that you plan to take a random sample from a population. You will use the value of a sample statistic to estimate the value of a population characteristic. You want the sampling variability of the sample statistic to be small so that the sample statistic is likely to be close to the value of the population characteristic.

When you increase the sample size, the sampling variability of the sample mean is decreased.

---

## Exit Ticket (5–8 minutes)

Name _____   Date_____

# Lesson 18:  Sampling Variability and the Effect of Sample Size

Exit Ticket

Suppose that you wanted to estimate the mean time per evening spent doing homework for students at your school. You decide to do this by taking a random sample of students from your school.  You will calculate the mean time spent doing homework for your sample.  You will then use your sample mean as an estimate of the population mean.

1.  The sample mean has "sampling variability."  Explain what this means.

2.  When you are using a sample statistic to estimate a population characteristic, do you want the sampling variability of the sample statistic to be large or small?  Explain why.

3.  Think about your estimate of the mean time spent doing homework for students at your school.  Given a choice of using a sample of size 20 or a sample of size 40, which should you choose?  Explain your answer.

## Exit Ticket Sample Solutions

Suppose that you wanted to estimate the mean time per evening spent doing homework for students at your school. You decide to do this by taking a random sample of students from your school. You will calculate the mean time spent doing homework for your sample. You will then use your sample mean as an estimate of the population mean.

1. The sample mean has "sampling variability." Explain what this means.

*There are many different possible samples of students at my school, and the value of the sample mean varies from sample to sample.*

2. When you are using a sample statistic to estimate a population characteristic, do you want the sampling variability of the sample statistic to be large or small? Explain why.

*You want the sampling variability of the sample statistic to be small because then you can expect the value of your sample statistic to be close to the value of the population characteristic that you are estimating.*

3. Think about your estimate of the mean time spent doing homework for students at your school. Given a choice of using a sample of size 20 or a sample of size 40, which should you choose? Explain your answer.

*I would use a sample of size 40, because then the sampling variability of the sample mean would be smaller than it would be for a sample of size 20.*

## Problem Set Sample Solutions

1. The owner of a new coffee shop is keeping track of how much each customer spends (in dollars). One hundred of these amounts are shown in the table below. These amounts will form the *population* for this question.

|   | 0 | 1 | 2 | 3 | 4 | 5 | 6 | 7 | 8 | 9 |
|---|---|---|---|---|---|---|---|---|---|---|
| 0 | 6.18 | 4.67 | 4.01 | 4.06 | 3.28 | 4.47 | 4.86 | 4.91 | 3.96 | 6.18 |
| 1 | 4.98 | 5.42 | 5.65 | 2.97 | 2.92 | 7.09 | 2.78 | 4.20 | 5.02 | 4.98 |
| 2 | 3.12 | 1.89 | 4.19 | 5.12 | 4.38 | 5.34 | 4.22 | 4.27 | 5.25 | 3.12 |
| 3 | 3.90 | 4.47 | 4.07 | 4.80 | 6.28 | 5.79 | 6.07 | 7.64 | 6.33 | 3.90 |
| 4 | 5.55 | 4.99 | 3.77 | 3.63 | 5.21 | 3.85 | 7.43 | 4.72 | 6.53 | 5.55 |
| 5 | 4.55 | 5.38 | 5.83 | 4.10 | 4.42 | 5.63 | 5.57 | 5.32 | 5.32 | 4.55 |
| 6 | 4.56 | 7.67 | 6.39 | 4.05 | 4.51 | 5.16 | 5.29 | 6.34 | 3.68 | 4.56 |
| 7 | 5.86 | 4.75 | 4.94 | 3.92 | 4.84 | 4.95 | 4.50 | 4.56 | 7.05 | 5.86 |
| 8 | 5.00 | 5.47 | 5.00 | 5.70 | 5.71 | 6.19 | 4.41 | 4.29 | 4.34 | 5.00 |
| 9 | 5.12 | 5.58 | 6.16 | 6.39 | 5.93 | 3.72 | 5.92 | 4.82 | 6.19 | 5.12 |

   a. Place the table of random digits in front of you. Select a starting point without looking at the page. Then, taking two digits at a time, select a random sample of size 5 from the population above. Write the 5 values in the space below. (For example, suppose you start at the third digit of row four of the random digit table. Taking two digits gives you 19. In the population above, go to the row labeled "1" and move across to the column labeled 9. This observation is 4.98, and that will be the first observation in your sample. Then continue in the random-digit table from the point you reached.)

   *For example, (starting in the random-digit table at the 8th digit in row 15)*

   6.10, 3.54, 9.85, 7.64, 7.23, 3.61, 4.31, 6.10, 9.85, 3.39

 Lesson 18: Sampling Variability and the Effect of Sample Size

213

Calculate the mean for your sample, showing your work. Round your answer to the nearest thousandth.

$$\frac{6.10 + 3.54 + 9.85 + 7.64 + 7.23 + 3.61 + 4.31 + 6.10 + 9.85 + 3.39}{10} = 6.162$$

b. Using the same approach as in part (a), select a random sample of size 20 from the population.

*For example, (continuing in the random-digit table from the point reached in Part (a))*

$5.24, 8.00, 4.98, 5.36, 3.80, 2.83, 4.76, 7.34, 4.78, 5.03, 9.85, 5.03, 5.22, 4.13, 4.71, 5.22, 2.39,$
$4.13, 5.45, 5.45$

Calculate the mean for your sample of size 20. Round your answer to the nearest thousandth.

$$\frac{5.24 + \cdots + 5.45}{20} = 5.235$$

c. Which of your sample means is likely to be the better estimate of the population mean? Explain your answer in terms of sampling variability.

*The sample mean from the sample of size 20 is likely to be the better estimate, since larger samples result in smaller sampling variability of the sample mean.*

2. Two dot plots are shown below. One of the dot plots shows the values of some sample means from random samples of size 5 from the population given in Problem 1. The other dot plot shows the values of some sample means from random samples of size 20 from the population given in Problem 1.

**Dot Plot A**

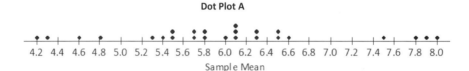

Sample Mean

**Dot Plot B**

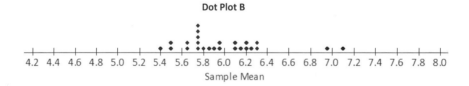

Sample Mean

Which dot plot is for sample means from samples of size 5, and which dot plot is for sample means from samples of size 20? Explain your reasoning.

The sample means from samples of size 5 are shown in Dot Plot  *A* .

*The variability is greater than in dot plot B.*

The sample means from samples of size 20 are shown in Dot Plot  *B* .

*The variability is smaller compared to dot plot A, which implies that the sample size was greater.*

3. You are going to use a random sample to estimate the mean travel time for getting to school for all the students in your grade. You will select a random sample of students from your grade. Explain why you would like the sampling variability of the sample mean to be *small*.

*I would like the sampling variability of the sample mean to be small because then it is likely that my sample mean will be close to the mean time for all students at the school.*

# Lesson 19: Understanding Variability when Estimating a Population Proportion

## Student Outcomes

- Students understand the term *sampling variability* in the context of estimating a population proportion.
- Students know that increasing the sample size decreases sampling variability.

## Lesson Notes

In a previous lesson, students investigated sampling variability in the sample mean for numerical data. In this lesson they will investigate sampling variability in the sample proportion for categorical data. Distinguishing between the sample mean for numerical data and the sample proportion for the categorical data is important. This lesson begins with students investigating these differences. It is important that students see the similarity between working with a proportion and what they did in the previous lessons when working with the mean; that is, as the sample size increases, the sampling variability of sample proportions decreases. The dot plots provide students a visual representation of sampling variability. Dot plots are also used to show the connection between sampling variability and sample size.

Before students start the exercises, prepare bags containing 100 cubes (or beads, or slips of paper, etc.) for each group of four students. Forty cubes should be red, and the remaining 60 cubes can be any other color (e.g., orange, blue, green, yellow). Also, prepare a number line (illustrated in Example 1) that is visible on the board or on poster paper.

## Classwork

As a warm-up activity, discuss how the data used to calculate a proportion (such as the proportion of gym members that are female) is different than the data used to determine a mean (such as the mean time people spend in the gym or the mean number of words in a children's literature book). Other examples of a proportion are: the proportion of people at a particular school who run a marathon, or the proportion of Great Lake perch that are 5 cm or less. Understanding what a proportion is and how the type of data used to calculate a proportion is different from the type of data they used to calculate a mean is important for the next couple of lessons. This difference is not always easy for students to see, so be deliberate in pointing this out with each new data scenario that is introduced.

Before students begin working with the example in their lesson, discuss the following statistical questions with students. Ask students what data they would collect to answer the questions, and what they would do with the data. Point out to the students those situations in which they would find the average or mean of the data and those situations in which they would find a proportion or a percent.

*Question 1:* How many hours of sleep do seventh graders get on a night in which there is school the next day?

Data on the number of hours students in a sample of seventh graders sleep would be collected. This would be numerical data and the mean of the numbers would be used to answer this statistical question.

**Question 2:** What proportion of the students in our school participate in band or orchestra?

A sample of students could be asked to indicate whether or not they participate in band or orchestra. This data would not be numerical. The proportion of the students participating in band for the sample would be used to answer this statistical question. (Point out that it would not be possible to calculate a mean based on this data.)

**Question 3:** What is the typical weight of a backpack for students at our school?

Data from a sample of students at the school on the weight of backpacks would be collected. This data would be numerical and the mean of the weights would be used to answer this statistical question.

**Question 4:** What is the likelihood that voters in a certain city will vote for building a new high school?

A sample of voters could be obtained. Each voter in the sample would be asked whether or not they would vote for building a new high school. This data is not numerical. The proportion of the voters who indicated they would vote "yes" for the sample would be used to answer this question.

After the warm-up, proceed with the example in the student lesson.

---

In a previous lesson, you selected several random samples from a population. You recorded values of a numerical variable. You then calculated the mean for each sample, saw that there was variability in the sample means, and created a distribution of sample means to better see the sampling variability. You then considered larger samples and saw that the variability in the distribution decreased when the sample size increases. In this lesson, you will use a similar process to investigate variability in sample proportions.

---

### Example 1 (10 minutes): Sample Proportion

Prior to the start of the lesson prepare the following:

- A bag containing 100 cubes (or beads, or slips of paper, etc.) for each group of four students. Forty should be red, and the remaining 60 cubes can be any other color.
- On the board draw a number line similar to the one shown below.

```
  0    0.1   0.2   0.3   0.4   0.5   0.6   0.7   0.8   0.9    1
```

Begin the lesson by saying, "Today we are going to learn something about the population of cubes in this bag. It is full of cubes. Based on your previous work, how could you learn something about the whole bag of cubes?" After this discussion, instruct each student to take a random sample of 10 cubes from the bag. This sample should be taken with replacement. You may want to begin by taking a sample as a demonstration.

---

**Example 1: Sample Proportion**

Your teacher will give your group a bag that contains colored cubes, some of which are red. With your classmates, you are going to build a distribution of sample proportions.

---

Each student should record the color of each draw, and then calculate the proportion of red in their sample of 10.

- How would you determine the sample proportion of red cubes?
    - $Sample\ proportion = \dfrac{number\ of\ successes\ in\ the\ sample}{total\ number\ of\ observations\ in\ the\ sample}$

Have students place post-it notes on the number line to build a dot plot. After all of the post-it notes have been placed, explain to students that the graph is the *sampling distribution* of the sample proportions.

## Exercises 1–6 (10 minutes)

Allow students to work with their groups on Exercises 1–6. Then discuss as a class.

As students discuss the answers to Exercises 4 and 5, note that the shape of the distribution of sample proportions is often mound shaped. For the variability students need to focus on the overall spread in the sample proportions and the main cluster around the center of the distribution.

An extension is to have the students calculate MAD for the distribution.

---

**Exercises 1–6**

1.  Each person in your group should randomly select a sample of 10 cubes from the bag. Record the data for your sample in the table below.

    *Students' tables will vary based their samples.*

    | Cube | Outcome (Color) |
    |------|-----------------|
    | 1 | Green |
    | 2 | Orange |
    | 3 | Red |
    | 4 | Red |
    | 5 | Blue |
    | 6 | Orange |
    | 7 | Red |
    | 8 | Green |
    | 9 | Blue |
    | 10 | Blue |

2.  What is the proportion of red cubes in your sample of 10?

    This value is called the sample proportion. The sample proportion is found by dividing the number of "successes" (in this example, the number of red cubes) by the total number of observations in the sample.

    *Student results will be around $0.4$. In this example, the sample proportion is $0.3$.*

---

3.  Write your sample proportion on a post-it note and place it on the number line that your teacher has drawn on the board. Place your note above the value on the number line that corresponds to your sample proportion.

The graph of all the students' sample proportions is called a sampling distribution of the sample proportions.

*This is an example of a dot plot of the sampling distribution.*

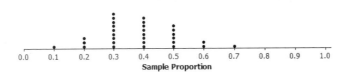

4.  Describe the shape of the distribution.

    *Nearly symmetrical distribution that is clustered around 0.4.*

5.  Describe the variability in the sample proportions.

    *The spread of the data is from 0.1 to 0.7. Much of the data clusters between 0.3 and 0.5.*

6.  Based on the distribution, answer the following:

    a.  What do you think is the population proportion?

        *Based on the dot plot, an estimate of the population proportion is approximately 0.4.*

    b.  How confident are you of your estimate?

        *Because there is a lot of variability from sample to sample (0.1 to 0.7), there is not a lot of confidence in the estimate.*

## Example 2 (10 minutes): Sampling Variability

This example parallels Example 1. Students now take a random sample of 30 cubes from the bag. This sample should be taken with replacement.

> **Example 2: Sampling Variability**
>
> What do you think would happen to the sampling distribution if everyone in class took a random sample of 30 cubes from the bag? To help answer this question you will repeat the random sampling you did in Exercise 1, except now you will draw a random sample of 30 cubes instead of 10.

Students record their sample results and find the sample proportion of red cubes in the bag. Again they should record the proportion on a post-it note.

Draw another number line on the board using the same scale as in Example 1.

## Exercises 7–15 (10 minutes)

Allow students to work with their groups on Exercises 7–10.  Then discuss the answers as a class.

---

**Exercises 7–15**

What do you think would happen to the sampling distribution if everyone in class took a random sample of 30 cubes from the bag?  To help answer this question you will repeat the random sampling you did in Exercise 1, except now you will draw a random sample of 30 cubes instead of 10.

7.  Take a random sample of 30 cubes from the bag.  Carefully record the outcome of each draw.

*Answers will vary.  An example follows:*

| 1 | 2 | 3 | 4 | 5 | 6 | 7 | 8 | 9 | 10 |
|---|---|---|---|---|---|---|---|---|---|
| Blue | Green | Red | Yellow | Blue | Red | Red | Green | Blue | Red |

| 11 | 12 | 13 | 14 | 15 | 16 | 17 | 18 | 19 | 20 |
|---|---|---|---|---|---|---|---|---|---|
| Green | Red | Red | Green | Yellow | Blue | Blue | Yellow | Red | Blue |

| 21 | 22 | 23 | 24 | 25 | 26 | 27 | 28 | 29 | 30 |
|---|---|---|---|---|---|---|---|---|---|
| Red | Green | Red | Red | Blue | Red | Yellow | Green | Green | Yellow |

8.  What is the proportion of red cubes in your sample of 30?

*Answers will vary.  In this example the sample proportion is $\frac{13}{30}$ or $0.433$.*

9.  Write your sample proportion on a post-it note and place the note on the number line that your teacher has drawn on the board.  Place your note above the value on the number line that corresponds to your sample proportion.

*An example of a dot plot:*

**Dot Plot of Sample Proportions for n=30**

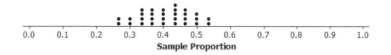

Sample Proportion

10.  Describe the shape of the distribution.

*Mound shaped, centered around $0.43$*

---

Students should work independently on Exercises 11–15.  Then discuss as a class.

11. Describe the variability in the sample proportions.

    *Spread of the data is from 0.25 to 0.55. Most of the data clusters between 0.35 and 0.5.*

12. Based on the distribution, answer the following:
    a.  What do you think is the population proportion?

        *Based on the dot plot, an estimate of the population proportion is approximately 0.43.*

    b.  How confident are you of your estimate?

        *Because there is a less variability from sample to sample (0.35 to 0.5), there is more confidence in the estimate.*

    c.  If you were taking a random sample of 30 cubes and determined the proportion that was red, do you think your sample proportion will be within 0.05 of the population proportion? Explain.

        *Answers depend on the dot plots prepared by students. If the dot plot in Exercise 9 is used as an example, note that only about half of the dots are between 0.35 to 0.45. There are several samples that had sample proportions that were farther away from the center than 0.05, so the sample proportion might not be within 0.05 of the population proportion.*

13. Compare the sampling distribution based on samples of size 10 to the sampling distribution based on samples of size 30.

    *Both distributions are mound shaped and center around 0.4. Variability is less in the sampling distribution of sample sizes of 30 versus sample sizes of 10.*

14. As the sample size increased from 10 to 30 describe what happened to the sampling variability of the sample proportions.

    *The sampling variability decreased as the sample size increased.*

15. What do you think would happen to the variability of the sample proportions if the sample size for each sample was 50 instead of 30? Explain.

    *The variability in the sampling distribution for samples of size 50 will be less than the variability of the sampling distribution for samples of size 30.*

MP.2

## Closing (5 minutes)

To highlight the Lesson Summary, ask students the following questions:

- How do you know if the data collected is summarized by a mean or by a proportion?
  - *Anticipate responses that indicate the different types of data collected and how the data is summarized. Examples similar to those discussed in the lesson would indicate an understanding of when a mean is calculated and when a proportion is calculated.*

- What would a dot plot of a sampling distribution of the sample proportion look like if proportions from many different random samples were used to create the plot?
  - *Anticipate responses that describe a dot plot in which each dot represents the sample proportion of one sample. The graph would show variability (or sampling variability) with a cluster around a value close to the value of the proportion for the population. Consider asking students to sketch a dot plot that they might get for one of the examples in which a proportion is used to answer the statistical question.*

- What happens to the sampling distribution as the sample size increases?

  □ *Anticipate responses that indicate the sampling variability (spread of the dots in the dot plot) decreases as the sample size increases.*

---

**Lesson Summary**

- The sampling distribution of the sample proportion is a graph of the sample proportions for many different samples.

- The mean of the sample proportions will be approximately equal to the value of the population proportion.

- As the sample size increases the sampling variability decreases.

---

**Exit Ticket (5 minutes)**

Name _____    Date_____

# Lesson 19:  Understanding Variability when Estimating a Population Proportion

A group of seventh graders took repeated samples of size 20 from a bag of colored cubes.  The dot plot below shows the sampling distribution of the sample proportion of blue cubes in the bag.

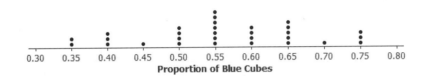

1.  Describe the shape of the distribution.

2.  Describe the variability of the distribution.

3.  Predict how the dot plot would look differently if the sample sizes had been 40 instead of 20.

## Exit Ticket Sample Solutions

A group of seventh graders took repeated samples of size 20 from a bag of colored cubes. The dot plot below shows the sampling distribution of the sample proportion of blue cubes in the bag.

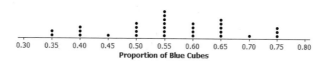

Proportion of Blue Cubes

1.  Describe the shape of the distribution.

    *Mound shaped, centered around* $0.6$.

2.  Describe the variability of the distribution.

    *The spread of the data is from* $0.35$ *to* $0.75$, *with much of the data between* $0.5$ *and* $0.65$.

3.  Predict how the dot plot would look differently if the sample sizes had been 40 instead of 20.

    *The variability will decrease as the sample size increases. The dot plot will be centered in a similar place but will be less spread out.*

## Problem Set Sample Solutions

1.  A class of seventh graders wanted to find the proportion of M&M's that are red. Each seventh grader took a random sample of 20 M&M's from a very large container of M&M's. Following is the proportion of red M&M's each student found.

| | | | | | | | | |
|---|---|---|---|---|---|---|---|---|
| 0.15 | 0 | 0.1 | 0.1 | 0.05 | 0.1 | 0.2 | 0.05 | 0.1 |
| 0.1 | 0.15 | 0.2 | 0 | 0.1 | 0.15 | 0.15 | 0.1 | 0.2 |
| 0.3 | 0.1 | 0.1 | 0.2 | 0.1 | 0.15 | 0.1 | 0.05 | 0.3 |

a.  Construct a dot plot of the sample proportions.

### Dot Plot of Proportion of Red M&Ms

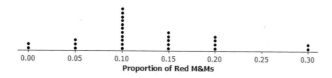

Proportion of Red M&Ms

b.  Describe the shape of the distribution.

    *Somewhat mounded shaped, slightly skewed to the right.*

c.   Describe the variability of the distribution.

*Spread of the data is from* $0.0$ *to* $0.3$. *Most of the data clusters between* $0.10$ *and* $0.20$.

    d.   Suppose the seventh grade students had taken random samples of size $50$. Describe how the sampling distribution would change from the one you constructed in part (a).

*The sampling variability would decrease.*

2.   A group of seventh graders wanted to estimate the proportion of middle school students who suffer from allergies. One group of seventh graders each took a random sample of $10$ middle school students, and another group of seventh graders each took a random sample of $40$ middle school students. Below are two sampling distributions of the sample proportions of middle school students who said that they suffer from allergies. Which dot plot is based on random samples of size $40$? How can you tell?

Dot Plot A

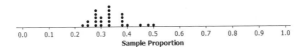

Dot Plot B

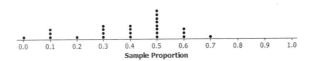

*Dot Plot A, because the variability of the distribution is less than the variability in Dot Plot B.*

3.   The nurse in your school district would like to study the proportion of middle school students who usually get at least eight hours of sleep on school nights. Suppose each student in your class plans on taking a random sample of $20$ middle school students from your district, and each calculates a sample proportion of students who said that they usually get at least eight hours of sleep on school nights.

    a.   Do you expect everyone in your class to get the same value for their sample proportion? Explain.

*No, we expect sample variability.*

    b.   Suppose each student in class increased the sample size from $20$ to $40$. Describe how you could reduce the sampling variability.

*I could reduce the sampling variability by using the larger sample size.*

# Lesson 20: Estimating a Population Proportion

## Student Outcome

- Students use data from a random sample to estimate a population proportion.

## Lesson Notes

In this lesson, students continue to work with random samples and the distribution of the sample proportions. The focus in this lesson is to derive the center of the sample proportions (or the mean of the sample proportions). Students begin to see how the distribution clusters around the mean of the distribution. This center is used to estimate the population proportion.

In preparation of this lesson, provide students or small groups of students the random-number table and the table of data for all 200 students in the middle school described in the exercises. Students will use the random-number table to select their random samples in the same way they used the random-number table in the previous lesson.

## Classwork

In a previous lesson, each student in your class selected a random sample from a population and calculated the sample proportion. It was observed that there was sampling variability in the sample proportions, and as the sample size increased, the variability decreased. In this lesson, you will investigate how sample proportions can be used to estimate population proportions.

### Example 1 (9 minutes): Mean of Sample Proportions

This example is similar to the data that students worked with in the previous lesson. The main idea is to have the students focus on the center of the distribution of sample proportions as an estimate for the population proportion. For some students the vocabulary can be problematic. Students are still learning the ideas behind samples and population.

Summarize the problems from the previous lesson by asking the following questions:

- How many samples are needed to calculate the sample proportion?
  - *The sample proportion is the result from one random sample.*
- How is the distribution of the sample proportions formed?
  - *The distribution of the sample proportions is a dot plot of the results from many randomly selected samples.*
- What is the population proportion?
  - *The population proportion is the actual value of the proportion of the population who would respond "yes" to the survey.*

**Example 1:  Mean of Sample Proportions**

A class of 30 seventh graders wanted to estimate the proportion of middle school students who were vegetarians. Each seventh grader took a random sample of 20 middle-school students. Students were asked the question, "Are you a vegetarian?" One sample of 20 students had three students who said that they were vegetarians. For this sample, the sample proportion is $\frac{3}{20}$ or $0.15$. Following are the proportions of vegetarians the seventh graders found in 30 samples. Each sample was of size 20 students. The proportions are rounded to the nearest hundredth.

| | | | | | | | | | |
|------|------|------|------|------|------|------|------|------|------|
| 0.15 | 0.10 | 0.15 | 0.00 | 0.10 | 0.15 | 0.10 | 0.10 | 0.05 | 0.20 |
| 0.25 | 0.15 | 0.25 | 0.25 | 0.30 | 0.20 | 0.10 | 0.20 | 0.05 | 0.10 |
| 0.10 | 0.30 | 0.15 | 0.05 | 0.25 | 0.15 | 0.20 | 0.10 | 0.20 | 0.15 |

## Exercises 1–9 (19 minutes)

Allow students to work in small groups on Exercises 1–9. Then discuss and confirm as a class.

**Exercises 1–9**

1.  The first student reported a sample proportion of $0.15$. Interpret this value in terms of the summary of the problem in the example.

    *Three of the 20 students surveyed responded that they were vegetarian.*

2.  Another student reported a sample proportion of $0$. Did this student do something wrong when selecting the sample of middle school students?

    *No, this means that none of the 20 students surveyed said that they were vegetarian.*

3.  Assume you were part of this seventh grade class and you got a sample proportion of $0.20$ from a random sample of middle school students. Based on this sample proportion, what is your estimate for the proportion of all middle school students who are vegetarians?

    *My estimate is $0.20$.*

4.  Construct a dot plot of the 30 sample proportions.

5.  Describe the shape of the distribution.

    *Nearly symmetrical or mound shaped centering at approximately $0.15$.*

6.  Using the 30 class results listed above, what is your estimate for the proportion of all middle school students who are vegetarians? Explain how you made this estimate.

    *About $0.15$. I chose this value because the sample proportions tend to cluster between $0.10$ and $0.15$, or $0.10$ and $0.20$.*

7. Calculate the mean of the 30 sample proportions. How close is this value to the estimate you made in Exercise 6?

*The mean of the 30 samples to the nearest thousandth is 0.153. The value is close to my estimate of 0.15, and if calculated to the nearest hundredth, they would be the same. (Most likely students will say between 0.10 and 0.15.)*

8. The proportion of all middle school students who are vegetarians is 0.15. This is the actual proportion for the entire population of middle school students used to select the samples. How the mean of the 30 sample proportions compares with the actual population proportion depends on the students' samples.

*In this case, the mean of the 30 sample proportions is very close to the actual population proportion.*

9. Do the sample proportions in the dot plot tend to cluster around the value of the population proportion? Are any of the sample proportions far away from 0.15? List the proportions that are far away from 0.15.

*They cluster around 0.15. The value of 0 and 0.30 are far away from the 0.15.*

## Example 2 (4 minutes): Estimating Population Proportion

This example asks students to work with data from a middle school of 200 students. Although the school is fictitious, the data were obtained from actual middle school students and are representative of middle school students' responses. A list of the entire 200 students' responses is provided at the end of the lesson. The data was collected from the website, http://www.amstat.org/censusatschool/. Details describing the Census at School project are also available on the website of the American Statistical Association, or http://www.amstat.org/

In this lesson, students are directed to analyze the last question summarized in the data file of the 200 students at Roosevelt Middle School. If students are more interested in one of the other questions listed, the exercise could be redirected or expanded to include analyzing the data from one of these questions.

---

**Example 2: Estimating Population Proportion**

Two hundred middle school students at Roosevelt Middle School responded to several survey questions. A printed copy of the responses the students gave to various questions is provided with this lesson.

The data are organized in columns and are summarized by the following table:

| Column Heading | Description |
|---|---|
| ID | Numbers from 1 to 200 |
| Travel to School | Method used to get to school<br>Walk, car, rail, bus, bike, skateboard, boat, other |
| Favorite Season | Summer, fall, winter, spring |
| Allergies | Yes or no |
| Favorite School Subject | Art, English, languages, social studies, history, geography, music, science, computers, math, PE, other |
| Favorite Music | Classical, Country, heavy metal, jazz, pop, punk rock, rap, reggae, R&B, rock and roll, techno, gospel, other |
| What superpower would you like? | Invisibility, super strength, telepathy, fly, freeze time |

The last column in the data file is based on the question: Which of the following superpowers would you most like to have? The choices were: invisibility, super-strength, telepathy, fly, or freeze time.

The class wants to determine the proportion of Roosevelt Middle School students who answered freeze time to the last question. You will use a sample of the Roosevelt Middle School population to estimate the proportion of the students who answered freeze time to the last question.

---

There are several options for obtaining random samples of 20 responses. It is anticipated that some classes can complete the exercise in the timeframe indicated; but it is also likely that other classes will require more time, which may require extending this lesson by another class period. One option is to provide each student a printed copy of the data file. A list of the data file in a table format is provided at the end of the student lesson and also at the end of the teacher notes. This option requires copying the data file for each student. A second option would be to provide small groups of students a copy of the data file, and allowing them to work in groups.

**MP.6** The standards for this lesson expect students will be involved in obtaining their own sample, and using the proportion derived from their sample to estimate the population proportion. By examining the distribution of sample proportions from many random samples, students see that sample proportions tend to cluster around the value of the population proportion. Students attend to precision in carefully describing how they use samples to describe the population.

The number of samples needed to illustrate this is a challenge. The more samples the class can generate, the more clearly the distribution of sample proportions will cluster around the value of the population proportion. For this lesson, a workable range would be between 20 to 30 samples.

Discuss how to obtain a random sample of size 20 from the 200 students represented in the data file. The student ID numbers should be used to select a student from the data file. The table of random digits that was used in previous lessons is provided in this lesson. Students drop their pencil on the random table and use the position of one end of the pencil (e.g., the eraser) as the starting point for generating 20 three-digit random numbers from 001 to 200. The ID numbers should be considered as three-digit numbers and used to obtain a random sample of 20 students. Students will read three digits in order from their starting point on the table as the student ID (e.g., $0 - 0 - 3$ is the selection of the student with ID number 3; $0 - 6 - 4$ is the selection of the student with ID number 64; $1 - 9 - 3$ is the student with ID 193). Any ID number formed in this way that is greater that 200 is simply disregarded, and students move on to form the next three-digit number from the random-number table. Indicate to students that they move to the top of the table if they reached the last digit in the table. If a number corresponding to a student that has already been selected into the sample appears again, students should ignore that number and move on to form another three-digit number.

After students obtain their 20 ID numbers, they connect the ID numbers to the students in the data file to generate a sample of 20 responses.

### Exercises 10–17 (14 minutes)

Let students work with their groups on Exercises 10–17. Then discuss answers as a class.

---

**Exercises 10–17**

A random sample of 20 student responses is needed. You are provided the random number table you used in a previous lesson. A printed list of the 200 Roosevelt Middle School students is also provided. In small groups, complete the following exercise:

10. Select a random sample of 20 student responses from the data file. Explain how you selected the random sample.

    *Generate 20 random numbers between 1 and 200. The random number chosen represents the ID number of the student. Go to that ID numbered row and record the outcome as a "yes" or "no" in the table regarding the freeze time response.*

---

11.  In the table below list the 20 responses for your sample.

*Answers will vary.  Below is one possible result.*

|    | Response |
|----|----------|
| 1  | *Yes*    |
| 2  | *No*     |
| 3  | *No*     |
| 4  | *No*     |
| 5  | *Yes*    |
| 6  | *No*     |
| 7  | *No*     |
| 8  | *No*     |
| 9  | *No*     |
| 10 | *Yes*    |
| 11 | *Yes*    |
| 12 | *No*     |
| 13 | *No*     |
| 14 | *No*     |
| 15 | *Yes*    |
| 16 | *No*     |
| 17 | *No*     |
| 18 | *No*     |
| 19 | *No*     |
| 20 | *No*     |

12.  Estimate the population proportion of students who responded "freeze time" by calculating the sample proportion of the 20 sampled students who responded "freeze time" to the question.

*Student answers will vary.  The sample proportion in the given example is $\frac{5}{20}$ or $0.25$.*

13.  Combine your sample proportion with other students' sample proportions and create a dot plot of the distribution of the sample proportions of students who responded "freeze time" to the question.

*An example is shown below.  Your class dot plot may differ somewhat from the one below, but the distribution should center at approximately $0.20$.  (Provide students this distribution of sample proportions if they were unable to obtain a distribution.)*

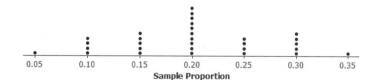

14.  By looking at the dot plot, what is the value of the proportion of the 200 Roosevelt Middle School students who responded "freeze time" to the question?

$0.20$

15.  Usually you will estimate the proportion of Roosevelt Middle School students using just a single sample proportion. How different was your sample proportion from your estimate based on the dot plot of many samples?

*Student answers will vary depending on their sample proportion.  For this example, the sample proportion is $0.25$, which is slightly greater than the $0.20$.*

16. Circle your sample proportion on the dot plot. How does your sample proportion compare with the mean of all the sample proportions?

*The mean of class distribution will vary from this example. The class distribution should center at approximately 0.20.*

17. Calculate the mean of all of the sample proportions. Locate the mean of the sample proportions in your dot plot; mark this position with an "$X$." How does the mean of the sample proportions compare with your sample proportion?

*Answers will vary based on the samples generated by students.*

---

### Lesson Summary

The sample proportion from a random sample can be used to estimate a population proportion. The sample proportion will not be exactly equal to the population proportion, but values of the sample proportion from random samples tend to cluster around the actual value of the population proportion.

---

**Exit Ticket (4 minutes)**

Name _____     Date_____

# Lesson 20:  Estimating a Population Proportion

Exit Ticket

Thirty seventh graders each took a random sample of 10 middle school students and asked each student whether or not they like pop music.  Then they calculated the proportion of students who like pop music for each sample.  The dot plot below shows the distribution of the sample proportions.

**Dot Plot of Sample Proportions for n=10**

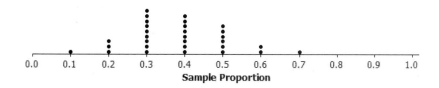

1.  There are three dots above 0.2.  What does each dot represent in terms of this scenario?

2.  Based on the dot plot, do you think the proportion of the middle school students at this school who like pop music is 0.6?  Explain why or why not.

## Exit Ticket Sample Solutions

Thirty seventh graders each took a random sample of 10 middle school students and asked each student whether or not they like pop music. Then they calculated the proportion of students who like pop music for each sample. The dot plot below shows the distribution of the sample proportions.

**Dot Plot of Sample Proportions for n=10**

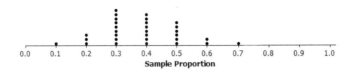

Sample Proportion

1. There are three dots above 0.2. What does each dot represent in terms of this scenario?

   *Each dot represents the survey results from one student.  0.2 means two students out of 10 said they like pop music.*

2. Based on the dot plot, do you think the proportion of the middle school students at this school who like pop music is 0.6? Explain why or why not

   *No. Based on the dot plot, 0.6 is not a likely proportion. The dots cluster at 0.3 to 0.5, and only a few dots were located at 0.6. An estimate of the proportion of students at this school who like pop music would be within the cluster of 0.3 to 0.5.*

## Problem Set Sample Solutions

1. A class of 30 seventh graders wanted to estimate the proportion of middle school students who played a musical instrument. Each seventh grader took a random sample of 25 middle school students and asked each student whether or not they played a musical instrument. Following are the sample proportions the seventh graders found in 30 samples.

   | | | | | | | | | | |
   |---|---|---|---|---|---|---|---|---|---|
   | 0.80 | 0.64 | 0.72 | 0.60 | 0.60 | 0.72 | 0.76 | 0.68 | 0.72 | 0.68 |
   | 0.72 | 0.68 | 0.68 | 0.76 | 0.84 | 0.60 | 0.80 | 0.72 | 0.76 | 0.80 |
   | 0.76 | 0.60 | 0.80 | 0.84 | 0.68 | 0.68 | 0.70 | 0.68 | 0.64 | 0.72 |

   a. The first student reported a sample proportion of 0.8. What does this value mean in terms of this scenario?

      *A sample proportion of 0.8 means 20 out of 25 answered "yes" to the survey.*

   b. Construct a dot plot of the 30 sample proportions.

   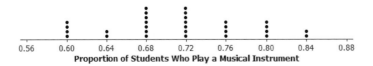

   Proportion of Students Who Play a Musical Instrument

c.   Describe the shape of the distribution.

*Nearly symmetrical. It centers at approximately* $0.72$.

d.   Describe the variability of the distribution.

*The spread of the distribution is from* $0.6$ *to* $0.84$.

e.   Using the 30 class sample proportions listed above, what is your estimate for the proportion of all middle school students who played a musical instrument?  Explain how you made this estimate.

*The mean of the 30 sample proportions is* $0.713$.

2.   Select another variable or column from the data file that is of interest.  Take a random sample of 30 students from the list and record the response to your variable of interest of each of the 30 students.

a.   Based on your random sample what is your estimate for the proportion of all middle school students?

*Student answers will vary depending on the column chosen.*

b.   If you selected a second random sample of 30, would you get the same sample proportion for the second random sample that you got for the first random sample?  Explain why or why not.

*No, it is very unlikely that you would get exactly the same result.  This is sampling variability—the value of a sample statistic will vary from one sample to another.*

## Table of Random Digits

| Row | | | | | | | | | | | | | | | | | | | | |
|---|---|---|---|---|---|---|---|---|---|---|---|---|---|---|---|---|---|---|---|---|
| 1 | 6 | 6 | 7 | 2 | 8 | 0 | 0 | 8 | 4 | 0 | 0 | 4 | 6 | 0 | 3 | 2 | 2 | 4 | 6 | 8 |
| 2 | 8 | 0 | 3 | 1 | 1 | 1 | 1 | 2 | 7 | 0 | 1 | 9 | 1 | 2 | 7 | 1 | 3 | 3 | 5 | 3 |
| 3 | 5 | 3 | 5 | 7 | 3 | 6 | 3 | 1 | 7 | 2 | 5 | 5 | 1 | 4 | 7 | 1 | 6 | 5 | 6 | 5 |
| 4 | 9 | 1 | 1 | 9 | 2 | 8 | 3 | 0 | 3 | 6 | 7 | 7 | 4 | 7 | 5 | 9 | 8 | 1 | 8 | 3 |
| 5 | 9 | 0 | 2 | 9 | 9 | 7 | 4 | 6 | 3 | 6 | 6 | 3 | 7 | 4 | 2 | 7 | 0 | 0 | 1 | 9 |
| 6 | 8 | 1 | 4 | 6 | 4 | 6 | 8 | 2 | 8 | 9 | 5 | 5 | 2 | 9 | 6 | 2 | 5 | 3 | 0 | 3 |
| 7 | 4 | 1 | 1 | 9 | 7 | 0 | 7 | 2 | 9 | 0 | 9 | 7 | 0 | 4 | 6 | 2 | 3 | 1 | 0 | 9 |
| 8 | 9 | 9 | 2 | 7 | 1 | 3 | 2 | 9 | 0 | 3 | 9 | 0 | 7 | 5 | 6 | 7 | 1 | 7 | 8 | 7 |
| 9 | 3 | 4 | 2 | 2 | 9 | 1 | 9 | 0 | 7 | 8 | 1 | 6 | 2 | 5 | 3 | 9 | 0 | 9 | 1 | 0 |
| 10 | 2 | 7 | 3 | 9 | 5 | 9 | 9 | 3 | 2 | 9 | 3 | 9 | 1 | 9 | 0 | 5 | 5 | 1 | 4 | 2 |
| 11 | 0 | 2 | 5 | 4 | 0 | 8 | 1 | 7 | 0 | 7 | 1 | 3 | 0 | 4 | 3 | 0 | 6 | 4 | 4 | 4 |
| 12 | 8 | 6 | 0 | 5 | 4 | 8 | 8 | 2 | 7 | 7 | 0 | 1 | 0 | 1 | 7 | 1 | 3 | 5 | 3 | 4 |
| 13 | 4 | 2 | 6 | 4 | 5 | 2 | 4 | 2 | 6 | 1 | 7 | 5 | 6 | 6 | 4 | 0 | 8 | 4 | 1 | 2 |
| 14 | 4 | 4 | 9 | 8 | 7 | 3 | 4 | 3 | 8 | 2 | 9 | 1 | 5 | 3 | 5 | 9 | 8 | 9 | 2 | 9 |
| 15 | 6 | 4 | 8 | 0 | 0 | 0 | 4 | 2 | 3 | 8 | 1 | 8 | 4 | 0 | 9 | 5 | 0 | 9 | 0 | 4 |
| 16 | 3 | 2 | 3 | 8 | 4 | 8 | 8 | 6 | 2 | 9 | 1 | 0 | 1 | 9 | 9 | 3 | 0 | 7 | 3 | 5 |
| 17 | 6 | 6 | 7 | 2 | 8 | 0 | 0 | 8 | 4 | 0 | 0 | 4 | 6 | 0 | 3 | 2 | 2 | 4 | 6 | 8 |
| 18 | 8 | 0 | 3 | 1 | 1 | 1 | 1 | 2 | 7 | 0 | 1 | 9 | 1 | 2 | 7 | 1 | 3 | 3 | 5 | 3 |
| 19 | 5 | 3 | 5 | 7 | 3 | 6 | 3 | 1 | 7 | 2 | 5 | 5 | 1 | 4 | 7 | 1 | 6 | 5 | 6 | 5 |
| 20 | 9 | 1 | 1 | 9 | 2 | 8 | 3 | 0 | 3 | 6 | 7 | 7 | 4 | 7 | 5 | 9 | 8 | 1 | 8 | 3 |
| 21 | 9 | 0 | 2 | 9 | 9 | 7 | 4 | 6 | 3 | 6 | 6 | 3 | 7 | 4 | 2 | 7 | 0 | 0 | 1 | 9 |
| 22 | 8 | 1 | 4 | 6 | 4 | 6 | 8 | 2 | 8 | 9 | 5 | 5 | 2 | 9 | 6 | 2 | 5 | 3 | 0 | 3 |
| 23 | 4 | 1 | 1 | 9 | 7 | 0 | 7 | 2 | 9 | 0 | 9 | 7 | 0 | 4 | 6 | 2 | 3 | 1 | 0 | 9 |
| 24 | 9 | 9 | 2 | 7 | 1 | 3 | 2 | 9 | 0 | 3 | 9 | 0 | 7 | 5 | 6 | 7 | 1 | 7 | 8 | 7 |
| 25 | 3 | 4 | 2 | 2 | 9 | 1 | 9 | 0 | 7 | 8 | 1 | 6 | 2 | 5 | 3 | 9 | 0 | 9 | 1 | 0 |
| 26 | 2 | 7 | 3 | 9 | 5 | 9 | 9 | 3 | 2 | 9 | 3 | 9 | 1 | 9 | 0 | 5 | 5 | 1 | 4 | 2 |
| 27 | 0 | 2 | 5 | 4 | 0 | 8 | 1 | 7 | 0 | 7 | 1 | 3 | 0 | 4 | 3 | 0 | 6 | 4 | 4 | 4 |
| 28 | 8 | 6 | 0 | 5 | 4 | 8 | 8 | 2 | 7 | 7 | 0 | 1 | 0 | 1 | 7 | 1 | 3 | 5 | 3 | 4 |
| 29 | 4 | 2 | 6 | 4 | 5 | 2 | 4 | 2 | 6 | 1 | 7 | 5 | 6 | 6 | 4 | 0 | 8 | 4 | 1 | 2 |
| 30 | 4 | 4 | 9 | 8 | 7 | 3 | 4 | 3 | 8 | 2 | 9 | 1 | 5 | 3 | 5 | 9 | 8 | 9 | 2 | 9 |
| 31 | 6 | 4 | 8 | 0 | 0 | 0 | 4 | 2 | 3 | 8 | 1 | 8 | 4 | 0 | 9 | 5 | 0 | 9 | 0 | 4 |
| 32 | 3 | 2 | 3 | 8 | 4 | 8 | 8 | 6 | 2 | 9 | 1 | 0 | 1 | 9 | 9 | 3 | 0 | 7 | 3 | 5 |
| 33 | 6 | 6 | 7 | 2 | 8 | 0 | 0 | 8 | 4 | 0 | 0 | 4 | 6 | 0 | 3 | 2 | 2 | 4 | 6 | 8 |
| 34 | 8 | 0 | 3 | 1 | 1 | 1 | 1 | 2 | 7 | 0 | 1 | 9 | 1 | 2 | 7 | 1 | 3 | 3 | 5 | 3 |
| 35 | 5 | 3 | 5 | 7 | 3 | 6 | 3 | 1 | 7 | 2 | 5 | 5 | 1 | 4 | 7 | 1 | 6 | 5 | 6 | 5 |
| 36 | 9 | 1 | 1 | 9 | 2 | 8 | 3 | 0 | 3 | 6 | 7 | 7 | 4 | 7 | 5 | 9 | 8 | 1 | 8 | 3 |
| 37 | 9 | 0 | 2 | 9 | 9 | 7 | 4 | 6 | 3 | 6 | 6 | 3 | 7 | 4 | 2 | 7 | 0 | 0 | 1 | 9 |
| 38 | 8 | 1 | 4 | 6 | 4 | 6 | 8 | 2 | 8 | 9 | 5 | 5 | 2 | 9 | 6 | 2 | 5 | 3 | 0 | 3 |
| 39 | 4 | 1 | 1 | 9 | 7 | 0 | 7 | 2 | 9 | 0 | 9 | 7 | 0 | 4 | 6 | 2 | 3 | 1 | 0 | 9 |
| 40 | 9 | 9 | 2 | 7 | 1 | 3 | 2 | 9 | 0 | 3 | 9 | 0 | 7 | 5 | 6 | 7 | 1 | 7 | 8 | 7 |

EUREKA MATH™ | Lesson 20:    Estimating a Population Proportion

| ID | Travel to School | Favorite Season | Allergies | Favorite School Subject | Favorite Music | Superpower |
|----|------------------|-----------------|-----------|-------------------------|----------------|------------|
| 1 | Car | Spring | Yes | English | Pop | Freeze time |
| 2 | Car | Summer | Yes | Music | Pop | Telepathy |
| 3 | Car | Summer | No | Science | Pop | Fly |
| 4 | Walk | Fall | No | Computers and technology | Pop | Invisibility |
| 5 | Car | Summer | No | Art | Country | Telepathy |
| 6 | Car | Summer | No | Physical education | Rap/Hip hop | Freeze time |
| 7 | Car | Spring | No | Physical education | Pop | Telepathy |
| 8 | Car | Winter | No | Art | Other | Fly |
| 9 | Car | Summer | No | Physical education | Pop | Fly |
| 10 | Car | Spring | No | Mathematics and statistics | Pop | Telepathy |
| 11 | Car | Summer | Yes | History | Rap/Hip hop | Invisibility |
| 12 | Car | Spring | No | Art | Rap/Hip hop | Freeze time |
| 13 | Bus | Winter | No | Computers and technology | Rap/Hip hop | Fly |
| 14 | Car | Winter | Yes | Social studies | Rap/Hip hop | Fly |
| 15 | Car | Summer | No | Art | Pop | Freeze time |
| 16 | Car | Fall | No | Mathematics and statistics | Pop | Fly |
| 17 | Bus | Winter | No | Science | Rap/Hip hop | Freeze time |
| 18 | Car | Spring | Yes | Art | Pop | Telepathy |
| 19 | Car | Fall | Yes | Science | Pop | Telepathy |
| 20 | Car | Summer | Yes | Physical education | Rap/Hip hop | Invisibility |
| 21 | Car | Spring | Yes | Science | Pop | Invisibility |
| 22 | Car | Winter | Yes | Mathematics and statistics | Country | Invisibility |
| 23 | Car | Summer | Yes | Art | Pop | Invisibility |
| 24 | Bus | Winter | Yes | Other | Pop | Telepathy |
| 25 | Bus | Summer | Yes | Science | Other | Fly |
| 26 | Car | Summer | No | Science | Pop | Fly |
| 27 | Car | Summer | Yes | Music | Pop | Telepathy |
| 28 | Car | Summer | No | Physical education | Country | Super strength |
| 29 | Car | Fall | Yes | Mathematics and statistics | Country | Telepathy |
| 30 | Car | Summer | Yes | Physical education | Rap/Hip hop | Telepathy |
| 31 | Boat | Winter | No | Computers and technology | Gospel | Invisibility |
| 32 | Car | Spring | No | Physical education | Pop | Fly |
| 33 | Car | Spring | No | Physical education | Pop | Fly |
| 34 | Car | Summer | No | Mathematics and statistics | Classical | Fly |
| 35 | Car | Fall | Yes | Science | Jazz | Telepathy |
| 36 | Car | Spring | No | Science | Rap/Hip hop | Telepathy |
| 37 | Car | Summer | No | Music | Country | Telepathy |
| 38 | Bus | Winter | No | Mathematics and statistics | Pop | Fly |
| 39 | Car | Spring | No | Art | Classical | Freeze time |
| 40 | Car | Winter | Yes | Art | Pop | Fly |
| 41 | Walk | Summer | Yes | Physical education | Rap/Hip hop | Fly |
| 42 | Bus | Winter | Yes | Physical education | Gospel | Invisibility |

Lesson 20:     Estimating a Population Proportion

| 43 | Bus | Summer | No | Art | Other | Invisibility |
| 44 | Car | Summer | Yes | Computers and technology | Other | Freeze time |
| 45 | Car | Fall | Yes | Science | Pop | Fly |
| 46 | Car | Summer | Yes | Music | Rap/Hip hop | Fly |
| 47 | Car | Spring | No | Science | Rap/Hip hop | Invisibility |
| 48 | Bus | Spring | No | Music | Pop | Telepathy |
| 49 | Car | Summer | Yes | Social studies | Techno/ Electronic | Telepathy |
| 50 | Car | Summer | Yes | Physical education | Pop | Telepathy |
| 51 | Car | Spring | Yes | Other | Other | Telepathy |
| 52 | Car | Summer | No | Art | Pop | Fly |
| 53 | Car | Summer | Yes | Other | Pop | Telepathy |
| 54 | Car | Summer | Yes | Physical education | Rap/Hip hop | Invisibility |
| 55 | Bus | Summer | Yes | Physical education | Other | Super strength |
| 56 | Car | Summer | No | Science | Rap/Hip hop | Invisibility |
| 57 | Car | Winter | No | Languages | Rap/Hip hop | Super strength |
| 58 | Car | Fall | Yes | English | Pop | Fly |
| 59 | Car | Winter | No | Science | Pop | Telepathy |
| 60 | Car | Summer | No | Art | Pop | Invisibility |
| 61 | Car | Summer | Yes | Other | Pop | Freeze time |
| 62 | Bus | Spring | No | Science | Pop | Fly |
| 63 | Car | Winter | Yes | Mathematics and statistics | Other | Freeze time |
| 64 | Car | Summer | No | Social studies | Classical | Fly |
| 65 | Car | Winter | Yes | Science | Pop | Telepathy |
| 66 | Car | Winter | No | Science | Rock and roll | Fly |
| 67 | Car | Summer | No | Mathematics and statistics | Rap/Hip hop | Super strength |
| 68 | Car | Fall | No | Music | Rock and roll | Super strength |
| 69 | Car | Spring | No | Other | Other | Invisibility |
| 70 | Car | Summer | Yes | Mathematics and statistics | Rap/Hip hop | Telepathy |
| 71 | Car | Winter | No | Art | Other | Fly |
| 72 | Car | Spring | Yes | Mathematics and statistics | Pop | Telepathy |
| 73 | Car | Winter | Yes | Computers and technology | Techno/ Electronic | Telepathy |
| 74 | Walk | Winter | No | Physical education | Techno/ Electronic | Fly |
| 75 | Walk | Summer | No | History | Rock and roll | Fly |
| 76 | Skateboard /Scooter/Ro llerblade | Winter | Yes | Computers and technology | Techno/ Electronic | Freeze time |
| 77 | Car | Spring | Yes | Science | Other | Telepathy |
| 78 | Car | Summer | No | Music | Rap/Hip hop | Invisibility |
| 79 | Car | Summer | No | Social studies | Pop | Invisibility |
| 80 | Car | Summer | No | Other | Rap/Hip hop | Telepathy |
| 81 | | Spring | Yes | History | Rap/Hip hop | Invisibility |
| 82 | Car | Summer | No | Art | Pop | Invisibility |

| 83 | Walk | Spring | No | Languages | Jazz | Super strength |
| 84 | Car | Fall | No | History | Jazz | Invisibility |
| 85 | Car | Summer | No | Physical education | Rap/Hip hop | Freeze time |
| 86 | Car | Spring | No | Mathematics and statistics | Pop | Freeze time |
| 87 | Bus | Spring | Yes | Art | Pop | Telepathy |
| 88 | Car | Winter | No | Mathematics and statistics | Other | Invisibility |
| 89 | Car | Summer | Yes | Physical education | Country | Telepathy |
| 90 | Bus | Summer | No | Computers and technology | Other | Fly |
| 91 | Car | Winter | No | History | Pop | Telepathy |
| 92 | Walk | Winter | No | Science | Classical | Telepathy |
| 93 | Bicycle | Summer | No | Physical education | Pop | Invisibility |
| 94 | Car | Summer | No | English | Pop | Telepathy |
| 95 | Car | Summer | Yes | Physical education | Pop | Fly |
| 96 | Car | Winter | No | Science | Other | Freeze time |
| 97 | Car | Winter | No | Other | Rap/Hip hop | Super strength |
| 98 | Car | Summer | Yes | Physical education | Rap/Hip hop | Freeze time |
| 99 | Car | Spring | No | Music | Classical | Telepathy |
| 100 | Car | Spring | Yes | Science | Gospel | Telepathy |
| 101 | Car | Summer | Yes | History | Pop | Super strength |
| 102 | Car | Winter | Yes | English | Country | Freeze time |
| 103 | Car | Spring | No | Computers and technology | Other | Telepathy |
| 104 | Car | Winter | No | History | Other | Invisibility |
| 105 | Car | Fall | No | Music | Pop | Telepathy |
| 106 | Car | Fall | No | Science | Pop | Telepathy |
| 107 | Car | Winter | No | Art | Heavy metal | Fly |
| 108 | Car | Spring | Yes | Science | Rock and roll | Fly |
| 109 | Car | Fall | Yes | Music | Other | Fly |
| 110 | Car | Summer | Yes | Social studies | Techno/ Electronic | Telepathy |
| 111 | Car | Spring | No | Physical education | Pop | Fly |
| 112 | Car | Summer | No | Physical education | Pop | Fly |
| 113 | Car | Summer | Yes | Social studies | Pop | Freeze time |
| 114 | Car | Summer | Yes | Computers and technology | Gospel | Freeze time |
| 115 | Car | Winter | Yes | Other | Rap/Hip hop | Telepathy |
| 116 | Car | Summer | Yes | Science | Country | Telepathy |
| 117 | Car | Fall | | Music | Country | Fly |
| 118 | Walk | Summer | No | History | Pop | Telepathy |
| 119 | Car | Spring | Yes | Art | Pop | Freeze time |
| 120 | Car | Fall | Yes | Physical education | Rap/Hip hop | Fly |
| 121 | Car | Spring | No | Music | Rock and roll | Telepathy |
| 122 | Car | Fall | No | Art | Pop | Invisibility |
| 123 | Car | Summer | Yes | Physical education | Rap/Hip hop | Fly |
| 124 | | Summer | No | Computers and technology | Pop | Telepathy |
| 125 | Car | Fall | No | Art | Pop | Fly |

| 126 | Bicycle | Spring | No | Science | Pop | Invisibility |
| 127 | Car | Summer | No | Social studies | Gospel | Fly |
| 128 | Bicycle | Winter | No | Social studies | Rap/Hip hop | Fly |
| 129 | Car | Summer | Yes | Mathematics and statistics | Pop | Invisibility |
| 130 | Car | Fall | Yes | Mathematics and statistics | Country | Telepathy |
| 131 | Car | Winter | Yes | Music | Gospel | Super strength |
| 132 | Rail (Train/Tram /Subway) | Fall | Yes | Art | Other | Fly |
| 133 | Walk | Summer | No | Social studies | Pop | Invisibility |
| 134 | Car | Summer | Yes | Music | Pop | Freeze time |
| 135 | Car | Winter | No | Mathematics and statistics | Pop | Telepathy |
| 136 | Car | Fall | Yes | Music | Pop | Telepathy |
| 137 | Car | Summer | Yes | Computers and technology | Other | Freeze time |
| 138 | Car | Summer | Yes | Physical education | Pop | Telepathy |
| 139 | Car | Summer | Yes | Social studies | Other | Telepathy |
| 140 | Car | Spring | Yes | Physical education | Other | Freeze time |
| 141 | Car | Fall | Yes | Science | Country | Telepathy |
| 142 | Car | Spring | Yes | Science | Pop | Invisibility |
| 143 | Car | Summer | No | Other | Rap/Hip hop | Freeze time |
| 144 | Car | Summer | No | Other | Other | Fly |
| 145 | Car | Summer | No | Languages | Pop | Freeze time |
| 146 | Car | Summer | Yes | Physical education | Pop | Telepathy |
| 147 | Bus | Winter | No | History | Country | Invisibility |
| 148 | Car | Spring | No | Computers and technology | Other | Telepathy |
| 149 | Bus | Winter | Yes | Science | Pop | Invisibility |
| 150 | Car | Summer | No | Social studies | Rap/Hip hop | Invisibility |
| 151 | Car | Summer | No | Physical education | Pop | Invisibility |
| 152 | Car | Summer | Yes | Physical education | Pop | Super strength |
| 153 | Car | Summer | No | Mathematics and statistics | Pop | Fly |
| 154 | Car | Summer | No | Art | Rap/Hip hop | Freeze time |
| 155 | Car | Winter | Yes | Other | Classical | Freeze time |
| 156 | Car | Summer | Yes | Computers and technology | Other | Telepathy |
| 157 | Car | Spring | No | Other | Pop | Freeze time |
| 158 | Car | Winter | Yes | Music | Country | Fly |
| 159 | Car | Winter | No | History | Jazz | Invisibility |
| 160 | Car | Spring | Yes | History | Pop | Fly |
| 161 | Car | Winter | Yes | Mathematics and statistics | Other | Telepathy |
| 162 | Car | Fall | No | Science | Country | Invisibility |
| 163 | Car | Winter | No | Science | Other | Fly |
| 164 | Car | Summer | No | Science | Pop | Fly |
| 165 | Skateboard /Scooter/ Rollerblade | Spring | Yes | Social studies | Other | Freeze time |
| 166 | Car | Winter | Yes | Art | Rap/Hip hop | Fly |

| 167 | Car | Summer | Yes | Other | Pop | Freeze time |
|-----|-----|--------|-----|-------|-----|-------------|
| 168 | Car | Summer | No | English | Pop | Telepathy |
| 169 | Car | Summer | No | Other | Pop | Invisibility |
| 170 | Car | Summer | Yes | Physical education | Techno/Electronic | Freeze time |
| 171 | Car | Summer | No | Art | Pop | Telepathy |
| 172 | Car | Summer | No | Physical education | Rap/Hip hop | Freeze time |
| 173 | Car | Winter | Yes | Mathematics and statistics | Other | Invisibility |
| 174 | Bus | Summer | Yes | Music | Pop | Freeze time |
| 175 | Car | Winter | No | Art | Pop | Fly |
| 176 | Car | Fall | No | Science | Rap/Hip hop | Fly |
| 177 | Car | Winter | Yes | Social studies | Pop | Telepathy |
| 178 | Car | Fall | No | Art | Other | Fly |
| 179 | Bus | Spring | No | Physical education | Country | Fly |
| 180 | Car | Winter | No | Music | Other | Telepathy |
| 181 | Bus | Summer | No | Computers and technology | Rap/Hip hop | Freeze time |
| 182 | Car | Summer | Yes | Physical education | Rap/Hip hop | Invisibility |
| 183 | Car | Summer | Yes | Music | Other | Telepathy |
| 184 | Car | Spring | No | Science | Rap/Hip hop | Invisibility |
| 185 | Rail (Train/Tram/Subway) | Summer | No | Physical education | Other | Freeze time |
| 186 | Car | Summer | Yes | Mathematics and statistics | Rap/Hip hop | Fly |
| 187 | Bus | Winter | Yes | Mathematics and statistics | Other | Super strength |
| 188 | Car | Summer | No | Mathematics and statistics | Other | Freeze time |
| 189 | Rail (Train/Tram/Subway) | Fall | Yes | Music | Jazz | Fly |
| 190 | Car | Summer | Yes | Science | Pop | Super strength |
| 191 | Car | Summer | Yes | Science | Techno/Electronic | Freeze time |
| 192 | Car | Spring | Yes | Physical education | Rap/Hip hop | Freeze time |
| 193 | Car | Summer | Yes | Physical education | Rap/Hip hop | Freeze time |
| 194 | Car | Winter | No | Physical education | Rap/Hip hop | Telepathy |
| 195 | Car | Winter | No | Music | Jazz | Freeze time |
| 196 | Walk | Summer | Yes | History | Country | Freeze time |
| 197 | Car | Spring | No | History | Rap/Hip hop | Freeze time |
| 198 | Car | Fall | Yes | Other | Pop | Freeze time |
| 199 | Car | Spring | Yes | Science | Other | Freeze time |
| 200 | Bicycle | Winter | Yes | Other | Rap/Hip hop | Freeze time |

Lesson 20:    Estimating a Population Proportion

GRADE

# Mathematics Curriculum

## Topic D:

# Comparing Populations

## 7.SP.B.3, 7.SP.B.4

| Focus Standard: | 7.SP.B.3 | Informally assess the degree of visual overlap of two numerical data distributions with similar variability, measuring the difference between the centers by expressing it as a multiple of a measure of variability. *For example, the mean height of players on the basketball team is 10 cm greater than the mean height of players on the soccer team, about twice the variability (mean absolute deviation) on either team; on a dot plot, the separation between the two distributions of heights is noticeable.* |
|---|---|---|
| | 7.SP.B.4 | Use measures of center and measures of variability for numerical data from random samples to draw informal comparative inferences about two populations. *For example, decide whether the words in a chapter of a seventh-grade science book are generally longer than the words in a chapter of a fourth-grade science book.* |
| **Instructional Days:** | 3 | |
| **Lesson 21:** | Why Worry About Sampling Variability? (E)[1] | |
| **Lessons 22–23:** | Using Sample Data to Decide if Two Population Means Are Different (P,P) | |

In Topic D, students learn to compare two populations with similar variability. They learn to consider sampling variability when deciding if there is evidence that population means or population proportions are actually different. In Lesson 21, students work with random samples from two different populations that have similar variability. They decide if there is "evidence" that the means of the two populations are actually different. In Lesson 22, students describe the difference in sample means from populations with similar variability by using a multiple of the measure of variability. They explore how big the difference in sample means would need to be in order to indicate a difference in population means. This lesson sets the stage for drawing informal conclusions about the difference between two populations' means from populations with similar variability. Lesson 23, again, uses random samples to draw informal inferences about the differences in means of two populations. Students work with examples in which there is a meaningful difference in population means and also with examples in which there is no evidence of a meaningful difference in population means.

---

[1] Lesson Structure Key: **P**-Problem Set Lesson, **M**-Modeling Cycle Lesson, **E**-Exploration Lesson, **S**-Socratic Lesson

# Lesson 21: Why Worry about Sampling Variability?

## Student Outcomes

▪ Students understand that a *meaningful* difference between two sample means is one that is greater than would have been expected due to just sampling variability.

## Lesson Notes

This lesson is the first of three lessons in which students are asked to compare the means of two populations. In this lesson, students will first consider random samples from two populations that have the same mean. Sampling variability must be considered when deciding if there is *evidence* that the two population means are actually the same or different, and this idea is developed throughout the exercises. Students then repeat the process by examining two populations that have different means. The investigation is presented in stages. Questions are posed in the stages that help students think about the sampling variability in different sample means. This lesson may take more than one class period.

## Classwork

*Preparation:* Organize the class in small groups. Before class, prepare three bags for each group, labeled $A$, $B$, and $C$. Each bag contains 100 numbers (written on 1-inch square pieces of cardstock, round counters, or precut foam squares, etc.) Bag $A$ and Bag $B$ both consist of the population of 100 numbers as follows: five 5's, five 6's, five 7's, five 8's … five 24's. Bag $C$ consists of the population of 100 numbers: five 1's, five 2's, five 3's, . . . , five 20's. Templates that can be used to produce the numbers for the bags are included at the end of this lesson.

> There are three bags, Bag $A$, Bag $B$, and Bag $C$, with 100 numbers in each bag. You and your classmates will investigate the population mean (the mean of all 100 numbers) in each bag. Each set of numbers has the same range. However, the population means of each set may or may not be the same. We will see who can uncover the mystery of the bags!

## Exercises 1–5 (8–10 minutes)

Before students begin this activity, ask: "How would you go about determining the mean of the numbers in each bag?" Anticipate students would indicate they would need to know all 100 numbers from each bag, and then use the numbers to determine the means. Explain that although knowing all the numbers would allow them to calculate the means, this activity introduces a way to estimate the means using random samples. This activity suggests how to estimate a mean when it is not possible or practical to know the values of the entire population.

The instructions for the investigation should be read aloud to the class. While each group needs only one bag of each population, each student is expected to draw a sample from each bag. Warn students that *before* they draw a new sample from a bag, *all* 100 numbers must be returned to the bag. Students should be allowed 5 minutes to take samples of ten numbers from Bag $A$, Bag $B$, and Bag $C$. Charts are provided to aid students in this process.

**Exercises 1–5**

1.  To begin your investigation, start by selecting a random sample of ten numbers from Bag $A$. Remember to mix the numbers in the bag first. Then, select one number from the bag. Do not put it back into the bag. Write the number in the chart below. Continue selecting one number at a time until you have selected ten numbers. Mix up the numbers in the bag between each selection.

| Selection | 1 | 2 | 3 | 4 | 5 | 6 | 7 | 8 | 9 | 10 |
|-----------|---|---|---|---|---|---|---|---|---|----|
| Bag $A$   |   |   |   |   |   |   |   |   |   |    |

   a.  Create a dot plot of your sample of ten numbers. Use a dot to represent each number in the sample.

   *Dot plot will vary based on the sample selected. One possible answer is shown here.*

   **Dotplot of Sample from Bag A**

   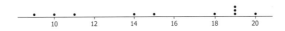

   b.  Do you think the mean of all the numbers in Bag $A$ might be 10? Why or why not?

   *Anticipate that students will indicate a mean of a sample from Bag $A$ is greater than 10. Responses depend on the students' samples and the resulting dot plots. In most cases, the dots will center around a value that is greater than 10 because the mean of the population is greater than 10.*

   c.  Based on the dot plot, what would you estimate the mean of the numbers in Bag $A$ to be? How did you make your estimate?

   *Answers will vary depending on students' samples. Anticipate that most students' estimates will correspond to roughly where the dots in the dot plot center. The population mean here is 14.5, so answers around 14 or 15 would be expected.*

   d.  Do you think your sample mean will be close to the population mean? Why or why not?

   *Students could answer "Yes," "No," or "I don't know." The goal of this question is to get students to think about the difference between a sample mean and the population mean.*

   e.  Is your sample mean the same as your neighbors' sample means? Why or why not?

   *No, when selecting a sample at random, different students will get different sets of numbers. This is sampling variability.*

2.  Repeat the process by selecting a random sample of ten numbers from Bag $B$.

| Selection | 1 | 2 | 3 | 4 | 5 | 6 | 7 | 8 | 9 | 10 |
|-----------|---|---|---|---|---|---|---|---|---|----|
| Bag $B$   |   |   |   |   |   |   |   |   |   |    |

a. Create a dot plot of your sample of ten numbers. Use a dot to represent each of the numbers in the sample.

*Dot plot will vary based on the sample selected. One possible answer is shown here.*

**Dotplot of Sample from Bag B**

b. Based on your dot plot, do you think the mean of the numbers in Bag $B$ is the same or different than the mean of the numbers in Bag $A$? Explain your thinking.

*Answer will vary as students will compare their center of the dot plot of the sample from Bag B to the center of the dot plot of the sample from Bag A. The centers will probably not be exactly the same; however, anticipate centers that are close to each other.*

3. Repeat the process once more by selecting a random sample of ten numbers from Bag $C$.

| Selection | 1 | 2 | 3 | 4 | 5 | 6 | 7 | 8 | 9 | 10 |
|---|---|---|---|---|---|---|---|---|---|---|
| Bag $C$ | | | | | | | | | | |

a. Create a dot plot of your sample of ten numbers. Use a dot to represent each of the numbers in the sample.

*Dot plot will vary based on the sample selected. One possible answer is shown here.*

**Dotplot of Sample from Bag C**

b. Based on your dot plot, do you think the mean of the numbers in Bag $C$ is the same or different than the mean of the numbers in Bag $A$? Explain your thinking.

*Anticipate that students will indicate that the center of the dot plot of the sample from Bag C is less than the center of the dot plot of the sample from Bag A. Because the population mean for Bag C is less than the population mean for Bag A, the center of the dot plot will usually be less for the sample from Bag C.*

4. Are your dot plots of the three bags the same as the dot plots of other students in your class? Why or why not?

*Dot plots will vary. Because different students generally get different samples when they select a sample from the bags, the dot plots will vary from student to student.*

Below are dot plots of all three samples drawn on the same scale. Notice how the samples from Bags $A$ and $B$ tend to center at approximately 14.5, while the sample from Bag $C$ centers around 10.5.

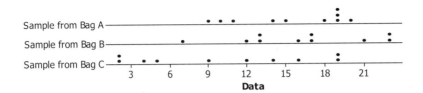

Sample from Bag A
Sample from Bag B
Sample from Bag C

3   6   9   12   15   18   21

**Data**

Next, students will calculate the means for each of the three samples. As students finish their calculation of the means, lead a discussion about why a student's sample means might differ from his or her neighbors' sample means. Point out that what they are observing is *sampling variability* – the chance variability that happens from one sample to another when repeated samples are taken from the same population.

5.  Calculate the mean of the numbers for each of the samples from Bag $A$, Bag, $B$, and Bag $C$.

| | Mean of the sample of numbers |
|---|---|
| Bag $A$ | |
| Bag $B$ | |
| Bag $C$ | |

*Answers will vary. For the samples shown in the dot plots above, the sample means are: Bag A 15.4, Bag B 16.2, Bag C 10.2.*

a.  Are the sample means you calculated the same as the sample means of other members of your class? Why or why not?

*No, when selecting a sample at random, you get different sets of numbers (again, sampling variability).*

b.  How do your sample means for Bag $A$ and for Bag $B$ compare?

*Students might answer that the mean for the sample from Bag A is larger, smaller, or equal to the mean for the sample from Bag B, depending on their samples. For the example, given above, the sample mean for Bag A is smaller than the sample mean for Bag B.*

c.  Calculate the difference of sample mean for Bag $A$ minus sample mean for Bag $B$ ($Mean_A - Mean_B$). Based on this difference, can you be sure which bag has the larger population mean? Why or why not?

*No, it is possible that you could get a sample mean that is larger than the population mean of Bag A, and then get a sample mean that is smaller than the population mean of Bag B, or vice versa.*

 **MP.3** Students begin to examine their samples and the samples of other students using the means and the dot plots. They begin to make conjectures about the populations based on their own samples and the samples of other students.

Creating dot plots of the means that each student calculated for each bag provides another way for students to formulate conjectures about the means of each bag. Have each student plot his or her sample mean for the sample from Bag $A$ on a class dot plot. Label this dot plot "Means of samples from Bag $A$." Then, construct class dot plots for the means from Bags $B$ and $C$. Anticipate that the dot plots for the means for Bags $A$ and $B$ will have similar distributions regarding center and spread. Anticipate that the dot plot for Bag $C$ will have similar spread to the dot plots of Bags $A$ or $B$; however, the center will be less than the center for Bags $A$ and $B$.

For a class of 30 students, the following dot plots were obtained. The dot plots for your class won't be exactly like these, but they will probably be similar.

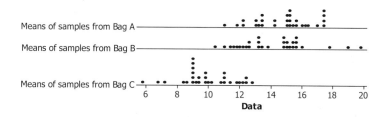

## Exercises 6–10 (8–10 minutes)

Continue the discussion with students as they work through the next set of exercises. Allow students to individually provide an answer to these questions. After a few minutes, organize a class discussion of each question.

---

**Exercises 6–10**

6.  Based on the class dot plots of the sample means, do you think the mean of the numbers in Bag $A$ and the mean of the numbers in Bag $B$ are different? Do you think the mean of the numbers in Bag $A$ and the mean of the numbers in Bag $C$ are different? Explain your answers.

    *At this stage of the lesson, students may suspect that Bags A and B are similar, and also that Bag C is different than the other two. This question sets the stage for the rest of this lesson.*

7.  Based on the difference between sample mean of Bag $A$ and the sample mean of Bag $B$, $(Mean_A - Mean_B)$, that you calculated in Exercise 5, do you think that the two populations (Bags $A$ and $B$) have different means or do you think that the two population means might be the same?

    *Answers will vary as the difference of the means will be based on each student's samples. Anticipate answers that indicate the difference in the sample means that the population means might be the same for differences that are close to 0. (Students learn later in this lesson that the populations of Bags A and B are the same, so most students will see differences that are not too far from 0.)*

8.  Based on this difference, can you be sure which bag has the larger population mean? Why or why not?

    *No, it is possible that you could get a sample mean that is larger than the population mean of Bag A and then get a sample mean that is smaller than the population mean of Bag B, or vice versa.*

9.  Is your difference in sample means the same as your neighbors' differences? Why or why not?

    *No. As the samples will vary due to sampling variability, so will the means of each sample.*

---

Next have the students plot their differences in sample means $(Mean_A - Mean_B)$ on a dot plot. A number line similar to the following may be drawn on the board or chart paper. If necessary, make the number line longer to include all of the differences.

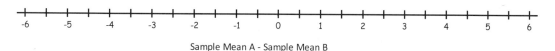

Sample Mean A - Sample Mean B

10. Plot your difference of the means, $(Mean_A - Mean_B)$, on a class dot plot. Describe the distribution of differences plotted on the graph. Remember to discuss center and spread.

*Provide each student an opportunity to place his or her difference of the means on this class dot plot. The distribution of the differences is expected to cluster around 0. One example for a class of 30 students is shown here.*

**Dotplot of Bag A mean - Bag B mean**

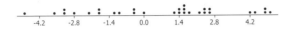

## Exercises 11–13 (5 minutes)

Reveal to students that the population of numbers in Bag $A$ is identical to the population of numbers in Bag $B$.

Use Exercises 11–13 for a class discussion of why the differences in sample means are not all equal to 0 even though the bags are the same. Point out that sampling variability will result in different means (and thus differences that are not all 0). Students should also think about why the differences in sample means may be relatively far from 0. (For example, why are some differences 3 or more?) In Exercise 12, point out that one reason for this is that a student could have randomly selected a sample from Bag $A$ in which the calculated sample mean was higher than the population mean of Bag $A$. Then the same student could have randomly selected a sample from Bag $B$ in which the calculated sample mean was lower than the population mean of Bag $B$. The difference in these sample means would then have a larger value. If there is an example provided by one of the students to illustrate this type of difference, use it to show that some larger differences in sample means are possible.

---

**Exercises 11–13**

11. Why are the differences in the sample means of Bag $A$ and Bag $B$ not always 0?

    *Discuss this question in the context of sampling variability. Students should comment on the fact that in order for the difference in sample means to be 0, the sample means must be the same value. This would rarely happen when selecting random samples.*

12. Does the class dot plot contain differences that were relatively far away from 0? If yes, why do you think this happened?

    *After students have a chance to respond to this question, discuss the points made above about why some larger differences might occur. For the dot plot given as an example above, some differences were as large as 4.*

13. Suppose you will take a sample from a new bag. How big would the difference in the sample mean for Bag $A$ and the sample mean for the new bag $(Mean_A - Mean_{new})$ have to be before you would be convinced that the population mean for the new bag is different than the population mean of Bag $A$? Use the class dot plot of the differences in sample means for Bags $A$ and $B$ (which have equal population means) to help you answer this question.

    *Students should recognize that the difference would need to be relatively far away from 0. They may give answers like "a difference of 5 (or larger)" or something similar. Remind students that the differences noted in the class dot plot are a result of sampling from bags that have the same numbers in them. As a result, you would expect students to suggest values that are greater than the values in the class dot plot.*

---

> The differences in the class dot plot occur because of sampling variability—the chance variability from one sample to another. In Exercise 13, you were asked about how great the difference in sample means would need to be before you have convincing evidence that one population mean is larger than another population mean. A *"meaningful"* difference between two sample means is one that is unlikely to have occurred by chance if the population means are equal. In other words, the difference is one that is greater than would have been expected just due to sampling variability.

How large the difference in sample means needs to be in order to say that it is unlikely to have occurred by chance depends on the context in which the data were collected, the sample size, and how much variability there is in the population, so it is not possible to give a general rule (such as "bigger than 5") that you can always use. In the lessons that follow, you will explore one method for determining if a difference in sample means is *meaningful* that depends on expressing the difference in sample means in terms of a measure of how much variability there is in the population.

### Exercise 14–16 (5–7 minutes)

Students examine similar questions with Bag $C$. After students develop their answers to the exercises, discuss their responses as a class.

---

**Exercises 14–17**

14. Calculate the sample mean of Bag $A$ minus the sample mean of Bag $C$, $(Mean_A - Mean_C)$.

   *Answers will vary as the samples collected by students will vary. Students might suspect, however, that you are setting them up for a discussion about populations that have different means. As a result, ask students what they think their difference is indicating about the populations of the two bags. For several students, this difference is larger than the difference they received for Bags A and B, and might suggest that the means of the bags are different. Not all students, however, will have differences that are noticeably different from what they obtained for bags A and B, and as a result, they will indicate that the bags could have the same or similar distribution of numbers.*

---

Students begin to reason quantitatively the meaning of the differences of the sample means. Encourage students to discuss what it means if the difference is positive. For those cases where the difference might have resulted in a negative difference, encourage students to indicate what that indicates about the two populations.

---

15. Plot your difference $(Mean_A - Mean_C)$ on a class dot plot.

   *Have each student or group of students place their differences on a class dot plot similar to what was developed for the dot plot of the difference of means in Bags A and B. Place the dot plots next to each other so that students can compare the centers and spread of each distribution. One example based on a class of 30 students is shown here. Notice that the differences for Bag A − Bag B center around 0, while the differences for Bag A − Bag C do not center around 0.*

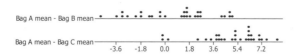

16. How do the centers of the class dot plots for $(Mean_A - Mean_B)$ and $(Mean_A - Mean_C)$ compare?

   *Students should recognize that the center of the second dot plot $(Mean_A - Mean_C)$ is shifted over to the right. Thus, it is not centered at 0, rather it is centered over a value that is larger than 0.*

---

## Exercise 17 (5 minutes)

> 17. Each bag has a population mean that is either $10.5$ or $14.5$. State what you think the population mean is for each bag. Explain your choice for each bag.
>
> *The population mean is $14.5$ for bags A and B, and $10.5$ for bag C. Students indicate their selections based on the class dot plots and the sample means they calculated in the exercises.*

## Closing (5 minutes)

Ask students to describe how this lesson involved *sampling variability*. Also, ask students to indicate what the comparison of the dot plots indicated about the population of the numbers in the bags.

Discuss the following Lesson Summary with students.

> Lesson Summary
>
> - Remember to think about sampling variability—the chance variability from sample to sample.
> - Beware of making decisions based just on the fact that two sample means are not equal.
> - Consider the distribution of the difference in sample means when making a decision.

## Exit Ticket (3–5 minutes)

Name _____    Date_____

# Lesson 21: Why Worry about Sampling Variability?

Exit Ticket

How is a *meaningful* difference in sample means different from a *non-meaningful* difference in sample means?  You may use what you saw in the dot plots of this lesson to help you answer this question.

## Exit Ticket Sample Solutions

How is a *meaningful* difference in sample means different from a *non-meaningful* difference in sample means? You may use what you saw in the dot plots of this lesson to help you answer this question.

*A meaningful difference in sample means is one that is not likely to have occurred by just chance if there is no difference in the population means. A meaningful difference in sample means would be one that is very far from 0 (or not likely to happen if the population means are equal). A non-meaningful difference in sample means would be one that is relatively close to 0, which indicates the population means are equal.*

*Note that how big this difference needs to be in order to be declared "meaningful" depends on the context, the sample size, and on the variability in the populations. This is explored in the next lesson.*

## Problem Set Sample Solutions

Below are three dot plots. Each dot plot represents the differences in sample means for random samples selected from two populations (Bag $A$ and Bag $B$). For each distribution, the differences were found by subtracting the sample means of Bag $B$ from the sample means of Bag $A$ (sample mean $A$ − sample mean $B$).

1. Does the graph below indicate that the population mean of $A$ is larger than the population mean of $B$? Why or why not?

   *No, since most of the differences are negative, it appears that the population mean of Bag $A$ is smaller than the population mean of Bag $B$.*

   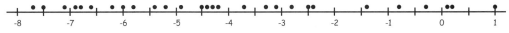

2. Use the graph above to estimate the difference in the population means $(A − B)$.

   *About $-4$, this is about the middle of the graph.*

3. Does the graph below indicate that the population mean of $A$ is larger than the population mean of $B$? Why or why not?

   *No, the dots are all centered around 0, meaning that the population means of Bag $A$ and Bag $B$ might be equal.*

   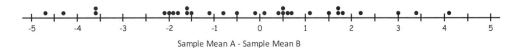

4. Does the graph below indicate that the population mean of $A$ is larger than the population mean of $B$? Why or why not?

   *Yes, the dots are near $1.5$. There is a small difference in the population means, but it is so small that it is difficult to detect. (Some students may answer this "No, the dots appear centered around 0". Problem 6 should cause students to rethink this answer.)*

   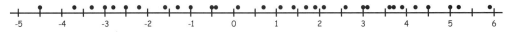

5.  In the above graph, how many differences are greater than 0?  How many differences are less than 0?  What might this tell you?

    *There are 18 dots greater than 0, and 12 dots less than 0. It tells me that there are more positive differences, which may mean that the population mean for Bag A is bigger than the population mean for Bag B.*

6.  In Problem 4, the population mean for Bag $A$ is really larger than the population mean for Bag $B$.  Why is it possible to still get so many negative differences in the graph?

    *It is possible to get so many negative values because the population mean of Bag A may only be a little bigger than the population mean of Bag B.*

**Template for Bags A and B**

| | | | | | | | | | |
|---|---|---|---|---|---|---|---|---|---|
| 5 | 5 | 5 | 5 | 5 | 6 | 6 | 6 | 6 | 6 |
| 7 | 7 | 7 | 7 | 7 | 8 | 8 | 8 | 8 | 8 |
| 9 | 9 | 9 | 9 | 9 | 10 | 10 | 10 | 10 | 10 |
| 11 | 11 | 11 | 11 | 11 | 12 | 12 | 12 | 12 | 12 |
| 13 | 13 | 13 | 13 | 13 | 14 | 14 | 14 | 14 | 14 |
| 15 | 15 | 15 | 15 | 15 | 16 | 16 | 16 | 16 | 16 |
| 17 | 17 | 17 | 17 | 17 | 18 | 18 | 18 | 18 | 18 |
| 19 | 19 | 19 | 19 | 19 | 20 | 20 | 20 | 20 | 20 |
| 21 | 21 | 21 | 21 | 21 | 22 | 22 | 22 | 22 | 22 |
| 23 | 23 | 23 | 23 | 23 | 24 | 24 | 24 | 24 | 24 |

**Template for Bag C**

| | | | | | | | | | |
|---|---|---|---|---|---|---|---|---|---|
| 1 | 1 | 1 | 1 | 1 | 2 | 2 | 2 | 2 | 2 |
| 3 | 3 | 3 | 3 | 3 | 4 | 4 | 4 | 4 | 4 |
| 5 | 5 | 5 | 5 | 5 | 6 | 6 | 6 | 6 | 6 |
| 7 | 7 | 7 | 7 | 7 | 8 | 8 | 8 | 8 | 8 |
| 9 | 9 | 9 | 9 | 9 | 10 | 10 | 10 | 10 | 10 |
| 11 | 11 | 11 | 11 | 11 | 12 | 12 | 12 | 12 | 12 |
| 13 | 13 | 13 | 13 | 13 | 14 | 14 | 14 | 14 | 14 |
| 15 | 15 | 15 | 15 | 15 | 16 | 16 | 16 | 16 | 16 |
| 17 | 17 | 17 | 17 | 17 | 18 | 18 | 18 | 18 | 18 |
| 19 | 19 | 19 | 19 | 19 | 20 | 20 | 20 | 20 | 20 |

**Lesson 22: Using Sample Data to Decide if Two Population Means Are Different**

## Student Outcomes

- Students express the difference in sample means as a multiple of a measure of variability.
- Students understand that a difference in sample means provides evidence that the population means are different if the difference is larger than what would be expected as a result of sampling variability alone.

## Lesson Notes

In previous lessons, seventh graders have been challenged to learn the very important roles that random sampling and sampling variability play in estimating a population mean. This lesson provides the groundwork for Lesson 23, in which students are asked to informally decide if two population means differ from each other using data from a random sample taken from each population. Implicit in the inference is sampling variability. Explicit in the inference is calculating sample means and the process of comparing them.

This lesson develops a measure for describing how far apart two sample means are from each other in terms of a multiple of some measure of variability. In this lesson, the MAD (mean absolute deviation) is used as the measure of variability. The next lesson builds on this work, introducing the idea of making informal inferences about the difference between two population means.

## Classwork

In previous lessons, you worked with one population. Many statistical questions involve comparing two populations. For example:

- On average, do boys and girls differ on quantitative reasoning?
- Do students learn basic arithmetic skills better with or without calculators?
- Which of two medications is more effective in treating migraine headaches?
- Does one type of car get better mileage per gallon of gasoline than another type?
- Does one type of fabric decay in landfills faster than another type?
- Do people with diabetes heal more slowly than people who do not have diabetes?

In this lesson, you will begin to explore how big of a difference there needs to be in sample means in order for the difference to be considered meaningful. The next lesson will extend that understanding to making informal inferences about population differences.

## Example 1 (2–3 Minutes)

Briefly summarize the data in the example. Note that KenKen is not important for this problem, any type of puzzle would do. It was mentioned only to possibly spark student interest.

---

**Example 1**

Tamika's mathematics project is to see whether boys or girls are faster in solving a KenKen-type puzzle. She creates a puzzle and records the following times that it took to solve the puzzle (in seconds) for a random sample of 10 boys from her school and a random sample of 11 girls from her school:

| | | | | | | | | | | | | Mean | MAD |
|---|---|---|---|---|---|---|---|---|---|---|---|---|---|
| Boys | 39 | 38 | 27 | 36 | 40 | 27 | 43 | 36 | 34 | 33 | | 35.3 | 4.04 |
| Girls | 41 | 41 | 33 | 42 | 47 | 38 | 41 | 36 | 36 | 32 | 46 | 39.4 | 3.96 |

---

## Exercises 1–3 (12–15 Minutes)

Work through the exercises one at a time as a class.

---

**Exercises 1–3**

1. On the same scale, draw dot plots for the boys' data and for the girls' data. Comment on the amount of overlap between the two dot plots. How are the dot plots the same, and how are they different?

   *The dot plots appear to have a considerable amount of overlap. The boys' data may be slightly skewed to the left, whereas the girls' are relatively symmetric.*

   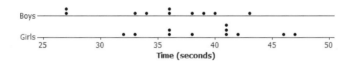

2. Compare the variability in the two data sets using the MAD (mean absolute deviation). Is the variability in each sample about the same? Interpret the MAD in the context of the problem.

   *The variability in each data set is about the same as measured by the mean absolute deviation (around 4 sec.). For boys and girls, a typical deviation from their respective mean times (35 for boys and 39 for girls) is about 4 sec.*

3. In the previous lesson, you learned that a difference between two sample means is considered to be meaningful if the difference is more than what you would expect to see just based on sampling variability. The difference in the sample means of the boys' times and the girls' times is 4.1 seconds (39.4 seconds−35.3 seconds). This difference is approximately 1 MAD.

---

Verify that the students understand the difference between the sample means is approximately 1 MAD. Boys and girls mean times differ by $39.4 - 35.3 = 4.1$ sec. The number of MADs that separate the sample means is $\frac{4.1}{4.04} = 1.01$, about 1 MAD.

---

a. If 4 sec. is used to approximate the values of 1 MAD (mean absolute deviation) for both boys and for girls, what is the interval of times that are within 1 MAD of the sample mean for boys?

   $35.3 + 4$ *sec. equals* $39.3$ *sec., and* $35.3 - 4$ *sec. equals* $31.4$ *sec. The interval of times that are within 1 MAD of the boys' mean time is approximately* $31.4$ *sec. to* $39.3$ *sec.*

---

> b.  Of the 10 sample means for boys, how many of them are within that interval?
>
> *Six of the sample means for boys are within the interval.*
>
> c.  Of the 11 sample means for girls, how many of them are within the interval you calculated in part (a)?
>
> *Five of the sample means for girls are within the interval.*
>
> d.  Based on the dot plots, do you think that the difference between the two sample means is a meaningful difference?  That is, are you convinced that the mean time for all girls at the school (not just this sample of girls) is different from the mean time for all boys at the school?  Explain your choice based on the dot plots.
>
> *Answers will vary.  Sample answer: I don't think that the difference is meaningful.  The dot plots overlap a lot and there is a lot of variability in the times for boys and the times for girls.*

Point out to students that here they are being asked to use just their own judgment to decide if the difference is meaningful, but later in this lesson a more objective criteria will be introduced.

**Note:** If some of your students prefer to use box plots rather than dot plots, then the more appropriate measure of center is the median rather than the mean, and the more appropriate measure of variability to use for measuring the separation of the medians would be the interquartile range (IQR), or the difference of quartile 3 and quartile 1.

### Example 2 (5–7 Minutes)

Read through the directions with students.  Each pair of students will need a stopwatch to record time.  Record the times for each group on the board.

An online stopwatch may be found at http://www.online-stopwatch.com.  Start the following activity by asking students, "How many of you would be better at guessing the length of a minute when it is quiet?  How many of you would be better at guessing a minute when people are talking?"

> **Example 2**
>
> How good are you at estimating a minute?  Work in pairs.  Flip a coin to determine which person in the pair will go first.  One of you puts your head down and raises your hand.  When your partner says "start," keep your head down and your hand raised.  When you think a minute is up, put your hand down.  Your partner will record how much time has passed.  Note that the room needs to be quiet.  Switch roles except this time you talk with your partner during the period when the person with his or her head down is indicating when they think a minute is up.  Note that the room will not be quiet.

### Exercises 4–7 (10–12 minutes)

Let students work with their partner on Exercises 4–7.  Then discuss and confirm as a class.

Suppose that the following data were collected:

> **Exercises 4–7**
>
> | Group | Estimate for a minute | | | | | | | | | | | | | |
> |-------|------|------|------|------|------|------|------|------|------|------|------|------|------|------|
> | Quiet | 58.1 | 56.9 | 60.1 | 56.6 | 56.4 | 54.7 | 64.5 | 62.5 | 58.6 | 55.6 | 61.7 | 58.0 | 55.4 | 63.8 |
> | Talking | 73.9 | 59.9 | 65.8 | 65.5 | 64.6 | 58.8 | 63.3 | 70.2 | 62.1 | 65.6 | 61.7 | 63.9 | 66.6 | 64.7 |

Use your class data to complete the following.

4.  Calculate the mean minute time for each group. Then find the difference between the "quiet" mean and the "talking" mean.

    *The mean of the "quiet" estimates is* $58.8$ *sec.*

    *The mean of the "talking" estimates is* $64.8$ *sec.*

    *The difference between the two means is* $64.8 - 58.8 = 6$ *sec.*

5.  On the same scale, draw dot plots of the two data distributions and discuss the similarities and differences in the two distributions.

    *The dot plots have quite a bit of overlap. The quiet group distribution is fairly symmetric; the talking group distribution is skewed somewhat to the right. The variability in each is about the same. The quiet group appears to be centered around* $60$ *sec. and the talking group appears to be centered around* $65$ *sec.*

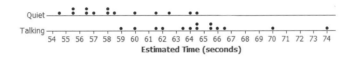

6.  Calculate the mean absolute deviation (MAD) for each data set. Based on the MADs, compare the variability in each sample. Is the variability about the same? Interpret the MADs in the context of the problem.

    *The MAD for the quiet distribution is* $2.68$ *sec.*

    *The MAD for the talking distribution is* $2.73$ *sec.*

    *The MAD measurements are about the same indicating that the variability in each data set is similar. In both groups, a typical deviation of students' minute estimates from their respective means is about* $2.7$ *sec.*

7.  Based on your calculations, is the difference in mean time estimates meaningful? Part of your reasoning should involve the number of MADs that separate the two sample means. Note that if the MADs differ, use the larger one in determining how many MADs separate the two means.

    *The number of MADs that separate the two sample means is* $\frac{6}{2.73} = 2.2$. *There is a meaningful difference between the means.*

Note:  In the next lesson, it is suggested that if the number of MADs that separate two means is 2 or more, then the separation is meaningful. (i.e., is due to more than sampling variability). The number of MADs separating two means is one way to informally gauge whether or not there is a meaningful difference in the population means. The suggestion of 2 MADs is not a hard and fast rule that provides a definitive estimate about the differences, but it represents a reasonable gauge based on both the variability observed in the distributions and the dot plots. Although a different number of MADs separating the means could have been used, providing a specific value (such as 2 MADs) is a reasonable starting point for discussing the differences in means. It also addresses the suggested measure as outlined in the standards (namely, 7.SP.B.3). It is interesting to note in this example that although the dot plots of the sample data overlap by quite a bit, it is still reasonable to think that the population means are different.

---

### Lesson Summary

Variability is a natural occurrence in data distributions. Two data distributions can be compared by describing how far apart their sample means are. The amount of separation can be measured in terms of how many MADs separate the means. (Note that if the two sample MADs differ, the larger of the two is used to make this calculation.)

---

**Exit Ticket (8–10 minutes)**

Name _____  Date_____

# Lesson 22: Using Sample Data to Decide if Two Population Means are Different

Exit Ticket

Suppose that Brett randomly sampled 12 tenth grade girls and boys in his school district and asked them for the number of minutes per day that they text. The data and summary measures follow.

| Gender | Number of Minutes of Texting | | | | | | | | | | | | mean | MAD |
|---|---|---|---|---|---|---|---|---|---|---|---|---|---|---|
| Girls | 98 | 104 | 95 | 101 | 98 | 107 | 86 | 92 | 96 | 107 | 88 | 95 | 97.3 | 5.3 |
| Boys | 66 | 72 | 65 | 60 | 78 | 82 | 63 | 56 | 85 | 79 | 68 | 77 | 70.9 | 7.9 |

1. Draw dot plots for the two data sets using the same numerical scales. Discuss the amount of overlap between the two dot plots that you drew and what it may mean in the context of the problem.

2. Compare the variability in the two data sets using the MAD. Interpret the result in the context of the problem.

3. From 1 and 2, does the difference in the two means appear to be meaningful? Explain.

## Exit Ticket Sample Solutions

Suppose that Brett randomly sampled 12 tenth grade girls and boys in his school district and asked them for the number of minutes per day that they text. The data and summary measures follow.

| Gender | Number of Minutes of Texting | | | | | | | | | | | | mean | MAD |
|---|---|---|---|---|---|---|---|---|---|---|---|---|---|---|
| Girls | 98 | 104 | 95 | 101 | 98 | 107 | 86 | 92 | 96 | 107 | 88 | 95 | 97.3 | 5.3 |
| Boys | 66 | 72 | 65 | 60 | 78 | 82 | 63 | 56 | 85 | 79 | 68 | 77 | 70.9 | 7.9 |

1.  Draw dot plots for the two data sets using the same numerical scales. Discuss the amount of overlap between the two dot plots that you drew and what it may mean in the context of the problem.

    *There is no overlap between the two data sets. This indicates that the sample means probably differ, with girls texting more than boys on average. The girls' data set is a little more compact than the boys, indicating that their measure of variability is smaller.*

    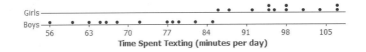

2.  Compare the variability in the two data sets using the MAD. Interpret the result in the context of the problem.

    *The MAD for the boys number of texting minutes is 7.9 min., which is higher than that for the girls, 5.3 min. This is not surprising as seen in the dot plots. The typical deviation from the mean of 70.9 is about 7.9 min. for boys. The typical deviation from the mean of 97.3 is about 5.3 min. for girls.*

3.  From 1 and 2, does the difference in the two means appear to be meaningful?

    *The difference in means is $97.3 - 70.9 = 26.4$ min. Using the larger MAD of 7.9 min., the means are separated by $\frac{26.4}{7.9} = 3.3$ MADs. Looking at the dot plots, it certainly seems as though a separation of more than 3 MADs is meaningful.*

## Problem Set Sample Solutions

Use this space to describe any specific details about the problem set for teacher reference.

1.  A school is trying to decide which reading program to purchase.

    a.  How many MADs separate the mean reading comprehension score for a standard program ($mean = 67.8$, $MAD = 4.6, n = 24$) and an activity-based program ($mean = 70.3, MAD = 4.5, n = 27$)?

        *The number of MADs that separate the sample mean reading comprehension score for a standard program and an activity-based program is $\frac{70.3 - 67.8}{4.6} = 0.54$, about half a MAD.*

    b.  What recommendation would you make based on this result?

        *The number of MADs that separate the programs is not large enough to indicate that one program is better than the other program based on mean scores. There is no noticeable difference in the two programs.*

2.  Does a football filled with helium go farther than one filled with air?  Two identical footballs were used, one filled with helium, and one filled with air to the same pressure.  Matt was chosen from your team to do the kicking.  You did not tell Matt which ball he was kicking.  The data (yards) follow.

| Air | 25 | 23 | 28 | 29 | 27 | 32 | 24 | 26 | 22 | 27 | 31 | 24 | 33 | 26 | 24 | 28 | 30 |
|-----|----|----|----|----|----|----|----|----|----|----|----|----|----|----|----|----|----|
| Helium | 24 | 19 | 25 | 25 | 22 | 24 | 28 | 31 | 22 | 26 | 24 | 23 | 22 | 21 | 21 | 23 | 25 |

|        | Mean | MAD |
|--------|------|-----|
| Air    | 27.0 | 2.59 |
| Helium | 23.8 | 2.07 |

a.  Calculate the difference between the sample mean distance for the football filled with air, and for the one filled with helium.

*The 17 air-filled balls had a mean of 27 yds. compared to 23.8 yds. for the 17 helium-filled balls, a difference of 3.2 yds.*

b.  On the same scale, draw dot plots of the two distributions and discuss the variability in each distribution.

*Based on the dot plots, it looks like the variability in the two distributions is about the same.*

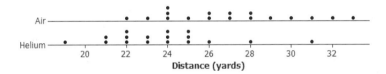

c.  Calculate the mean absolute deviations (MAD) for each distribution.  Based on the MADs, compare the variability in each distribution.  Is the variability about the same?  Interpret the MADs in the context of the problem.

*The MAD is 2.59 yds. for the air-filled balls and 2.07 yds. for the helium filled balls.  The typical deviation from the mean of 27.0 is about 2.59 yds. for the air filled balls.  The typical deviation from the mean of 23.8 is about 2.07 yds. for the helium filled balls.  There is a slight difference in variability.*

d.  Based on your calculations, is the difference in mean distance meaningful?  Part of your reasoning should involve the number of MADs that separate the sample means.  Note that if the MADs differ, use the larger one in determining how many MADs separate the two means.

*There is a separation of $\frac{3.2}{2.59} = 1.2$ MADs.  There is no meaningful distance between the means.*

3.  Suppose that your classmates were debating about whether going to college is really worth it.  Based on the following data of annual salaries (rounded to the nearest thousands of dollars) for college graduates and high school graduates with no college experience, does it appear that going to college is indeed worth the effort?  The data are from people in their second year of employment.

| College Grad | 41 | 67 | 53 | 48 | 45 | 60 | 59 | 55 | 52 | 52 | 50 | 59 | 44 | 49 | 52 |
|--------------|----|----|----|----|----|----|----|----|----|----|----|----|----|----|----|
| High School Grad | 23 | 33 | 36 | 29 | 25 | 43 | 42 | 38 | 27 | 25 | 33 | 41 | 29 | 33 | 35 |

a.  Calculate the difference between the sample mean salary for college graduates and for high school graduates.

*The 15 college graduates had a mean salary of $52,400, compared to $32,800 for the 15 high school graduates, a difference of $19,600.*

b.   On the same scale, draw dot plots of the two distributions and discuss the variability in each distribution.

*Based on the dot plots, the variability of the two distributions appears to be about the same.*

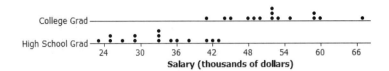

c.   Calculate the mean absolute deviations (MAD) for each distribution. Based on the MADs, compare the variability in each distribution. Is the variability about the same? Interpret the MADs in the context of the problem.

*The MAD is 5.15 for college graduates, and 5.17 for high school graduates. The typical deviation from the mean of 52.4 is about 5.15 (or $5,150) for college graduates. The typical deviation from the mean of 32.8 is about 5.17 ($5,170) for high school graduates. The variability in the two distributions is nearly the same.*

d.   Based on your calculations, is going to college worth the effort? Part of your reasoning should involve the number of MADs that separate the sample means.

*There is a separation of $\dfrac{19.6}{5.17} = 3.79$ MADs. There is a meaningful difference between the population means. Going to college is worth the effort.*

# Lesson 23: Using Sample Data to Decide if Two Population Means Are Different

## Student Outcomes

- Students use data from random samples to draw informal inferences about the difference in population means.

## Lesson Notes

Statistics is about making decisions concerning populations. In many settings, this involves deciding whether two population means are similar or different. More formal procedures are presented at the high school level and beyond. This lesson builds on the previous lesson to present an informal inferential procedure. The bottom line is that if two sample means are separated by at least 2 MADs, then it is reasonable to conclude that there is a difference between the two corresponding population means. Recall that in Lesson 21, a difference between two sample means was called "meaningful" if it is greater than what would have been expected just due to sampling variability. This is an indication that the corresponding population means are not equal.

In the previous lesson, students calculated how many MADs separated two sample means. This lesson extends that procedure to informal inference by introducing a way to decide if the two sample means are different enough to be confident in deciding that the population means differ.

## Classwork

> In the previous lesson, you described how far apart the means of two data sets are in terms of the MAD (mean absolute deviation), a measure of variability. In this lesson, you will extend that idea to informally determine when two sample means computed from random samples are far enough apart from each other so to imply that the population means also differ in a "meaningful" way. Recall that a "meaningful" difference between two means is a difference that is greater than would have been expected just due to sampling variability.

## Example 1 (3 minutes): Texting

The purpose of this example is to take students through the informal inferential process focusing on the concept and not on calculation. Introduce the scenario presented in the example. The statistical question is whether there is a difference on average in recalling *real* words or *fake* words. Linda randomly selected 28 students from her district's middle school students and assigned 14 of them (chosen at random from the 28) a list of 20 real words and assigned the other half a list of 20 fake words.

> **Example 1: Texting**
>
> With texting becoming so popular, Linda wanted to determine if middle school students memorize *real* words more or less easily than *fake* words. For example, real words are "food," "car," "study," "swim;" whereas fake words are "stk," "fonw," "cqur," "ttnsp." She randomly selected 28 students from all middle school students in her district and gave half of them a list of 20 real words and the other half a list of 20 fake words.

## Exercises 1–3 (5–7 minutes)

Give students a moment to think about the questions and then discuss as a class.

---

**Exercises 1–3**

1. How do you think Linda might have randomly selected 28 students from all middle school students in her district?

   *Random selection is done in an attempt to obtain students to represent all middle school students in the district. Linda would need to number all middle school students and use a random device to generate 28 numbers. One device to generate integers is http://www.rossmanchance.com/applets/RandomGen/GenRandom01.htm. Note that if there are duplicates, additional random numbers need to be generated. A second way to generate the random selections is using the random-number table from previous lessons.*

2. Why do you think Linda selected the students for her study randomly? Explain.

   *Linda randomly assigned her chosen 28 students to two groups of 14 each. Random assignment is done to help assure that groups are similar to each other.*

3. She gave the selected students one minute to memorize their list after which they were to turn the list over and after two minutes write down all the words that they could remember. Afterwards, they calculated the number of correct "words" that they were able to write down. Do you think a penalty should be given for an incorrect "word" written down? Explain your reasoning.

   *Answers will vary. Either position is acceptable. The purpose is to get students to take a position and argue for it.*

---

MP.3

## Exercises 4–7 (8–10 Minutes)

Let students work independently, and compare answers with a neighbor. Then discuss Exercises 6 and 7 as a class.

---

**Exercises 4–7**

Suppose the data (number of correct words recalled) she collected were:

For students given the real words list: $8, 11, 12, 8, 4, 7, 9, 12, 12, 9, 14, 11, 5, 10$.
For students given the fake words list: $3, 5, 4, 4, 4, 7, 11, 9, 7, 7, 1, 3, 3, 7$.

4. On the same scale, draw dot plots for the two data sets.

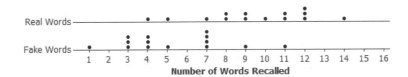

5. From looking at the dot plots, write a few sentences comparing the distribution of the number of correctly recalled real words with the distribution of number of correctly recalled fake words. In particular, comment on which type of word, if either, that students recall better. Explain.

   *There is a considerable amount of overlap between data from the two random samples. The distribution of number of real words recalled is somewhat skewed to the left; the distribution of number of fake words recalled is fairly symmetric. The real words distribution appears to be centered around 9 or 10, whereas the fake words distribution appears to be centered around 6. Whether the separation between 9 and 6 is meaningful remains to be seen. If it is meaningful, then the mean number of real words recalled is greater than the mean number of fake words recalled.*

---

**6.** Linda made the following calculations for the two data sets:

| | Mean | MAD |
|---|---|---|
| Real words recalled | 9.43 | 2.29 |
| Fake words recalled | 5.36 | 2.27 |

In the previous lesson, you calculated the number of MADs that separated two sample means. You used the larger MAD to make this calculation if the two MADs were not the same. How many MADs separate the mean number of real words recalled and the mean number of fake words recalled for the students in the study?

*The difference between the two means is* $9.43 - 5.36 = 4.07$*. The larger of the two MADs is* $2.29$*. The number of MADs that separate the two means is* $\frac{4.07}{2.29} = 1.78$*.*

**7.** In the last lesson, our work suggested that if the number of MADs that separate the two sample means is 2 or more, then it is reasonable to conclude that not only do the means differ in the samples, but that the means differ in the populations as well. If the number of MADs is less than 2, then you can conclude that the difference in the sample means might just be sampling variability and that there may not be a meaningful difference in the population means. Using these criteria, what can Linda conclude about the difference in population means based on the sample data that she collected? Be sure to express your conclusion in the context of this problem.

*Since* $1.78$ *is below the suggested 2 MADs, Linda would conclude that the average number of real words that all middle school students in her district would recall might be the same as the average number of fake words that they would recall.*

Ask your students if they are surprised by her conclusion. Some may say that they are surprised because 9 is bigger than 5. Be sure that those students understand the concept of sampling variability. Their response might be a mathematical one in comparing two numbers. Other students may not be surprised, as they can see from the dot plots that there is quite a bit of overlap.

## Example 2 (2 minutes)

This example has a population database in Excel consisting of four numerical variables: Texting is in column 2, the number of minutes per day students text (whole number); ReacTime is in column 3, the number of seconds that it takes students to respond to a computer screen stimulus (two decimal places); Homework is in column 4, the total number of hours per week that students spend on doing homework both in school and at home (one decimal place); and Sleep is in column 5, the number of hours per night that students sleep (one decimal place). Display the data file on an overhead projector and be sure students understand the data located in each column.

---

**Example 2**

Ken, an eighth grade student, was interested in doing a statistics study involving sixth grade and eleventh grade students in his school district. He conducted a survey on four numerical variables and two categorical variables (grade level and gender). His Excel population database for the 265 sixth graders and 175 eleventh graders in his district has the following description:

| Column | Name | Description |
|---|---|---|
| 1 | ID | ID numbers are from 1 through 440<br>   1 – 128 Sixth grade females<br>129 – 265 Sixth grade males<br>266 – 363 Eleventh grade females<br>364 – 440 Eleventh grade males |
| 2 | Texting | Number of minutes per day text (whole number) |
| 3 | ReacTime | Time in seconds to respond to a computer screen stimulus (two decimal places) |
| 4 | Homework | Total number of hours per week spend on doing homework (one decimal place) |
| 5 | Sleep | Number of hours per night sleep (one decimal place) |

---

## Exercise 8 (3 minutes)

Discuss the question in this exercise with students.

---

**Exercise 8**

8.  Ken decides to base his study on a random sample of 20 sixth graders and a random sample of 20 eleventh graders. The sixth graders have IDs 1–265 and the eleventh graders are numbered 266–440. Advise him on how to randomly sample 20 sixth graders and 20 eleventh graders from his data file.

*Advise Ken to use a random-number generator such as http://www.rossmanchance.com/applets/RandomGen/GenRandom01.html. If access to this site is limited, students can use the random-number table used in previous lessons. Choose 20 different random integers from 1–265 for the sixth grade participants and 20 different random integers from 266–440 for the eleventh graders.*

---

## Exercises 9–13 (15–20 minutes)

You may want students to generate their own random integers. This exercise provides random numbers for both grade levels so that all of your students are on the same page for this exercise.

It is easier to use the data file if the random numbers are ordered. The data file (an EXCEL spreadsheet) can be obtained from the Engage New York website, or from a printed copy located at the end of this lesson. Organize students in small groups and provide each group a copy of the data file.

---

**Exercises 9–13**

Suppose that from a random-number generator:

The random ID numbers for his 20 sixth graders are:
231 15 19 206 86 183 233 253 142 36 195 139 75 210 56 40 66 114 127 9

The random ID numbers for his 20 eleventh graders are:
391 319 343 426 307 360 289 328 390 350 279 283 302 287 269 332 414 267 428

9.  For each set, find the homework hours data from the population database that corresponds to these randomly selected ID numbers.

*Sixth graders IDs ordered:*
9 15 19 36 40 56 66 75 86 114 127 139 142 183 195 206 210 231 233 253

*Their data are:*
8.4 7.2 9.2 7.9 9.3 7.6 6.5 6.7 8.2 7.7 7.7 4.8 5.4 4.7 7.1 6.6 2.9 8.5 6.1 8.6

*Eleventh graders IDs ordered:*
267 269 279 280 283 287 289 302 307 319 328 332 343 350 360 390 391 414 426 428

*Their data are:*
9.5 9.8 10.9 11.8 10.0 12.0 9.1 11.8 10.0 8.3 10.6 9.5 9.8 10.8 10.7 9.4 13.2 7.3 11.6 10.3

10. On the same scale, draw dot plots for the two sample data sets.

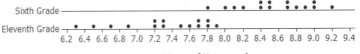

---

11. From looking at the dot plots, list some observations comparing the number of hours per week that sixth graders spend on doing homework and the number of hours per week that eleventh graders spend on doing homework.

*There is some overlap between the data for the two random samples. The sixth grade distribution may be slightly skewed to the left. The eleventh grade distribution is fairly symmetric. The mean number of homework hours for sixth graders appears to be around 7 hours, whereas that for the eleventh graders is around 10.*

12. Calculate the mean and MAD for each of the data sets. How many MADs separate the two sample means? (Use the larger MAD to make this calculation if the sample MADs are not the same.)

| | Mean (hr.) | MAD (hr.) |
|---|---|---|
| Sixth Grade | 7.055 | 1.274 |
| Eleventh Grade | 10.32 | 1.052 |

*The number of MADs that separate the two means is $\frac{10.32-7.055}{1.274} = 2.56$.*

13. Ken recalled Linda suggesting that if the number of MADs is greater than or equal to 2 then it would be reasonable to think that the population of all sixth grade students in his district and the population of all eleventh grade students in his district have different means. What should Ken conclude based on his homework study?

*Since 2.56 is greater than 2, it is reasonable to conclude that on average eleventh graders spend more time doing homework per week than do sixth graders.*

Note: Some students may ask why they have to do the sampling when they have the population database. Since they have all the information, why not just calculate the two population means and be done with it? Good question. Perhaps the best answer is that they are learning a procedure. In practice, populations are often very large and observations for the subjects are not known. If you think that your students are bothered by having a population database, then you may decide not to give them access to it, and provide them just the data corresponding to the given random numbers. A minor rewrite of the following problem set would be needed in that case.

## Closing (4 minutes)

Discuss the Lesson Summary with students.

---

**Lesson Summary**

To determine if the mean value of some numerical variable differs for two populations, take random samples from each population. It is very important that the samples be random samples. If the number of MADs that separate the two sample means is 2 or more, then it is reasonable to think that the populations have different means. Otherwise, the population means are considered to be the same.

---

## Exit Ticket (5 minutes)

Name _____   Date_____

# Lesson 23: Using Sample Data to Decide if Two Population Means are Different

1. Do eleventh grade males text more per day than eleventh grade females do? To answer this question, two randomly selected samples were obtained from the Excel data file used in this lesson. Indicate how 20 randomly selected eleventh grade females would be chosen for this study. Indicate how 20 randomly selected eleventh grade males would be chosen.

2. Two randomly selected samples (one of eleventh grade females, and one of eleventh grade males) were obtained from the database. The results are indicated below:

| | Mean number of minutes per day texting | MAD (minutes) |
|---|---|---|
| Eleventh grade females | 102.55 | 1.31 |
| Eleventh grade males | 100.32 | 1.12 |

Is there a meaningful difference in the number of minutes per day that eleventh grade females and males text? Explain your answer.

## Exit Ticket Sample Solutions

1.  Do eleventh grade males text more per day than eleventh grade females? To answer this question, two randomly selected samples were obtained from the Excel data file used in this lesson. Indicate how 20 randomly selected eleventh grade females would be chosen for this study. Indicate how 20 randomly selected eleventh grade males would be chosen.

    *To pick 20 females, 20 randomly selected numbers from 266–363 would be generated from a random-number generator or from a random-number table. Duplicates would be disregarded and a new number would be generated. To pick 20 males, 20 randomly selected numbers from 264–440 would be generated. Again, duplicates would be disregarded, and a new number would be generated.*

2.  Two randomly selected samples (one of eleventh grade females, and one of eleventh grade males) were obtained from the database. The results are indicated below:

    |  | Mean number of minutes per day texting | MAD (minutes) |
    |---|---|---|
    | Eleventh grade females | 102.55 | 1.31 |
    | Eleventh grade males | 100.32 | 1.12 |

    Is there a meaningful difference in the number of minutes per day eleventh grade females and males text? Explain your answer.

    *The difference in the means is 102.55 min. −100.32 min., or 2.23 min. (to the nearest hundredth of a minute). Divide this by 1.31 minutes or the MAD for females (the larger of the two MADs): $\frac{2.23}{1.31} = 1.70$ min. to the nearest hundredth of a minute. This difference is less than 2 MADs, and therefore, the difference in the male and female number of minutes per day of texting is not a meaningful difference.*

## Problem Set Sample Solutions

1.  Based on Ken's population database, compare the amount of sleep that sixth grade females get on average to the amount of sleep that eleventh grade females get on average.

    Find the data for 15 sixth grade females based on the following random ID numbers:
    65  1  67  101  106  87  85  95  120  4  64  74 102  31  128

    Find the data for 15 eleventh grade females based on the following random ID numbers:
    348  313  297  351  294  343  275  354  311  328  274  305  288  267  301

    *This problem compares the amount of sleep that sixth grade females get on average to the amount of sleep that eleventh grade females get on average.*

    *Random numbers are provided for students. Provide students access to the data file (a printed copy or access to the file at the website), or if that is not possible, provide them the following values to use in the remaining questions:*

    *Sixth grade females number of hours of sleep per night:*
    *8.2  7.8  8.0  8.1  8.7  9.0  8.9  8.7  8.4  9.0  8.4  8.5  8.8  8.5  9.2*

    *Eleventh grade females number of hours of sleep per night:*
    *6.9  7.8  7.2  7.9  7.8  6.7  7.6  7.3  7.7  7.3  6.5  7.7  7.2  6.3  7.5*

2. On the same scale, draw dot plots for the two sample data sets.

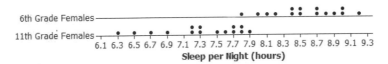

3. Looking at the dot plots, list some observations comparing the number of hours per week that sixth graders spend on doing homework and the number of hours per week that eleventh graders spend on doing homework.

   *There is a small amount of overlap between the data sets for the two random samples. The distribution of sixth grade hours of sleep is symmetric, whereas that for eleventh graders is skewed to the left. It appears that the mean number of hours of sleep for sixth grade females is around 8.5 and the mean number for eleventh grade females is around 7.3 or so. Whether or not the difference is meaningful depends on the amount of variability that separates them.*

4. Calculate the mean and MAD for each of the data sets. How many MADs separate the two sample means? (Use the larger MAD to make this calculation if the sample MADs are not the same.)

   |                        | Mean (hr.) | MAD (hr.) |
   |------------------------|------------|-----------|
   | Sixth Grade Females    | 8.55       | 0.33      |
   | Eleventh Grade Females | 7.29       | 0.40      |

   *The number of MADs that separate the two means is* $\dfrac{8.55 - 7.29}{0.4} = 3.15.$

5. Recall that if the number of MADs in the difference of two sample means is greater than or equal to 2, then it would be reasonable to think that the population means are different. Using this guideline, what can you say about the average number of hours of sleep per night for all sixth grade females in the population compared to all eleventh grade females in the population?

   *Since 3.15 is well above the criteria of 2 MADs, it can be concluded that on average sixth grade females get more sleep per night than do eleventh grade females.*

**Copy of the Excel student data file:**

| ID Number | Texting | ReacTime | Homework | Sleep |
|-----------|---------|----------|----------|-------|
| 1 | 99 | 0.33 | 9.0 | 8.2 |
| 2 | 69 | 0.39 | 8.6 | 7.5 |
| 3 | 138 | 0.36 | 6.1 | 8.7 |
| 4 | 100 | 0.40 | 7.9 | 7.8 |
| 5 | 116 | 0.28 | 5.1 | 8.8 |
| 6 | 112 | 0.38 | 6.5 | 7.9 |
| 7 | 79 | 0.35 | 6.5 | 8.8 |
| 8 | 111 | 0.41 | 8.8 | 8.5 |
| 9 | 115 | 0.49 | 8.4 | 8.4 |
| 10 | 82 | 0.43 | 8.7 | 8.8 |
| 11 | 136 | 0.46 | 7.2 | 8.4 |
| 12 | 112 | 0.51 | 8.3 | 9.0 |
| 13 | 101 | 0.42 | 7.0 | 8.8 |
| 14 | 89 | 0.38 | 5.6 | 8.3 |
| 15 | 120 | 0.35 | 7.2 | 8.2 |
| 16 | 144 | 0.36 | 3.9 | 8.8 |
| 17 | 131 | 0.26 | 9.0 | 8.9 |
| 18 | 126 | 0.39 | 7.0 | 8.5 |
| 19 | 118 | 0.37 | 9.2 | 8.7 |
| 20 | 83 | 0.34 | 7.4 | 8.6 |
| 21 | 120 | 0.20 | 4.5 | 8.7 |
| 22 | 114 | 0.38 | 6.0 | 8.6 |
| 23 | 90 | 0.25 | 7.0 | 8.4 |
| 24 | 116 | 0.36 | 5.8 | 8.4 |
| 25 | 108 | 0.36 | 8.9 | 8.1 |
| 26 | 89 | 0.31 | 8.4 | 8.8 |
| 27 | 124 | 0.44 | 6.3 | 8.3 |
| 28 | 121 | 0.32 | 5.2 | 8.0 |
| 29 | 104 | 0.30 | 6.7 | 8.1 |
| 30 | 110 | 0.39 | 7.8 | 8.1 |
| 31 | 119 | 0.36 | 8.5 | 8.0 |
| 32 | 113 | 0.40 | 5.3 | 9.4 |
| 33 | 106 | 0.36 | 5.7 | 8.6 |
| 34 | 119 | 0.33 | 5.9 | 8.4 |
| 35 | 129 | 0.38 | 6.2 | 9.0 |
| 36 | 95 | 0.44 | 7.9 | 8.3 |
| 37 | 126 | 0.41 | 7.2 | 8.6 |
| 38 | 106 | 0.26 | 7.1 | 8.5 |

| 39 | 116 | 0.34 | 4.9 | 8.4 |
| 40 | 107 | 0.35 | 9.3 | 8.1 |
| 41 | 108 | 0.48 | 8.1 | 8.6 |
| 42 | 97 | 0.40 | 7.1 | 8.8 |
| 43 | 97 | 0.27 | 4.2 | 8.3 |
| 44 | 100 | 0.24 | 6.2 | 8.9 |
| 45 | 123 | 0.50 | 8.1 | 8.6 |
| 46 | 94 | 0.39 | 5.2 | 8.3 |
| 47 | 87 | 0.37 | 8.0 | 8.3 |
| 48 | 93 | 0.42 | 9.7 | 8.1 |
| 49 | 117 | 0.39 | 7.9 | 8.3 |
| 50 | 94 | 0.36 | 6.9 | 8.9 |
| 51 | 124 | 0.29 | 8.1 | 8.4 |
| 52 | 116 | 0.44 | 4.9 | 8.2 |
| 53 | 137 | 0.25 | 9.1 | 8.3 |
| 54 | 123 | 0.30 | 3.8 | 8.7 |
| 55 | 122 | 0.21 | 6.6 | 8.8 |
| 56 | 92 | 0.41 | 7.6 | 8.6 |
| 57 | 101 | 0.36 | 8.5 | 8.2 |
| 58 | 111 | 0.34 | 6.5 | 8.7 |
| 59 | 126 | 0.31 | 5.6 | 8.6 |
| 60 | 81 | 0.33 | 7.3 | 8.4 |
| 61 | 118 | 0.35 | 8.4 | 8.3 |
| 62 | 113 | 0.37 | 7.7 | 8.4 |
| 63 | 114 | 0.49 | 6.9 | 8.6 |
| 64 | 124 | 0.34 | 8.5 | 8.1 |
| 65 | 90 | 0.48 | 6.6 | 8.7 |
| 66 | 99 | 0.39 | 6.5 | 8.5 |
| 67 | 155 | 0.40 | 6.5 | 9.0 |
| 68 | 77 | 0.45 | 4.9 | 8.3 |
| 69 | 79 | 0.23 | 7.7 | 8.6 |
| 70 | 91 | 0.43 | 6.7 | 8.9 |
| 71 | 107 | 0.42 | 6.9 | 8.6 |
| 72 | 112 | 0.39 | 6.3 | 8.6 |
| 73 | 147 | 0.34 | 7.8 | 8.3 |
| 74 | 126 | 0.52 | 9.2 | 8.9 |
| 75 | 106 | 0.27 | 6.7 | 8.0 |
| 76 | 86 | 0.21 | 6.1 | 8.4 |
| 77 | 111 | 0.29 | 7.4 | 8.0 |
| 78 | 118 | 0.41 | 5.2 | 8.8 |
| 79 | 105 | 0.28 | 5.9 | 8.3 |

| 80 | 108 | 0.40 | 7.0 | 8.6 |
|---|---|---|---|---|
| 81 | 131 | 0.32 | 7.5 | 8.5 |
| 82 | 86 | 0.34 | 7.4 | 8.4 |
| 83 | 87 | 0.27 | 7.7 | 8.6 |
| 84 | 116 | 0.34 | 9.6 | 9.1 |
| 85 | 102 | 0.42 | 5.6 | 8.7 |
| 86 | 105 | 0.41 | 8.2 | 8.2 |
| 87 | 96 | 0.33 | 4.9 | 8.4 |
| 88 | 90 | 0.20 | 7.5 | 8.6 |
| 89 | 109 | 0.39 | 6.0 | 8.5 |
| 90 | 103 | 0.36 | 5.7 | 9.1 |
| 91 | 98 | 0.26 | 5.5 | 8.4 |
| 92 | 103 | 0.41 | 8.2 | 8.0 |
| 93 | 110 | 0.39 | 9.2 | 8.4 |
| 94 | 101 | 0.39 | 8.8 | 8.6 |
| 95 | 101 | 0.40 | 6.1 | 9.0 |
| 96 | 98 | 0.22 | 7.0 | 8.6 |
| 97 | 105 | 0.43 | 8.0 | 8.5 |
| 98 | 110 | 0.33 | 8.0 | 8.8 |
| 99 | 104 | 0.45 | 9.6 | 8.5 |
| 100 | 119 | 0.29 | 7.2 | 9.1 |
| 101 | 120 | 0.48 | 6.7 | 8.4 |
| 102 | 109 | 0.36 | 6.8 | 8.5 |
| 103 | 105 | 0.29 | 8.6 | 8.6 |
| 104 | 110 | 0.29 | 7.1 | 8.9 |
| 105 | 116 | 0.41 | 6.7 | 8.8 |
| 106 | 114 | 0.25 | 9.0 | 8.8 |
| 107 | 111 | 0.35 | 9.9 | 8.6 |
| 108 | 137 | 0.37 | 6.1 | 7.9 |
| 109 | 104 | 0.37 | 7.3 | 9.0 |
| 110 | 95 | 0.33 | 8.3 | 8.0 |
| 111 | 117 | 0.29 | 7.5 | 8.8 |
| 112 | 100 | 0.30 | 2.9 | 9.3 |
| 113 | 105 | 0.24 | 6.1 | 8.5 |
| 114 | 102 | 0.41 | 7.7 | 8.0 |
| 115 | 109 | 0.32 | 5.8 | 8.2 |
| 116 | 108 | 0.36 | 6.6 | 8.3 |
| 117 | 111 | 0.43 | 8.2 | 8.5 |
| 118 | 115 | 0.20 | 5.7 | 8.9 |
| 119 | 89 | 0.34 | 5.1 | 8.1 |
| 120 | 94 | 0.29 | 9.2 | 8.5 |

| 121 | 105 | 0.26 | 7.3 | 8.6 |
| 122 | 143 | 0.42 | 5.8 | 8.1 |
| 123 | 129 | 0.29 | 6.8 | 8.8 |
| 124 | 118 | 0.31 | 6.2 | 8.7 |
| 125 | 129 | 0.38 | 9.1 | 8.4 |
| 126 | 96 | 0.25 | 6.7 | 8.5 |
| 127 | 95 | 0.27 | 7.7 | 8.3 |
| 128 | 43 | 0.22 | 6.3 | 9.2 |
| 129 | 64 | 0.23 | 3.7 | 8.5 |
| 130 | 38 | 0.41 | 5.2 | 9.1 |
| 131 | 46 | 0.35 | 7.1 | 8.3 |
| 132 | 56 | 0.32 | 4.7 | 8.9 |
| 133 | 41 | 0.45 | 7.2 | 8.8 |
| 134 | 55 | 0.27 | 4.1 | 8.5 |
| 135 | 57 | 0.40 | 7.5 | 8.6 |
| 136 | 66 | 0.41 | 6.9 | 8.0 |
| 137 | 62 | 0.39 | 3.8 | 8.5 |
| 138 | 49 | 0.36 | 6.4 | 8.5 |
| 139 | 33 | 0.21 | 4.8 | 9.0 |
| 140 | 38 | 0.34 | 5.6 | 8.5 |
| 141 | 75 | 0.19 | 5.0 | 8.2 |
| 142 | 68 | 0.25 | 5.4 | 8.4 |
| 143 | 60 | 0.47 | 7.4 | 9.1 |
| 144 | 63 | 0.33 | 4.4 | 8.2 |
| 145 | 48 | 0.28 | 7.1 | 8.2 |
| 146 | 49 | 0.36 | 2.7 | 8.5 |
| 147 | 72 | 0.44 | 5.6 | 7.6 |
| 148 | 54 | 0.51 | 4.3 | 8.7 |
| 149 | 65 | 0.38 | 7.7 | 8.5 |
| 150 | 72 | 0.40 | 3.4 | 9.1 |
| 151 | 51 | 0.16 | 4.9 | 8.4 |
| 152 | 64 | 0.16 | 4.4 | 8.5 |
| 153 | 43 | 0.34 | 0.1 | 8.8 |
| 154 | 57 | 0.38 | 4.4 | 8.2 |
| 155 | 72 | 0.51 | 3.2 | 8.4 |
| 156 | 37 | 0.46 | 5.3 | 8.6 |
| 157 | 50 | 0.33 | 4.1 | 8.2 |
| 158 | 41 | 0.46 | 4.5 | 8.9 |
| 159 | 63 | 0.40 | 5.0 | 8.7 |
| 160 | 51 | 0.33 | 5.4 | 7.9 |
| 161 | 57 | 0.51 | 4.9 | 8.6 |

Lesson 23:   Using Sample Data to Decide if Two Population Means Are Different

| 162 | 51 | 0.24 | 1.7 | 8.4 |
| 163 | 73 | 0.32 | 5.6 | 8.6 |
| 164 | 51 | 0.37 | 4.0 | 8.5 |
| 165 | 52 | 0.36 | 5.8 | 8.3 |
| 166 | 52 | 0.34 | 4.6 | 8.1 |
| 167 | 63 | 0.34 | 4.1 | 8.1 |
| 168 | 76 | 0.30 | 6.1 | 8.2 |
| 169 | 56 | 0.40 | 5.7 | 8.5 |
| 170 | 47 | 0.33 | 5.0 | 8.2 |
| 171 | 44 | 0.41 | 5.0 | 8.3 |
| 172 | 60 | 0.33 | 7.7 | 8.4 |
| 173 | 36 | 0.39 | 6.5 | 8.8 |
| 174 | 52 | 0.30 | 5.4 | 8.2 |
| 175 | 53 | 0.27 | 5.6 | 8.2 |
| 176 | 60 | 0.35 | 6.0 | 8.6 |
| 177 | 48 | 0.43 | 3.6 | 8.6 |
| 178 | 63 | 0.49 | 0.2 | 8.2 |
| 179 | 76 | 0.42 | 5.9 | 8.9 |
| 180 | 58 | 0.34 | 7.3 | 8.3 |
| 181 | 51 | 0.43 | 6.4 | 8.7 |
| 182 | 38 | 0.33 | 4.9 | 8.5 |
| 183 | 46 | 0.17 | 4.7 | 8.3 |
| 184 | 53 | 0.34 | 6.4 | 8.7 |
| 185 | 60 | 0.38 | 6.1 | 8.7 |
| 186 | 71 | 0.23 | 6.9 | 8.2 |
| 187 | 54 | 0.41 | 2.9 | 8.3 |
| 188 | 61 | 0.44 | 5.8 | 8.4 |
| 189 | 62 | 0.35 | 3.9 | 8.9 |
| 190 | 55 | 0.15 | 4.8 | 8.0 |
| 191 | 57 | 0.22 | 4.1 | 8.2 |
| 192 | 43 | 0.41 | 7.5 | 8.5 |
| 193 | 51 | 0.34 | 2.4 | 8.6 |
| 194 | 34 | 0.55 | 3.5 | 8.4 |
| 195 | 38 | 0.43 | 7.1 | 8.8 |
| 196 | 49 | 0.38 | 3.5 | 8.3 |
| 197 | 57 | 0.30 | 3.6 | 8.5 |
| 198 | 53 | 0.37 | 5.2 | 9.1 |
| 199 | 51 | 0.36 | 5.1 | 8.2 |
| 200 | 59 | 0.38 | 3.6 | 8.7 |
| 201 | 35 | 0.44 | 4.0 | 8.0 |
| 202 | 73 | 0.32 | 3.0 | 8.3 |

| 203 | 68 | 0.37 | 2.7 | 8.4 |
|-----|----|------|-----|-----|
| 204 | 31 | 0.36 | 4.6 | 8.6 |
| 205 | 40 | 0.33 | 9.0 | 8.3 |
| 206 | 60 | 0.36 | 6.6 | 8.5 |
| 207 | 66 | 0.44 | 4.2 | 8.5 |
| 208 | 47 | 0.22 | 4.5 | 8.7 |
| 209 | 56 | 0.30 | 4.8 | 8.6 |
| 210 | 72 | 0.36 | 2.9 | 8.8 |
| 211 | 68 | 0.50 | 6.6 | 8.3 |
| 212 | 45 | 0.37 | 7.3 | 8.5 |
| 213 | 58 | 0.17 | 4.9 | 9.0 |
| 214 | 64 | 0.34 | 3.2 | 8.6 |
| 215 | 66 | 0.34 | 2.5 | 8.4 |
| 216 | 49 | 0.29 | 5.0 | 8.3 |
| 217 | 83 | 0.39 | 2.5 | 8.8 |
| 218 | 73 | 0.33 | 3.6 | 8.4 |
| 219 | 52 | 0.34 | 3.3 | 8.8 |
| 220 | 56 | 0.28 | 8.8 | 8.7 |
| 221 | 58 | 0.32 | 5.6 | 8.3 |
| 222 | 53 | 0.40 | 5.9 | 8.1 |
| 223 | 50 | 0.23 | 4.4 | 8.4 |
| 224 | 43 | 0.34 | 3.9 | 8.7 |
| 225 | 50 | 0.39 | 4.4 | 8.0 |
| 226 | 44 | 0.31 | 4.4 | 8.4 |
| 227 | 59 | 0.36 | 6.0 | 9.1 |
| 228 | 41 | 0.35 | 3.2 | 8.4 |
| 229 | 53 | 0.29 | 6.6 | 8.7 |
| 230 | 49 | 0.37 | 5.7 | 8.3 |
| 231 | 42 | 0.22 | 8.5 | 8.6 |
| 232 | 48 | 0.34 | 3.9 | 8.2 |
| 233 | 60 | 0.31 | 6.1 | 8.8 |
| 234 | 56 | 0.50 | 2.6 | 8.5 |
| 235 | 43 | 0.25 | 5.3 | 8.9 |
| 236 | 67 | 0.32 | 6.1 | 8.8 |
| 237 | 43 | 0.24 | 8.6 | 8.8 |
| 238 | 41 | 0.46 | 5.1 | 8.7 |
| 239 | 66 | 0.45 | 4.9 | 8.3 |
| 240 | 44 | 0.52 | 4.1 | 8.7 |
| 241 | 70 | 0.43 | 6.6 | 8.8 |
| 242 | 63 | 0.38 | 7.9 | 8.4 |
| 243 | 47 | 0.24 | 3.9 | 8.3 |

| | | | | |
|---|---|---|---|---|
| 244 | 52 | 0.38 | 5.4 | 8.8 |
| 245 | 49 | 0.47 | 4.2 | 8.4 |
| 246 | 45 | 0.31 | 8.1 | 8.8 |
| 247 | 46 | 0.37 | 1.9 | 8.3 |
| 248 | 19 | 0.31 | 6.5 | 8.3 |
| 249 | 63 | 0.40 | 6.1 | 8.5 |
| 250 | 64 | 0.35 | 5.8 | 8.1 |
| 251 | 63 | 0.34 | 6.7 | 8.5 |
| 252 | 68 | 0.46 | 6.9 | 8.5 |
| 253 | 48 | 0.43 | 8.6 | 8.7 |
| 254 | 43 | 0.38 | 4.4 | 8.3 |
| 255 | 50 | 0.32 | 4.6 | 8.7 |
| 256 | 76 | 0.31 | 4.0 | 8.3 |
| 257 | 64 | 0.39 | 5.7 | 8.6 |
| 258 | 38 | 0.29 | 6.4 | 8.0 |
| 259 | 90 | 0.30 | 7.0 | 8.6 |
| 260 | 37 | 0.39 | 4.8 | 8.8 |
| 261 | 58 | 0.37 | 6.5 | 8.0 |
| 262 | 42 | 0.27 | 4.5 | 8.6 |
| 263 | 58 | 0.37 | 6.0 | 8.3 |
| 264 | 42 | 0.42 | 7.2 | 8.8 |
| 265 | 66 | 0.33 | 12.6 | 8.8 |
| 266 | 116 | 0.44 | 8.7 | 7.5 |
| 267 | 76 | 0.43 | 9.5 | 6.9 |
| 268 | 125 | 0.46 | 9.9 | 6.8 |
| 269 | 128 | 0.41 | 9.8 | 6.6 |
| 270 | 128 | 0.37 | 11.3 | 7.1 |
| 271 | 125 | 0.44 | 6.7 | 7.9 |
| 272 | 80 | 0.49 | 10.6 | 7.1 |
| 273 | 110 | 0.48 | 9.9 | 7.2 |
| 274 | 135 | 0.41 | 9.8 | 7.8 |
| 275 | 136 | 0.45 | 8.9 | 7.2 |
| 276 | 142 | 0.43 | 10.2 | 8.0 |
| 277 | 120 | 0.48 | 10.2 | 7.5 |
| 278 | 109 | 0.43 | 10.1 | 7.1 |
| 279 | 109 | 0.50 | 10.9 | 7.5 |
| 280 | 111 | 0.35 | 11.8 | 7.4 |
| 281 | 101 | 0.49 | 8.5 | 7.8 |
| 282 | 98 | 0.50 | 11.6 | 7.2 |
| 283 | 91 | 0.56 | 10.0 | 7.3 |
| 284 | 151 | 0.50 | 7.7 | 6.7 |

| 285 | 82 | 0.48 | 14.0 | 7.5 |
| 286 | 107 | 0.48 | 9.5 | 7.5 |
| 287 | 83 | 0.40 | 12.0 | 7.2 |
| 288 | 91 | 0.40 | 9.2 | 7.9 |
| 289 | 127 | 0.40 | 9.1 | 7.6 |
| 290 | 115 | 0.42 | 11.6 | 6.8 |
| 291 | 118 | 0.40 | 9.8 | 7.3 |
| 292 | 89 | 0.42 | 10.8 | 7.0 |
| 293 | 100 | 0.46 | 11.6 | 7.3 |
| 294 | 97 | 0.39 | 8.5 | 7.8 |
| 295 | 110 | 0.36 | 11.1 | 7.7 |
| 296 | 88 | 0.40 | 9.0 | 6.7 |
| 297 | 103 | 0.47 | 11.7 | 6.7 |
| 298 | 82 | 0.49 | 10.7 | 7.5 |
| 299 | 87 | 0.41 | 8.1 | 7.4 |
| 300 | 130 | 0.39 | 9.8 | 7.6 |
| 301 | 116 | 0.42 | 9.6 | 7.6 |
| 302 | 96 | 0.42 | 11.8 | 7.1 |
| 303 | 122 | 0.39 | 7.9 | 7.1 |
| 304 | 70 | 0.38 | 11.1 | 7.4 |
| 305 | 116 | 0.47 | 8.8 | 7.3 |
| 306 | 122 | 0.48 | 9.0 | 7.0 |
| 307 | 109 | 0.45 | 10.0 | 7.3 |
| 308 | 114 | 0.50 | 10.1 | 7.3 |
| 309 | 62 | 0.47 | 11.4 | 7.3 |
| 310 | 120 | 0.51 | 9.5 | 6.3 |
| 311 | 130 | 0.38 | 10.5 | 7.7 |
| 312 | 92 | 0.47 | 11.8 | 7.3 |
| 313 | 81 | 0.55 | 7.9 | 7.3 |
| 314 | 82 | 0.51 | 10.1 | 7.7 |
| 315 | 102 | 0.48 | 10.9 | 6.5 |
| 316 | 113 | 0.43 | 10.2 | 7.9 |
| 317 | 119 | 0.43 | 8.0 | 6.8 |
| 318 | 108 | 0.48 | 8.9 | 7.0 |
| 319 | 130 | 0.53 | 8.3 | 7.3 |
| 320 | 111 | 0.50 | 9.9 | 6.6 |
| 321 | 132 | 0.50 | 11.5 | 6.8 |
| 322 | 110 | 0.47 | 10.8 | 7.1 |
| 323 | 95 | 0.49 | 10.4 | 7.5 |
| 324 | 137 | 0.29 | 9.8 | 7.5 |
| 325 | 98 | 0.53 | 11.5 | 7.0 |

| 326 | 124 | 0.55 | 10.2 | 6.6 |
| 327 | 146 | 0.36 | 10.2 | 7.5 |
| 328 | 126 | 0.51 | 10.6 | 6.5 |
| 329 | 124 | 0.53 | 9.4 | 7.6 |
| 330 | 99 | 0.47 | 8.7 | 7.7 |
| 331 | 100 | 0.51 | 9.5 | 7.9 |
| 332 | 101 | 0.45 | 9.5 | 7.1 |
| 333 | 113 | 0.37 | 9.4 | 7.8 |
| 334 | 139 | 0.42 | 8.9 | 7.1 |
| 335 | 105 | 0.38 | 8.7 | 7.4 |
| 336 | 113 | 0.45 | 10.7 | 7.3 |
| 337 | 104 | 0.45 | 9.6 | 7.2 |
| 338 | 117 | 0.48 | 10.3 | 7.3 |
| 339 | 132 | 0.43 | 10.9 | 7.7 |
| 340 | 100 | 0.44 | 11.8 | 6.8 |
| 341 | 109 | 0.40 | 8.1 | 7.2 |
| 342 | 95 | 0.39 | 9.7 | 7.4 |
| 343 | 139 | 0.39 | 9.8 | 7.7 |
| 344 | 140 | 0.47 | 8.9 | 7.3 |
| 345 | 110 | 0.48 | 12.1 | 7.2 |
| 346 | 97 | 0.56 | 11.5 | 8.2 |
| 347 | 98 | 0.49 | 11.2 | 6.9 |
| 348 | 146 | 0.44 | 10.0 | 7.2 |
| 349 | 92 | 0.47 | 12.0 | 6.5 |
| 350 | 128 | 0.43 | 10.8 | 7.7 |
| 351 | 156 | 0.50 | 11.4 | 6.3 |
| 352 | 134 | 0.39 | 9.1 | 8.2 |
| 353 | 110 | 0.44 | 7.6 | 6.6 |
| 354 | 104 | 0.45 | 12.4 | 7.5 |
| 355 | 98 | 0.54 | 11.0 | 7.1 |
| 356 | 120 | 0.50 | 10.5 | 7.3 |
| 357 | 140 | 0.50 | 10.6 | 6.9 |
| 358 | 130 | 0.53 | 10.7 | 7.4 |
| 359 | 115 | 0.45 | 10.1 | 7.1 |
| 360 | 159 | 0.41 | 10.7 | 7.5 |
| 361 | 114 | 0.43 | 9.9 | 6.9 |
| 362 | 128 | 0.46 | 9.3 | 7.0 |
| 363 | 96 | 0.49 | 7.6 | 7.5 |
| 364 | 61 | 0.49 | 12.0 | 6.7 |
| 365 | 60 | 0.46 | 8.2 | 7.6 |
| 366 | 51 | 0.50 | 7.8 | 7.6 |

| 367 | 61 | 0.49 | 7.9 | 7.1 |
| 368 | 46 | 0.57 | 7.5 | 6.5 |
| 369 | 60 | 0.44 | 8.0 | 7.7 |
| 370 | 53 | 0.36 | 12.5 | 7.1 |
| 371 | 55 | 0.45 | 10.7 | 7.3 |
| 372 | 59 | 0.38 | 9.0 | 7.0 |
| 373 | 61 | 0.38 | 9.3 | 6.9 |
| 374 | 69 | 0.57 | 10.4 | 7.4 |
| 375 | 63 | 0.51 | 10.6 | 7.0 |
| 376 | 62 | 0.48 | 12.8 | 7.1 |
| 377 | 57 | 0.49 | 7.8 | 7.4 |
| 378 | 70 | 0.40 | 11.2 | 6.9 |
| 379 | 31 | 0.46 | 9.2 | 6.6 |
| 380 | 70 | 0.41 | 10.8 | 6.8 |
| 381 | 66 | 0.39 | 10.9 | 7.8 |
| 382 | 62 | 0.51 | 9.8 | 6.3 |
| 383 | 75 | 0.50 | 9.6 | 7.2 |
| 384 | 58 | 0.34 | 9.1 | 7.2 |
| 385 | 50 | 0.47 | 11.3 | 7.3 |
| 386 | 73 | 0.44 | 9.1 | 7.4 |
| 387 | 61 | 0.37 | 10.8 | 7.2 |
| 388 | 48 | 0.48 | 8.1 | 6.8 |
| 389 | 54 | 0.52 | 12.4 | 7.6 |
| 390 | 63 | 0.52 | 9.4 | 7.1 |
| 391 | 69 | 0.35 | 13.2 | 7.1 |
| 392 | 71 | 0.39 | 12.0 | 7.6 |
| 393 | 44 | 0.40 | 10.6 | 7.4 |
| 394 | 60 | 0.42 | 11.8 | 6.8 |
| 395 | 79 | 0.37 | 9.4 | 7.9 |
| 396 | 38 | 0.50 | 9.9 | 7.3 |
| 397 | 80 | 0.57 | 11.0 | 7.0 |
| 398 | 54 | 0.46 | 8.8 | 6.9 |
| 399 | 74 | 0.44 | 10.8 | 7.6 |
| 400 | 37 | 0.40 | 9.3 | 7.8 |
| 401 | 69 | 0.47 | 9.8 | 7.1 |
| 402 | 54 | 0.47 | 9.6 | 7.3 |
| 403 | 68 | 0.42 | 10.1 | 8.1 |
| 404 | 49 | 0.56 | 8.9 | 7.2 |
| 405 | 55 | 0.45 | 6.1 | 7.2 |
| 406 | 64 | 0.43 | 10.2 | 6.9 |
| 407 | 83 | 0.41 | 7.3 | 6.6 |

| 408 | 36 | 0.46 | 11.5 | 7.3 |
| 409 | 44 | 0.43 | 11.0 | 6.7 |
| 410 | 65 | 0.44 | 11.1 | 7.0 |
| 411 | 77 | 0.39 | 12.1 | 7.7 |
| 412 | 33 | 0.44 | 6.9 | 7.1 |
| 413 | 45 | 0.47 | 8.2 | 6.9 |
| 414 | 70 | 0.53 | 7.3 | 7.2 |
| 415 | 77 | 0.44 | 8.9 | 7.2 |
| 416 | 53 | 0.46 | 9.2 | 6.8 |
| 417 | 60 | 0.49 | 11.0 | 7.4 |
| 418 | 86 | 0.44 | 5.8 | 7.8 |
| 419 | 49 | 0.55 | 8.4 | 7.2 |
| 420 | 50 | 0.45 | 12.3 | 6.5 |
| 421 | 64 | 0.41 | 10.7 | 7.2 |
| 422 | 57 | 0.45 | 7.0 | 7.1 |
| 423 | 56 | 0.40 | 12.1 | 6.5 |
| 424 | 41 | 0.45 | 12.7 | 7.2 |
| 425 | 50 | 0.50 | 8.3 | 6.8 |
| 426 | 63 | 0.45 | 11.6 | 7.4 |
| 427 | 44 | 0.43 | 7.7 | 7.1 |
| 428 | 51 | 0.42 | 10.3 | 7.5 |
| 429 | 51 | 0.50 | 8.7 | 7.1 |
| 430 | 54 | 0.43 | 8.4 | 7.2 |
| 431 | 45 | 0.44 | 7.0 | 6.9 |
| 432 | 65 | 0.46 | 10.5 | 7.5 |
| 433 | 60 | 0.45 | 7.4 | 7.1 |
| 434 | 52 | 0.42 | 4.1 | 7.1 |
| 435 | 50 | 0.49 | 11.2 | 6.9 |
| 436 | 61 | 0.52 | 10.7 | 6.7 |
| 437 | 42 | 0.43 | 9.2 | 6.8 |
| 438 | 42 | 0.50 | 11.4 | 6.8 |
| 439 | 66 | 0.47 | 7.9 | 7.2 |
| 440 | 65 | 0.43 | 9.2 | 7.1 |

Name _____     Date _____

Round all decimal answers to the nearest hundredth.

1.  You and a friend decide to conduct a survey at your school to see whether students are in favor of a new dress code policy. Your friend stands at the school entrance and asks the opinions of the first 100 students who come to campus on Monday. You obtain a list of all students at the school and randomly select 60 to survey.

    a.  Your friend finds 34% of his sample in favor of the new dress code policy, but you find only 16%. Which do you believe is more likely to be representative of the school population? Explain your choice.

    b.  Suppose 25% of the students at the school are in favor of the new dress code policy. Below is a dot plot of the proportion of students who favor the new dress code for each of 100 different random samples of 50 students at the school.

    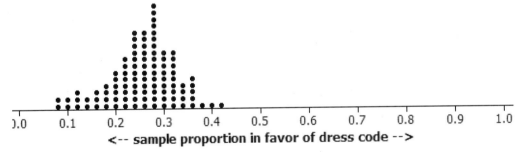

    If you were to select a random sample of 50 students and ask them if they favor the new dress code, do you think that your sample proportion will be within 0.05 of the population proportion? Explain.

c.  Suppose ten people each take a simple random sample of 100 students from the school and calculate the proportion in the sample who favors the new dress code.  On the dot plot axis below, place 10 values that you think are most believable for the proportions you could obtain.

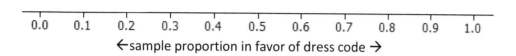

←sample proportion in favor of dress code →

Explain your reasoning.

2.  Students in a random sample of 57 students were asked to measure their hand-spans (distance from outside of thumb to outside of little finger when the hand is stretched out as far as possible).  The graphs below show the results for the males and females.

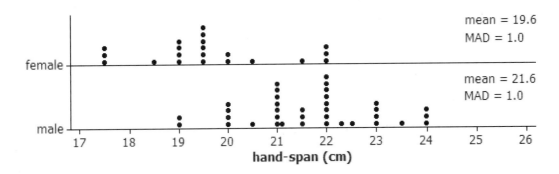

mean = 19.6
MAD = 1.0

mean = 21.6
MAD = 1.0

a.  Based on these data, do you think there is a difference between the population mean hand-span for males and the population mean hand-span for females?  Justify your answer.

b.  The same students were asked to measure their heights, with the results shown below.

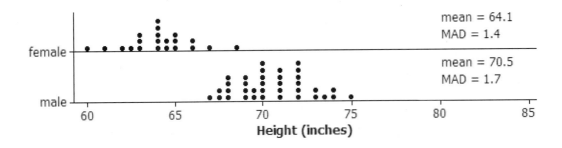

mean = 64.1
MAD = 1.4

mean = 70.5
MAD = 1.7

Are these height data more or less convincing of a difference in the population mean height than the hand span data are of a difference in population mean handspan?  Explain.

3.  A student purchases a bag of "mini" chocolate chip cookies, and, after opening the bag, finds one cookie that does not contain any chocolate chips!  The student then wonders how unlikely it is to randomly find a cookie with no chocolate chips for this brand.

    a.  Based on the bag of 30 cookies, estimate the probability of this company producing a cookie with no chocolate chips.

    b.  Suppose the cookie company claims that 90% of all cookies it produces contain chocolate chips. Explain how you could simulate randomly selecting 30 cookies (one bag) from such a population to determine how many of the sampled cookies do not contain chocolate chips.  Explain the details of your method so it could be carried out by another person.

    c.  Now explain how you could use simulation to estimate the probability of obtaining a bag of 30 cookies with exactly one cookie with no chocolate chips.

d. If 90% of the cookies made by this company contain chocolate chips, then the actual probability of obtaining a bag of 30 cookies with one chipless cookie equals 0.143. Based on this result, would you advise this student to complain to the company about finding one cookie with no chocolate chips in her bag of 30? Explain.

## A Progression Toward Mastery

| Assessment Task Item | | STEP 1<br>Missing or incorrect answer and little evidence of reasoning or application of mathematics to solve the problem. | STEP 2<br>Missing or incorrect answer but evidence of some reasoning or application of mathematics to solve the problem. | STEP 3<br>A correct answer with some evidence of reasoning or application of mathematics to solve the problem, or an incorrect answer with substantial evidence of solid reasoning or application of mathematics to solve the problem. | STEP 4<br>A correct answer supported by substantial evidence of solid reasoning or application of mathematics to solve the problem. |
|---|---|---|---|---|---|
| 1 | a<br><br>7.SP.A.1 | Student answer based on personal experience and does not use information from problem stem. | Student believes neither sample will be representative because of the small sample sizes. | Student indicates the friend because the sample size is larger or indicates the student but without justification. | Student indicates that only the second method, based on the random sampling method, is likely to produce a representative result for the school population. |
| | b<br><br>7.SP.A.2 | Student answer does not make use of dot plot. | Student focuses only on all dots being less than 0.50. | Student guarantees that any proportion will fall between 0.15 and 0.35 or focuses only on 0.20 and 0.30. | Student notes that most of the random samples fall between 0.25 and 0.35, so the chance of the next sample proportion being in that range is high. |
| | c<br><br>7.SP.A.2 | Student does not attempt problem or does not have ten dots. | Student distribution is similar to (b) but with fewer dots. | Student distribution indicates less variability than (b) but student does not explain reasoning. | Student distribution indicates less variability based on the larger sample size producing results that tend to fall closer to the population proportion. |
| 2 | a<br><br>7.SP.B.3 | Student answer based on personal experience and does not use information from problem stem. | Student only focuses on the sample sizes that are unequal, so no comparison can be made. OR Student focuses only on how irregular looking the distributions are. | Student only discusses amount of overlap in distributions or the centers of the distributions (the means) with no consideration of variability; reasoning is not complete. | Student measures difference in centers of distributions as a multiple of MAD. |

| | | | | | |
|---|---|---|---|---|---|
| | **b**<br><br>7.SP.B.4 | Student answer based on personal experience and does not use information from problem stem. | Student focuses only on sample size or how data were collected. | Student examines amount of overlap in distributions using mean, but makes no consideration of variability. <u>OR</u> Student cannot reconcile the different MAD values. | Student measures difference in centers of distributions as a multiple of MAD. Student could also discuss how bulk of distributions does not overlap at all (especially compared to previous question). |
| **3** | **a**<br><br>7.SP.C.5<br>7.SP.C.6 | Student does not provide an estimate of a probability. | Student uses context/intuition to estimate the probability rather than the given information. | Student makes a statement about how unusual the outcome is, but does not give a numerical estimate; or, reports "1." | Student reports 1/30. |
| | **b**<br><br>7.SP.C.8 | Student does not provide meaningful instructions for carrying out a simulation. | Student description is very generic and not specific to this problem. | Student explains part of the simulation (e.g., how to represent 90%) but description is either incomplete (e.g., does not draw 30 cookies) or not sufficiently detailed that it could be implemented by another person. | Student explains how to set up a simulation (e.g., random digits, to represent 90% with $1 =$ no chocolate chips, everything else $=$ chocolate chips) and how to select 30 one-digit numbers. |
| | **c**<br><br>7.SP.C.8 | Student does not include instructions for carrying out a simulation. | Student response does not differ from (b) or only differs in looking for one cookie vs. no cookie without chocolate chips. | Student focuses on one cookie with no chocolate chips, but does not clearly indicate replication of the chance experiment a large number of times. | Student clearly describes repeating process in (b) a large number of times and looking at the proportion of "bags" with exactly one chocolate chip. |
| | **d**<br><br>7.SP.C.7 | Student does not make use of 0.143 in making a decision. | Student only comments that 30 is a small sample size, so it is difficult to make a decision. | Student discusses how her bag could have happened by chance, but does not tie to 0.143. <u>OR</u> Student considers 0.143 a small value and evidence that her bag would be unusual if the company's claim was true. | Student states that the purchased bag is within the expected sampling chance variability and supports this conclusion by stating that 0.143 is not a small number. |

Name _____     Date _____

1. You and a friend decide to conduct a survey at your school to see whether students are in favor of a new dress code policy. Your friend stands at the school entrance and asks the opinions of the first 100 students to come into campus on Monday. You obtain a list of all students at the school and randomly select 60 to survey.

   a. Your friend finds 34% of his sample in favor of the new dress code policy but you find only 16%. Which do you believe is more likely to be representative of the school population? Explain your choice.

   > My students were randomly sampled instead of only the early arrivers so my result should be more representative.

   b. Suppose 25% of the students at the school are in favor of the new dress code policy. Below is a dot plot of the proportion of students who favor the new dress code for each of 100 different random samples of 50 students at the school.

   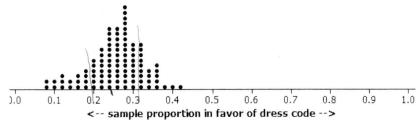

   If you were to select a random sample of 50 students and ask them if they favor the new dress code, do you think that your sample proportion will be within 0.05 of the population proportion? Explain.

   > A little more than half of these 100 samples are between .25 and .30 so there is a good chance but a value like 0.10 should be even better.

c. Suppose ten people each take a simple random sample of 100 students from the school and calculate the proportion in the sample who favors the new dress code. On the dot plot axis below, place 10 values that you think are most believable for the proportions you could obtain.

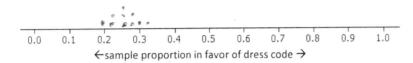

0.0   0.1   0.2   0.3   0.4   0.5   0.6   0.7   0.8   0.9   1.0

← sample proportion in favor of dress code →

Explain your reasoning.

Values will still center around 0.25 but will tend to be much closer together than in (b) where samples only had 50 students.

2. Students in a random sample of 57 students were asked to measure their hand-spans (distance from outside of thumb to outside of little finger when the hand is stretched out as far as possible). The graphs below show the results for the males and females.

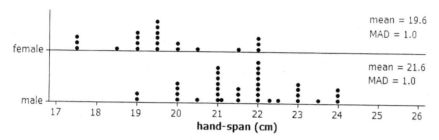

a. Based on these data, do you think there is a difference between the population mean hand-span for males and the population mean hand-span for females? Justify your answer.

*Yes! The male hand-spans tend to be larger for males. All but two males are at least 20 cm but less than 50% of the female hands are that large*

b. The same students were asked asked to measure their heights, with the results shown below.

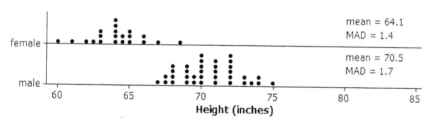

Are these height data more or less convincing of a difference in the population mean height than the hand span data are of a difference in population mean handspan? Explain.

*Even more convincing because there is even less overlap between the two distributions.*

3. A student purchases a bag of "mini" chocolate chip cookies and among those 30 cookies she finds one cookie that does not contain any chocolate chips! She then wonders how unlikely it is to randomly find a cookie with no chips for this brand.

a. Based on her bag of 30 cookies, estimate the probability this company producing a cookie with no chips.

$1/30 \approx .0333$

b. Suppose the cookie company claims that 90% of all cookies it produces contain chips. Explain how you could simulate randomly selecting 30 cookies (one bag) from such a population to determine how many of the sampled cookies do not contain chips. Explain the details of your method so it could be carried out by another person.

Have a bag of 100 chips, 90 of them red. Pull out a chip and record its color and put it back. Do this 30 times and count how many are not red (say blue).

c. Now explain how you could use simulation to estimate the probability of obtaining a bag of 30 cookies with exactly one cookie with no chips.

Repeat the above process from (b) many, many times, like 1000. See what proportion of those 1000 bags had exactly one blue chip. That number over 1000 is your estimate of the probability of a bag of 30 cookies with one chipless cookie.

d. If 90% of cookies made by this company contain chips, then the actual probability of obtaining a bag of 30 cookies with one chipless cookie equals 0.143. Based on this result, would you advise this student to complain to the company about finding one cookie with no chips in her bag of 30? Explain.

No, that is not that small of a probability. I would not find that value convincing that this didn't just happen to her randomly.

# Mathematics Curriculum

## Student Material

# Lesson 1:  Chance Experiments

Have you ever heard a weatherman say there is a 40% chance of rain tomorrow or a football referee tell a team there is a 50/50 chance of getting a head on a coin toss to determine which team starts the game?  These are probability statements.  In this lesson, you are going to investigate probability and how likely it is that some events will occur.

### Example 1:  Spinner Game

Suppose you and your friend will play a game using the spinner shown here:

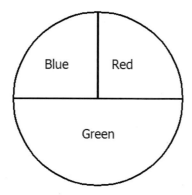

Rules of the game:

1.  Decide who will go first.
2.  Each person picks a color.  Both players cannot pick the same color.
3.  Each person takes a turn spinning the spinner and recording what color the spinner stops on.  The winner is the person whose color is the first to happen 10 times.

Play the game and remember to record the color the spinner stops on for each spin.

**Exercises 1–4**

1. Which color was the first to occur 10 times?

2. Do you think it makes a difference who goes first to pick a color?

3. Which color would you pick to give you the best chance of winning the game?  Why would you pick that color?

4. Below are three different spinners.  If you pick green for your color, which spinner would give you the best chance to win?  Give a reason for your answer.

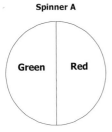

Spinner A

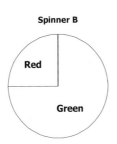

Spinner B

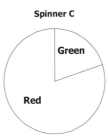

Spinner C

## Example 2:  What is Probability?

**Probability** is about how likely it is that an event will happen.  A probability is indicated by a number between 0 and 1.  Some events are certain to happen, while others are impossible.  In most cases, the probability of an event happening is somewhere between certain and impossible.

For example, consider a bag that contains only red balls.  If you were to select one ball from the bag, you are certain to pick a red one.  We say that an event that is certain to happen has a probability of 1.  If we were to reach into the same bag of balls, it is impossible to select a yellow ball.  An impossible event has a probability of 0.

| Description | Example | Explanation |
|---|---|---|
| Some events are impossible.  These events have a probability of 0. | You have a bag with two green cubes and you select one at random.  Selecting a blue cube is an impossible event. | There is no way to select a blue cube if there are no blue cubes in the bag. |
| Some events are certain.  These events have a probability of 1. | You have a bag with two green cubes and you select one at random.  Selecting a green cube is a certain event. | You will always get a green cube if there are only green cubes in the bag. |
| Some events are classified as equally likely to happen or to not happen.  These events have a probability of $\frac{1}{2}$. | You have a bag with one blue cube and one red cube, and you randomly pick one.  Selecting a blue cube is equally likely to happen or to not happen. |  |
| Some events are more likely to happen than to not happen.  These events will have a probability that is greater than 0.5.  These events could be described as "likely" to occur. | If you have a bag that contains eight blue cubes and two red cubes, and you select one at random, it is likely that you will get a blue cube. | Even though it is not certain that you will get a blue cube, a blue cube would be selected most of the time because there are many more blue cubes than red cubes. |
| Some events are less likely to happen than to not happen.  These events will have a probability that is less than 0.5.  These events could be described as "unlikely" to occur. | If you have a bag that contains eight blue cubes and two red cubes, and you select one at random, it is unlikely that you will get a red cube. | Even though it is not impossible to get a red cube, a red cube would not be selected very often because there are many more blue cubes than red cubes. |

The figure below shows the probability scale.

**Probability Scale**

| 0 | | 1/2 | | 1 |
|---|---|---|---|---|
| Impossible | Unlikely | Equally Likely to Occur or Not Occur | Likely | Certain |

## Exercises 5–8

5.  Decide where each event would be located on the scale above.  Place the letter for each event on the appropriate place on the probability scale.

Event:

    A.  You will see a live dinosaur on the way home from school today.

    B.  A solid rock dropped in the water will sink.

    C.  A round disk with one side red and the other side yellow will land yellow side up when flipped.

    D.  A spinner with four equal parts numbered 1–4 will land on the 4 on the next spin.

    E.  Your name will be drawn when a name is selected randomly from a bag containing the names of all of the students in your class.

    F.  A red cube will be drawn when a cube is selected from a bag that has five blue cubes and five red cubes.

    G.  The temperature outside tomorrow will be $-250$ degrees.

6.  Design a spinner so that the probability of green is 1.

7. Design a spinner so that the probability of green is 0.

8. Design a spinner with two outcomes in which it is equally likely to land on the red and green parts.

**Exercises 9–10**

An event that is impossible has probability 0 and will never occur, no matter how many observations you make. This means that in a long sequence of observations, it will occur 0% of the time. An event that is certain, has probability 1 and will always occur. This means that in a long sequence of observations, it will occur 100% of the time.

9. What do you think it means for an event to have a probability of $\frac{1}{2}$?

10. What do you think it means for an event to have a probability of $\frac{1}{4}$?

**Lesson Summary**

- **Probability** is a measure of how likely it is that an event will happen.
- A probability is a number between 0 and 1.
- The probability scale is:

**Probability Scale**

| 0 | | 1/2 | | 1 |
|---|---|---|---|---|
| Impossible | Unlikely | Equally Likely to Occur or Not Occur | Likely | Certain |

**Problem Set**

1. Match each spinner below with the words Impossible, Unlikely, Equally likely to occur or not occur, Likely and Certain to describe the chance of the spinner landing on black.

**Spinner A**

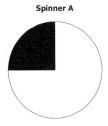

**Spinner B**

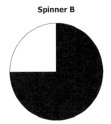

**Spinner C**

**Spinner D**

**Spinner E**

2. Decide if each of the following events is Impossible, Unlikely, Equally likely to occur or not occur, Likely, or Certain to occur.

   a. A vowel will be picked when a letter is randomly selected from the word "lieu."

   b. A vowel will be picked when a letter is randomly selected from the word "math."

   c. A blue cube will be drawn from a bag containing only five blue and five black cubes.

   d. A red cube will be drawn from a bag of 100 red cubes.

   e. A red cube will be drawn from a bag of 10 red and 90 blue cubes.

3. A shape will be randomly drawn from the box shown below. Decide where each event would be located on the probability scale. Then, place the letter for each event on the appropriate place on the probability scale.

   Event:

   A. A circle is drawn.

   B. A square is drawn.

   C. A star is drawn.

   D. A shape that is not a square is drawn.

**Probability Scale**

| 0 | | 1/2 | | 1 |
|---|---|---|---|---|
| Impossible | Unlikely | Equally Likely to Occur or Not Occur | Likely | Certain |

4.  Color the cubes below so that it would be equally likely to choose a blue or yellow cube.

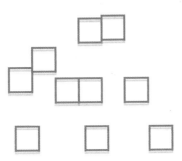

5.  Color the cubes below so that it would be likely but not certain to choose a blue cube from the bag.

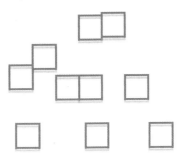

6.  Color the cubes below so that it would be unlikely but not impossible to choose a blue cube from the bag.

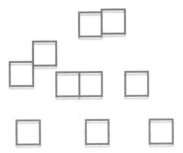

7.  Color the cubes below so that it would be impossible to choose a blue cube from the bag.

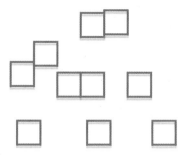

# Lesson 2: Estimating Probabilities by Collecting Data

### Example 1: Carnival Game

At the school carnival, there is a game in which students spin a large spinner. The spinner has four equal sections numbered 1–4 as shown below. To play the game, a student spins the spinner twice and adds the two numbers that the spinner lands on. If the sum is greater than or equal to 5, the student wins a prize.

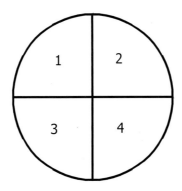

### Exercises 1–8

You and your partner will play this game 15 times. Record the outcome of each spin in the table below.

| Turn | 1st Spin Results | 2nd Spin Results | Sum |
|------|------------------|------------------|-----|
| 1    |                  |                  |     |
| 2    |                  |                  |     |
| 3    |                  |                  |     |
| 4    |                  |                  |     |
| 5    |                  |                  |     |
| 6    |                  |                  |     |
| 7    |                  |                  |     |
| 8    |                  |                  |     |
| 9    |                  |                  |     |
| 10   |                  |                  |     |
| 11   |                  |                  |     |
| 12   |                  |                  |     |
| 13   |                  |                  |     |
| 14   |                  |                  |     |
| 15   |                  |                  |     |

1. Out of the 15 turns how many times was the sum greater than or equal to 5?

2. What sum occurred most often?

3. What sum occurred least often?

4. If students played a lot of games, what proportion of the games played will they win?  Explain your answer.

5. Name a sum that would be impossible to get while playing the game.

6. What event is certain to occur while playing the game?

When you were spinning the spinner and recording the outcomes, you were performing a **chance experiment**.  You can use the results from a chance experiment to estimate the probability of an event.  In the example above, you spun the spinner 15 times and counted how many times the sum was greater than or equal to 5.  An estimate for the probability of a sum greater than or equal to 5 is:

$$P(sum \geq 5) = \frac{Number\ of\ observed\ occurrences\ of\ the\ event}{Total\ number\ of\ observations}$$

7. Based on your experiment of playing the game, what is your estimate for the probability of getting a sum of 5 or more?

8. Based on your experiment of playing the game, what is your estimate for the probability of getting a sum of exactly 5?

### Example 2:  Animal Crackers

A student brought a very large jar of animal crackers to share with students in class.  Rather than count and sort all the different types of crackers, the student randomly chose 20 crackers and found the following counts for the different types of animal crackers.

| | |
|---|---|
| Lion | 2 |
| Camel | 1 |
| Monkey | 4 |
| Elephant | 5 |
| Zebra | 3 |
| Penguin | 3 |
| Tortoise | 2 |
| | Total 20 |

The student can now use that data to find an estimate for the probability of choosing a zebra from the jar by dividing the observed number of zebras by the total number of crackers selected.  The estimated probability of picking a zebra is $\frac{3}{20}$, or 0.15 or 15%. This means that an estimate of the proportion of the time a zebra will be selected is 0.15, or 15% of the time.  This could be written as $P(zebra) = 0.15$, or the probability of selecting a zebra is 0.15.

**Exercises 9–15**

If a student were to randomly select a cracker from the large jar:

9.  What is your estimate for the probability of selecting a lion?

10. What is the estimate for the probability of selecting a monkey?

11. What is the estimate for the probability of selecting a penguin or a camel?

12. What is the estimate for the probability of selecting a rabbit?

13. Is there the same number of each animal cracker in the large jar?  Explain your answer.

14. If the student were to randomly select another 20 animal crackers, would the same results occur?  Why or why not?

15. If there are 500 animal crackers in the jar, how many elephants are in the jar?  Explain your answer.

Lesson Summary

An estimate for finding the probability of an event occurring is

$$P(event\ occurring) = \frac{Number\ of\ observed\ occurrences\ of\ the\ event}{Total\ number\ of\ observations}$$

Problem Set

1.  Play a game using the two spinners below. Spin each spinner once, and then multiply the outcomes together. If the result is less than or equal to 8, you win the game. Play the game 15 times, and record your results in the table below.

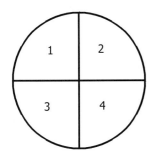 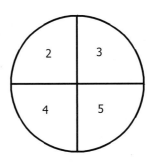

| Turn | 1$^{st}$ Spin Results | 2$^{nd}$ Spin Results | Product |
|------|------|------|------|
| 1 | | | |
| 2 | | | |
| 3 | | | |
| 4 | | | |
| 5 | | | |
| 6 | | | |
| 7 | | | |
| 8 | | | |
| 9 | | | |
| 10 | | | |
| 11 | | | |
| 12 | | | |
| 13 | | | |
| 14 | | | |
| 15 | | | |

2.  a.  What is your estimate for the probability of getting a product of 8 or less?

    b.  What is your estimate for the probability of getting a product more than 8?

    c.  What is your estimate for the probability of getting a product of exactly 8?

    d.  What is the most likely product for this game?

    e.  If you played this game another 15 times, will you get the exact same results?  Explain.

3.  A seventh grade student surveyed students at her school.  She asked them to name their favorite pet.  Below is a bar graph showing the results of the survey.

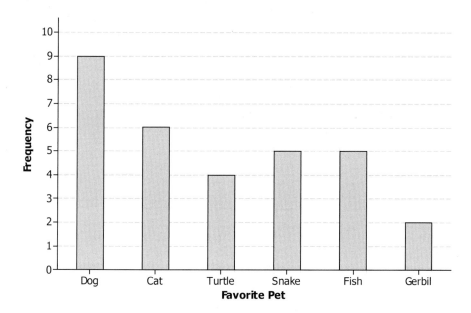

Use the results from the survey to answer the following questions.

a.  How many students answered the survey question?

b.  How many students said that a snake was their favorite pet?

Now suppose a student will be randomly selected and asked what his or her favorite pet is.

c.  What is your estimate for the probability of that student saying that a dog is their favorite pet?

d.  What is your estimate for the probability of that student saying that a gerbil is their favorite pet?

e.  What is your estimate for the probability of that student saying that a frog is their favorite pet?

4.  A seventh grade student surveyed 25 students at her school.  She asked them how many hours a week they spend playing a sport or game outdoors.  The results are listed in the table below.

| Number of hours | Tally | Frequency |
|---|---|---|
| 0 | \| \| \| | 3 |
| 1 | \| \| \| \| | 4 |
| 2 | ⊬⊬ | 5 |
| 3 | ⊬⊬ \| \| | 7 |
| 4 | \| \| \| | 3 |
| 5 |  | 0 |
| 6 | \| \| | 2 |
| 7 |  | 0 |
| 8 | \| | 1 |

a.  Draw a dot plot of the results

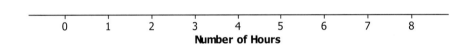

Suppose a student will be randomly selected.

b.  What is your estimate for the probability of that student answering 3 hours?

c.  What is your estimate for the probability of that student answering 8 hours?

d.  What is your estimate for the probability of that student answering 6 or more hours?

e.  What is your estimate for the probability of that student answering 3 or less hours?

f.  If another 25 students were surveyed do you think they will give the exact same results?  Explain your answer.

g.  If there are 200 students at the school, what is your estimate for the number of students who would say they play a sport or game outdoors 3 hours per week?  Explain your answer.

5.  A student played a game using one of the spinners below. The table shows the results of 15 spins. Which spinner did the student use? Give a reason for your answer.

| Spin | Results |
|------|---------|
| 1    | 1       |
| 2    | 1       |
| 3    | 2       |
| 4    | 3       |
| 5    | 1       |
| 6    | 2       |
| 7    | 3       |
| 8    | 2       |
| 9    | 2       |
| 10   | 1       |
| 11   | 2       |
| 12   | 2       |
| 13   | 1       |
| 14   | 3       |
| 15   | 1       |

Spinner A

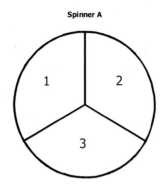

Spinner B

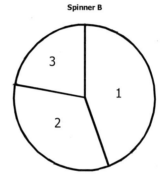

Spinner C

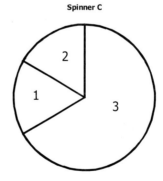

# Lesson 3: Chance Experiments with Equally Likely Outcomes

Classwork

### Example 1

Jamal, a 7$^{th}$ grader, wants to design a game that involves tossing paper cups. Jamal tosses a paper cup five times and records the outcome of each toss. An **outcome** is the result of a single trial of an experiment.

Here are the results of each toss:

Jamal noted that the paper cup could land in one of three ways: on its side, right side up, or upside down. The collection of these three outcomes is called the *sample space* of the experiment. The **sample space** of an experiment is the set of all possible outcomes of that experiment.

For example, the sample space when flipping a coin is Heads, Tails.

The sample space when drawing a colored cube from a bag that has 3 red, 2 blue, 1 yellow, and 4 green cubes is red, blue, yellow, green.

### Exercises 1–6

For each of the following chance experiments, list the sample space (i.e., all the possible outcomes).

1. Drawing a colored cube from a bag with 2 green, 1 red, 10 blue, and 3 black.

2. Tossing an empty soup can to see how it lands.

3.  Shooting a free-throw in a basketball game.

4.  Rolling a number cube with the numbers 1– 6 on its faces.

5.  Selecting a letter from the word:  probability

6.  Spinning the spinner:

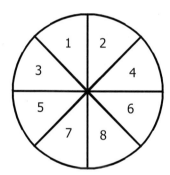

## Example 2:  Equally Likely Outcomes

The sample space for the paper cup toss was on its side, right side up, and upside down.  Do you think each of these outcomes has the same chance of occurring?  If they do, then they are equally likely to occur.

The outcomes of an experiment are equally likely to occur when the probability of each outcome is equal.

You and your partner toss the paper cup 30 times and record in a table the results of each toss.

| Toss | Outcome |
|------|---------|
| 1 | |
| 2 | |
| 3 | |
| 4 | |
| 5 | |
| 6 | |
| 7 | |
| 8 | |
| 9 | |
| 10 | |
| 11 | |
| 12 | |
| 13 | |
| 14 | |
| 15 | |
| 16 | |
| 17 | |
| 18 | |
| 19 | |
| 20 | |
| 21 | |
| 22 | |
| 23 | |
| 24 | |
| 25 | |
| 26 | |
| 27 | |
| 28 | |
| 29 | |
| 30 | |

Lesson 3:       Chance Experiments with Equally Likely Outcomes

**Exercises 7–12**

7.  Using the results of your experiment, what is your estimate for the probability of a paper cup landing on its side?

8.  Using the results of your experiment, what is your estimate for the probability of a paper cup landing upside down?

9.  Using the results of your experiment, what is your estimate for the probability of a paper cup landing right side up?

10. Based on your results, do you think the three outcomes are equally likely to occur?

11. Using the spinner below, answer the following questions.

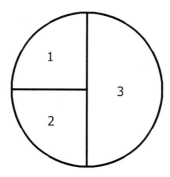

   a.   Are the events spinning and landing on 1 or a 2 equally likely?

b.   Are the events spinning and landing on 2 or 3 equally likely?

c.   How many times do you predict the spinner to land on each section after 100 spins?

12.  Draw a spinner that has 3 sections that are equally likely to occur when the spinner is spun.  How many times do you think the spinner will land on each section after 100 spins?

> **Lesson Summary:**
>
> An **outcome** is the result of a single observation of an experiment.
>
> The **sample space** of an experiment is the set of all possible outcomes of that experiment.
>
> The outcomes of an experiment are **equally likely** to occur when the probability of each outcome is equal.
>
> Suppose a bag of crayons contains 10 green, 10 red, 10 yellow, 10 orange, and 10 purple pieces of crayons. If one crayon is selected from the bag and the color is noted, the *outcome* is the color that will be chosen. The *sample space* will be the colors: green, red, yellow, orange, and purple. Each color is *equally likely* to be selected because each color has the same chance of being chosen.

## Problem Set

1. For each of the following chance experiments, list the sample space (all the possible outcomes).

   a. Rolling a 4-sided die with the numbers 1– 4 on the faces of the die.

   b. Selecting a letter from the word: mathematics.

   c. Selecting a marble from a bag containing 50 black marbles and 45 orange marbles.

   d. Selecting a number from the even numbers from 2– 14, inclusive.

   e. Spinning the spinner below:

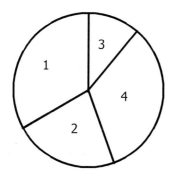

2.  For each of the following decide if the two outcomes listed are equally likely to occur.  Give a reason for your answer.

    a.  Rolling a 1 or a 2 when a 6-sided number cube with the numbers 1– 6 on the faces of the cube is rolled.

    b.  Selecting the letter *a* or *k* from the word:  take.

    c.  Selecting a black or an orange marble from a bag containing 50 black and 45 orange marbles.

    d.  Selecting a 4 or an 8 from the even numbers from 2– 14, inclusive.

    e.  Landing on a 1 or 3 when spinning the spinner below.

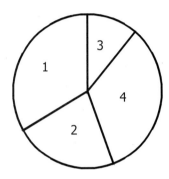

3.  Color the cubes below so that it would be equally likely to choose a blue or yellow cube.

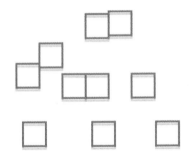

4.  Color the cubes below so that it would be more likely to choose a blue than a yellow cube.

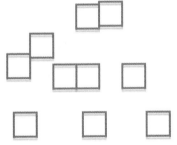

5. You are playing a game using the spinner below. The game requires that you spin the spinner twice. For example, one outcome could be Yellow on 1ˢᵗ spin and Red on 2ⁿᵈ spin. List the sample space (all the possible outcomes) for the two spins.

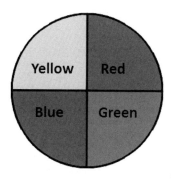

6. List the sample space for the chance experiment of flipping a coin twice.

# Lesson 4:  Calculating Probabilities for Chance Experiments with Equally Likely Outcomes

## Classwork

### Example 1:  Theoretical Probability

In a previous lesson, you saw that to find an estimate of the probability of an event for a chance experiment you divide:

$$P(event) = \frac{Number\ of\ observed\ occurrences\ of\ the\ event}{Total\ number\ of\ observations}$$

Your teacher has a bag with some cubes colored yellow, green, blue, and red.  The cubes are identical except for their color.  Your teacher will conduct a chance experiment by randomly drawing a cube with replacement from the bag.  Record the outcome of each draw in the table below.

| Trial | Outcome |
|:-----:|:--------|
| 1 | |
| 2 | |
| 3 | |
| 4 | |
| 5 | |
| 6 | |
| 7 | |
| 8 | |
| 9 | |
| 10 | |
| 11 | |
| 12 | |
| 13 | |
| 14 | |
| 15 | |
| 16 | |
| 17 | |
| 18 | |
| 19 | |
| 20 | |

**Exercises 1–6**

1.  Based on the 20 trials, estimate for the probability of
    a.  choosing a yellow cube.




    b.  choosing a green cube.




    c.  choosing a red cube.




    d.  choosing a blue cube.




2.  If there are 40 cubes in the bag, how many cubes of each color are in the bag?  Explain.




3.  If your teacher were to randomly draw another 20 cubes one at a time and with replacement from the bag, would you see exactly the same results?  Explain.

4.  Find the fraction of each color of cubes in the bag.

    Yellow

    Green

    Red

    Blue

Each fraction is the **theoretical probability** of choosing a particular color of cube when a cube is randomly drawn from the bag.

When all the possible outcomes of an experiment are equally likely, the probability of each outcome is

$$P(outcome) = \frac{1}{Number\ of\ possible\ outcomes}.$$

An event is a collection of outcomes, and when the outcomes are equally likely, the theoretical probability of an event can be expressed as

$$P(event) = \frac{Number\ of\ favorable\ outcomes}{Number\ of\ possible\ outcomes}.$$

The theoretical probability of drawing a blue cube is

$$P(blue) = \frac{Number\ of\ blue\ cubes}{Total\ number\ of\ cubes} = \frac{10}{40}.$$

5.  Is each color equally likely to be chosen? Explain your answer.

6.  How do the theoretical probabilities of choosing each color from Exercise 4 compare to the experimental probabilities you found in Exercise 1?

### Example 2

An experiment consisted of flipping a nickel and a dime. The first step in finding the theoretical probability of obtaining a head on the nickel and a head on the dime is to list the sample space. For this experiment, the sample space is shown below.

| Nickel | Dime |
|--------|------|
| H | H |
| H | T |
| T | H |
| T | T |

If the counts are fair, these outcomes are equally likely, so the probability of each outcome is $\frac{1}{4}$.

| Nickel | Dime | Probability |
|--------|------|-------------|
| H | H | $\frac{1}{4}$ |
| H | T | $\frac{1}{4}$ |
| T | H | $\frac{1}{4}$ |
| T | T | $\frac{1}{4}$ |

The probability of two heads is $\frac{1}{4}$ or $P(two\ heads) = \frac{1}{4}$.

### Exercises 7–10

7.  Consider a chance experiment of rolling a number cube.

   a.   What is the sample space? List the probability of each outcome in the sample space.

   b.   What is the probability of rolling an odd number?

   c.   What is the probability of rolling a number less than 5?

8.  Consider an experiment of randomly selecting a letter from the word:  number.

a.  What is the sample space?   List the probability of each outcome in the sample space.

b.  What is the probability of selecting a vowel?

c.  What is the probability of selecting the letter z?

9.  Consider an experiment of randomly selecting a cube from a bag of 10 cubes.

a.  Color the cubes below so that the probability of selecting a blue cube is $\frac{1}{2}$.

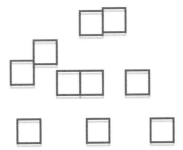

b.   Color the cubes below so that the probability of selecting a blue cube is $\frac{4}{5}$.

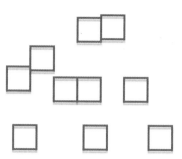

10.  Students are playing a game that requires spinning the two spinners shown below.  A student wins the game if both spins land on Red.  What is the probability of winning the game?  Remember to first list the sample space and the probability of each outcome in the sample space.  There are eight possible outcomes to this chance experiment.

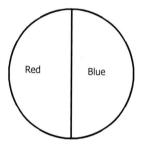    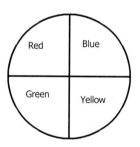

**Lesson Summary**

When all the possible outcomes of an experiment are equally likely, the probability of each outcome is

$$P(outcome) = \frac{1}{Number\ of\ possible\ outcomes}.$$

An event is a collection of outcomes, and when all outcomes are equally likely, the theoretical probability of an event can be expressed as

$$P(event) = \frac{Number\ of\ favorable\ outcomes}{Number\ of\ possible\ outcomes}.$$

**Problem Set**

1.  In a seventh grade class of 28 students, there are 16 girls and 12 boys. If one student is randomly chosen to win a prize, what is the probability that a girl is chosen?

2.  An experiment consists of spinning the spinner once.

    a.  Find the probability of landing on a 2.
    b.  Find the probability of landing on a 1.
    c.  Is landing in each section of the spinner equally likely to occur? Explain.

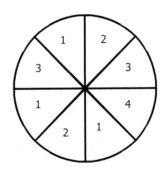

3.  An experiment consists of randomly picking a square section from the board shown below.
    a.  Find the probability of choosing a triangle.
    b.  Find the probability of choosing a star.
    c.  Find the probability of choosing an empty square.
    d.  Find the probability of choosing a circle.

4. Seventh graders are playing a game where they randomly select two integers from 0– 9, inclusive, to form a two-digit number. The same integer might be selected twice.

   a. List the sample space for this chance experiment. List the probability of each outcome in the sample space.

   b. What is the probability that the number formed is between 90 and 99, inclusive?

   c. What is the probability that the number formed is evenly divisible by 5?

   d. What is the probability that the number formed is a factor of 64?

5. A chance experiment consists of flipping a coin and rolling a number cube with the numbers 1– 6 on the faces of the cube.

   a. List the sample space of this chance experiment. List the probability of each outcome in the sample space.

   b. What is the probability of getting a head on the coin and the number 3 on the number cube?

   c. What is the probability of getting a tail on the coin and an even number on the number cube?

6. A chance experiment consists of spinning the two spinners below.

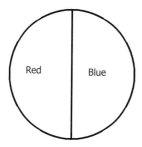
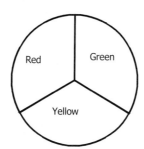

   a. List the sample space and the probability of each outcome.

   b. Find the probability of the event of getting a red on the first spinner and a red on the second spinner.

   c. Find the probability of a red on at least one of the spinners.

# Lesson 5: Chance Experiments with Outcomes that are Not Equally Likely

## Classwork

In previous lessons, you learned that when the outcomes in a sample space are equally likely, the probability of an event is the number of outcomes in the event divided by the number of outcomes in the sample space. However, when the outcomes in the sample space are *not* equally likely, we need to take a different approach.

### Example 1

When Jenna goes to the farmer's market she usually buys bananas. The numbers of bananas she might buy and their probabilities are shown in the table below.

| Number of Bananas | 0 | 1 | 2 | 3 | 4 | 5 |
|---|---|---|---|---|---|---|
| Probability | 0.1 | 0.1 | 0.1 | 0.2 | 0.2 | 0.3 |

a. What is the probability that Jenna buys exactly 3 bananas?

b. What is the probability that Jenna doesn't buy any bananas?

c. What is the probability that Jenna buys more than 3 bananas?

d. What is the probability that Jenna buys at least 3 bananas?

e. What is the probability that Jenna doesn't buy exactly 3 bananas?

Notice that the probabilities in the table add to 1. $(0.1 + 0.1 + 0.1 + 0.2 + 0.2 + 0.3 = 1)$ This is always true; when we add up the probabilities of all the possible outcomes, the result is always 1. So, taking 1 and subtracting the probability of the event gives us the probability of something NOT occurring.

## Exercises 1–2

Jenna's husband, Rick, is concerned about his diet. On any given day, he eats 0, 1, 2, 3, or 4 servings of fruit and vegetables. The probabilities are given in the table below.

| Number of Servings of Fruit and Vegetables | 0 | 1 | 2 | 3 | 4 |
|---|---|---|---|---|---|
| Probability | 0.08 | 0.13 | 0.28 | 0.39 | 0.12 |

1.  On a given day, find the probability that Rick eats:

    a.  Two servings of fruit and vegetables.

    b.  More than two servings of fruit and vegetables.

    c.  At least two servings of fruit and vegetables.

2.  Find the probability that Rick does not eat exactly two servings of fruit and vegetables.

## Example 2

Luis works in an office, and the phone rings occasionally. The possible numbers of phone calls he receives in an afternoon and their probabilities are given in the table below.

| Number of Phone Calls | 0 | 1 | 2 | 3 | 4 |
|---|---|---|---|---|---|
| Probability | $\frac{1}{6}$ | $\frac{1}{6}$ | $\frac{2}{9}$ | $\frac{1}{3}$ | $\frac{1}{9}$ |

a.  Find the probability that Luis receives 3 or 4 phone calls.

b.   Find the probability that Luis receives fewer than 2 phone calls.

c.   Find the probability that Luis receives 2 or fewer phone calls.

d.   Find the probability that Luis does not receive 4 phone calls.

**Exercises 3–6**

When Jenna goes to the farmer's market, she also usually buys some broccoli.  The possible number of heads of broccoli that she buys and the probabilities are given in the table below.

| Number of Heads of Broccoli | 0 | 1 | 2 | 3 | 4 |
|---|---|---|---|---|---|
| Probability | $\frac{1}{12}$ | $\frac{1}{6}$ | $\frac{5}{12}$ | $\frac{1}{4}$ | $\frac{1}{12}$ |

Find the probability that Jenna:

3.   buys exactly 3 heads of broccoli.

4.   does not buy exactly 3 heads of broccoli.

5.   buys more than 1 head of broccoli.

6.   buys at least 3 heads of broccoli.

### Exercises 7–10

The diagram below shows a spinner designed like the face of a clock. The sectors of the spinner are colored red (R), blue (B), green (G), and yellow (Y).

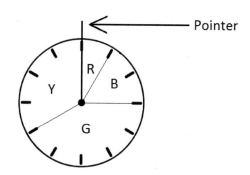

Spin the pointer, and award the player a prize according to the color on which the pointer stops.

7.  Writing your answers as fractions in lowest terms, find the probability that the pointer stops on:

a.  red:


b.  blue:


c.  green:


d.  yellow:

8.  Complete the table of probabilities below.

| Color | Red | Blue | Green | Yellow |
|-------|-----|------|-------|--------|
| Probability | | | | |

9.  Find the probability that the pointer stops in either the blue region or the green region.

10. Find the probability that the pointer does not stop in the green region.

**Lesson Summary**

In a probability experiment where the outcomes are not known to be equally likely, the formula for the probability of an event does not necessarily apply:

$$P(event) = \frac{Number\ of\ outcomes\ in\ the\ event}{Number\ of\ outcomes\ in\ the\ sample\ space}.$$

For example:

- To find the probability that the score is greater than 3, add the probabilities of all the scores that are greater than 3.
- To find the probability of not getting a score of 3, calculate $1 - (the\ probability\ of\ getting\ a\ 3)$.

**Problem Set**

1. The "Gator Girls" are a soccer team. The possible numbers of goals the Gator Girls will score in a game and their probabilities are shown in the table below.

| Number of Goals | 0 | 1 | 2 | 3 | 4 |
|---|---|---|---|---|---|
| Probability | 0.22 | 0.31 | 0.33 | 0.11 | 0.03 |

Find the probability that the Gator girls

a. score more than two goals.

b. score at least two goals.

c. do not score exactly 3 goals.

2. The diagram below shows a spinner. The pointer is spun, and the player is awarded a prize according to the color on which the pointer stops.

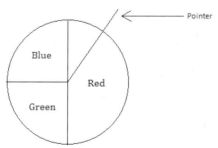

a. What is the probability that the pointer stops in the red region?

b. Complete the table below showing the probabilities of the three possible results.

| Color | Red | Green | Blue |
|---|---|---|---|
| Probability | | | |

c. Find the probability that the pointer stops on green or blue.

d. Find the probability that the pointer does not stop on green.

3. Wayne asked every student in his class how many siblings (brothers and sisters) they had. Survey results are shown in the table below. (Wayne included himself in the results.)

| Number of Siblings | 0 | 1 | 2 | 3 | 4 |
|---|---|---|---|---|---|
| Number of Students | 4 | 5 | 14 | 6 | 3 |

(Note: The table tells us that 4 students had no siblings, 5 students had one sibling, 14 students had two siblings, and so on.)

a. How many students are there in Wayne's class?

b. What is the probability that a randomly selected student does not have any siblings? Write your answer as a fraction in lowest terms.

c. The table below shows the possible number of siblings and the probabilities of each number. Complete the table by writing the probabilities as fractions in lowest terms.

| Number of Siblings | 0 | 1 | 2 | 3 | 4 |
|---|---|---|---|---|---|
| Probability | | | | | |

d. Writing your answers as fractions in lowest terms, find the probability that the student:

i. has fewer than two siblings.

ii. has two or fewer siblings.

iii. does not have exactly one sibling.

# Lesson 6: Using Tree Diagrams to Represent a Sample Space and to Calculate Probabilities

Suppose a girl attends a preschool where the students are studying primary colors. To help teach calendar skills, the teacher has each student maintain a calendar in his or her cubby. For each of the four days that the students are covering primary colors in class, each student gets to place a colored dot on his/her calendar: blue, yellow, or red. When the four days of the school week have passed (Monday–Thursday), what might the young girl's calendar look like?

One outcome would be four blue dots if the student chose blue each day. But consider that the first day (Monday) could be blue, and the next day (Tuesday) could be yellow, and Wednesday could be blue, and Thursday could be red. Or, maybe Monday and Tuesday could be yellow, Wednesday could be blue, and Thursday could be red. Or, maybe Monday, Tuesday, and Wednesday could be blue, and Thursday could be red …

As hard to follow as this seems now, we have only mentioned 3 of the 81 possible outcomes in terms of the four days of colors! Listing the other 78 outcomes would take several pages! Rather than listing outcomes in the manner described above (particularly when the situation has multiple stages, such as the multiple days in the case above), we often use a *tree diagram* to display all possible outcomes visually. Additionally, when the outcomes of each stage are the result of a chance experiment, tree diagrams are helpful for computing probabilities.

## Classwork

### Example 1: Two Nights of Games

Imagine that a family decides to play a game each night. They all agree to use a tetrahedral die (i.e., a four-sided pyramidal die where each of four possible outcomes is equally likely—see image on page 9) each night to randomly determine if they will play a board game ($B$) or a card game ($C$). The tree diagram mapping the possible overall outcomes over two consecutive nights will be developed below.

To make a tree diagram, first present all possibilities for the first stage. (In this case, Monday.)

*Monday*     *Tuesday*     *Outcome*


*B*



*C*

Then, from *each* branch of the first stage, attach all possibilities for the second stage (Tuesday).

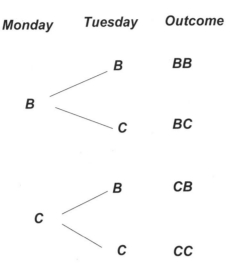

Note: If the situation has more than two stages, this process would be repeated until all stages have been presented.

a.   If "$BB$" represents two straight nights of board games, what does "$CB$" represent?

b.   List the outcomes where exactly one board game is played over two days.  How many outcomes were there?

**Example 2: Two Nights of Games (with Probabilities)**

In the example above, each night's outcome is the result of a chance experiment (rolling the tetrahedral die). Thus, there is a probability associated with each night's outcome.

By multiplying the probabilities of the outcomes from each stage, we can obtain the probability for each "branch of the tree." In this case, we can figure out the probability of each of our four outcomes: $BB$, $BC$, $CB$, and $CC$.

For this family, a card game will be played if the die lands showing a value of 1 and a board game will be played if the die lands showing a value of 2, 3, or 4. This makes the probability of a board game ($B$) on a given night 0.75.

| Monday | Tuesday | Outcome | |
|---|---|---|---|
| | **B** | **BB** | $(0.75)(0.75) = 0.5625$ |
| **B** | 0.75 | | |
| 0.75 | **C** | **BC** | $(0.75)(0.25) = 0.1875$ |
| | 0.25 | | |
| | **B** | **CB** | |
| **C** | 0.75 | | |
| 0.25 | **C** | **CC** | |
| | 0.25 | | |

a.  The probabilities for two of the four outcomes are shown. Now, compute the probabilities for the two remaining outcomes.

b.  What is the probability that there will be exactly one night of board games over the two nights?

### Exercises 1–3:  Two Children

Two friends meet at a grocery store and remark that a neighboring family just welcomed their second child.  It turns out that both children in this family are girls, and they are not twins.  One of the friends is curious about what the chances are of having 2 girls in a family's first 2 births.  Suppose that for each birth the probability of a "boy" birth is 0.5 and the probability of a "girl" birth is also 0.5.

1.  Draw a tree diagram demonstrating the four possible birth outcomes for a family with 2 children (no twins).  Use the symbol "$B$" for the outcome of "boy" and "$G$" for the outcome of "girl."  Consider the first birth to be the "first stage."  (Refer to Example 1 if you need help getting started.)

2.  Write in the probabilities of each stage's outcome to the tree diagram you developed above, and determine the probabilities for each of the 4 possible birth outcomes for a family with 2 children (no twins).

3.  What is the probability of a family having 2 girls in this situation?  Is that greater than or less than the probability of having exactly 1 girl in 2 births?

---

**Lesson Summary**

Tree diagrams can be used to organize outcomes in the sample space for chance experiments that can be thought of as being performed in multiple stages. Tree diagrams are also useful for computing probabilities of events with more than one outcome.

---

## Problem Set

1. Imagine that a family of three (Alice, Bill, and Chester) plays bingo at home every night. Each night, the chance that any one of the three players will win is $\frac{1}{3}$.

   a. Using "$A$" for Alice wins, "$B$" for Bill wins, and "$C$" for Chester wins, develop a tree diagram that shows the nine possible outcomes for two consecutive nights of play.

   b. Is the probability that "Bill wins both nights" the same as the probability that "Alice wins the first night and Chester wins the second night"? Explain.

2. According to the Washington, DC Lottery's website for its "Cherry Blossom Doubler" instant scratch game, the chance of winning a prize on a given ticket is about 17%. Imagine that a person stops at a convenience store on the way home from work every Monday and Tuesday to buy a "scratcher" ticket to play the game.

   (Source: http://dclottery.com/games/scratchers/1223/cherry-blossom-doubler.aspx accessed May 27, 2013).

   a. Develop a tree diagram showing the four possible outcomes of playing over these two days. Call stage 1 "Monday" and use the symbols "$W$" for a winning ticket and "$L$" for a non-winning ticket.

   b. What is the chance that the player will not win on Monday but will win on Tuesday?

   c. What is the chance that the player will win at least once during the two-day period?

Image of Tetrahedral Die

Source: http://commons.wikimedia.org/wiki/File:4-sided_dice_250.jpg

# Lesson 7: Calculating Probabilities of Compound Events

A previous lesson introduced *tree diagrams* as an effective method of displaying the possible outcomes of certain multistage chance experiments. Additionally, in such situations, tree diagrams were shown to be helpful for computing probabilities.

In those previous examples, diagrams primarily focused on cases with two stages. However, the basic principles of tree diagrams can apply to situations with more than two stages.

## Classwork

### Example 1: Three Nights of Games

Recall a previous example where a family decides to play a game each night, and they all agree to use a four-sided die in the shape of a pyramid (where each of four possible outcomes is equally likely) each night to randomly determine if the game will be a board ($B$) or a card ($C$) game. The tree diagram mapping the possible overall outcomes over two consecutive nights was as follows:

$$
\begin{array}{ccc}
\textit{Monday} & \textit{Tuesday} & \textit{Outcome} \\
\end{array}
$$

|  |  |  |
|---|---|---|
| | B | BB |
| B | | |
| | C | BC |
| | B | CB |
| C | | |
| | C | CC |

But how would the diagram change if you were interested in mapping the possible overall outcomes over three consecutive nights? To accommodate this additional "third stage," you would take steps similar to what you did before. You would attach all possibilities for the third stage (Wednesday) to each branch of the previous stage (Tuesday).

| Monday | Tuesday | Wednesday | Outcome |
|--------|---------|-----------|---------|
|        |         | B         | BBB     |
|        | B       | C         | BBC     |
| B      |         | B         | BCB     |
|        | C       | C         | BCC     |
|        | B       | B         | CBB     |
| C      |         | C         | CBC     |
|        | C       | B         | CCB     |
|        |         | C         | CCC     |

### Exercises 1–3

1. If "$BBB$" represents three straight nights of board games, what does "$CBB$" represent?

2. List all outcomes where exactly two board games were played over three days. How many outcomes were there?

3. There are eight possible outcomes representing the three nights. Are the eight outcomes representing the three nights equally likely? Why or why not?

## Example 2:  Three Nights of Games (with Probabilities)

In the example above, each night's outcome is the result of a chance experiment (rolling the four-sided die).  Thus, there is a probability associated with each night's outcome.

By multiplying the probabilities of the outcomes from each stage, you can obtain the probability for each "branch of the tree."  In this case, you can figure out the probability of each of our eight outcomes.

For this family, a card game will be played if the die lands showing a value of 1, and a board game will be played if the die lands showing a value of 2, 3, or 4.  This makes the probability of a board game ($B$) on a given night 0.75.

Let's use a tree to examine the probabilities of the outcomes for the three days.

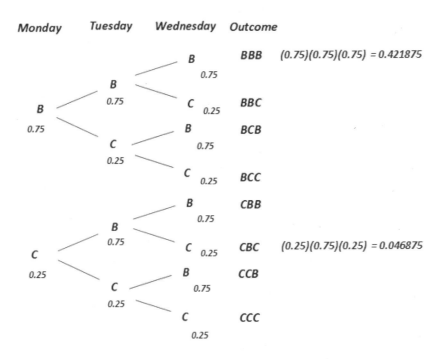

## Exercises 4–6

4.  Probabilities for two of the eight outcomes are shown.  Calculate the approximate probabilities for the remaining six outcomes.

5. What is the probability that there will be exactly two nights of board games over the three nights?

6. What is the probability that the family will play at least one night of card games?

### Exercises 7–10:  Three Children

A neighboring family just welcomed their third child.  It turns out that all 3 of the children in this family are girls, and they are not twins.  Suppose that for each birth the probability of a "boy" birth is 0.5, and the probability of a "girl" birth is also 0.5.  What are the chances of having 3 girls in a family's first 3 births?

7. Draw a tree diagram showing the eight possible birth outcomes for a family with 3 children (no twins).  Use the symbol "$B$" for the outcome of "boy" and "$G$" for the outcome of "girl."  Consider the first birth to be the "first stage."  (Refer to Example 1 if you need help getting started.)

8. Write in the probabilities of each stage's outcomes in the tree diagram you developed above, and determine the probabilities for each of the eight possible birth outcomes for a family with 3 children (no twins).

9. What is the probability of a family having 3 girls in this situation?  Is that greater than or less than the probability of having exactly 2 girls in 3 births?

10. What is the probability of a family of 3 children having at least 1 girl?

Problem Set

1.   According to the Washington, DC Lottery's website for its "Cherry Blossom Doubler" instant scratch game, the chance of winning a prize on a given ticket is about 17%.  Imagine that a person stops at a convenience store on the way home from work every Monday, Tuesday, and Wednesday to buy a "scratcher" ticket and plays the game.

   (Source:  http://dclottery.com/games/scratchers/1223/cherry-blossom-doubler.aspx accessed May 27, 2013)

   a.   Develop a tree diagram showing the eight possible outcomes of playing over these three days.  Call stage one "Monday," and use the symbols "$W$" for a winning ticket and "$L$" for a non-winning ticket.

   b.   What is the probability that the player will not win on Monday but will win on Tuesday and Wednesday?

   c.   What is the probability that the player will win at least once during the 3-day period?

2.   A survey company is interested in conducting a statewide poll prior to an upcoming election.  They are only interested in talking to registered voters.

   Imagine that 55% of the registered voters in the state are male, and 45% are female.  Also, consider that the distribution of ages may be different for each group.  In this state, 30% of male registered voters are age 18–24, 37% are age 25–64, and 33% are 65 or older.  32% of female registered voters are age 18–24, 26% are age 25–64, and 42% are 65 or older.

   The following tree diagram describes the distribution of registered voters.  The probability of selecting a male registered voter age 18–24 is 0.165.

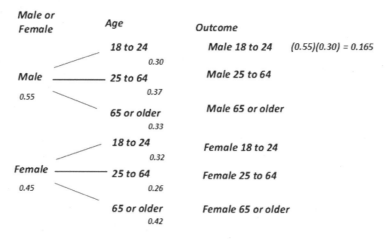

   a.   What is the chance that the polling company will select a registered female voter age 65 or older?

   b.   What is the chance that the polling company will select any registered voter age 18–24?

# Lesson 8:  The Difference Between Theoretical Probabilities and Estimated Probabilities

Did you ever watch the beginning of a Super Bowl game?  After the traditional handshakes, a coin is tossed to determine which team gets to kick-off first.  Whether or not you are a football fan, the toss of a fair coin is often used to make decisions between two groups.

## Classwork

### Example 1:  Why a Coin?

Coins were discussed in previous lessons of this module.  What is special about a coin?  In most cases, a coin has two different sides:  a head side ("heads") and a tail side ("tails").  The sample space for tossing a coin is {heads, tails}.  If each outcome has an equal chance of occurring when the coin is tossed, then the probability of getting heads is $\frac{1}{2}$, or 0.5.  The probability of getting tails is also 0.5.  Note that the sum of these probabilities is 1.

The probabilities formed using the sample space and what we know about coins are called the theoretical probabilities. Using observed relative frequencies is another method to estimate the probabilities of heads or tails.  A relative frequency is the proportion derived from the number of the observed outcomes of an event divided by the total number of outcomes.  Recall from earlier lessons that a relative frequency can be expressed as a fraction, a decimal, or a percent. Is the estimate of a probability from this method close to the theoretical probability?  The following exercise investigates how relative frequencies can be used to estimate probabilities.

### Exercises 1–9

Beth tosses a coin 10 times and records her results.  Here are the results from the 10 tosses:

| Toss | 1 | 2 | 3 | 4 | 5 | 6 | 7 | 8 | 9 | 10 |
|---|---|---|---|---|---|---|---|---|---|---|
| Result | H | H | T | H | H | H | T | T | T | H |

The total number of heads divided by the total number of tosses is the relative frequency of heads.  It is the proportion of the time that heads occurred on these tosses.  The total number of tails divided by the total number of tosses is the relative frequency of tails.

1. Beth started to complete the following table as a way to investigate the relative frequencies. For each outcome, the total number of tosses increased. The total number of heads or tails observed so far depends on the outcome of the current toss. Complete this table for the 10 tosses recorded above.

| Toss | Outcome | Total number of Heads so far | Relative frequency of heads so far (to the nearest hundredth) | Total number of tails so far | Relative frequency of tails so far (to the nearest hundredth) |
|---|---|---|---|---|---|
| 1 | H | 1 | $\frac{1}{1} = 1$ | 0 | $\frac{0}{1} = 0$ |
| 2 | H | 2 | $\frac{2}{2} = 1$ | 0 | $\frac{0}{2} = 0$ |
| 3 | T | 2 | $\frac{2}{3} = 0.67$ | 1 | $\frac{1}{3} = 0.33$ |
| 4 | | | | | |
| 5 | | | | | |
| 6 | | | | | |
| 7 | | | | | |
| 8 | | | | | |
| 9 | | | | | |
| 10 | | | | | |

2. What is the sum of the relative frequency of heads and the relative frequency of tails for each row of the table?

3. Beth's results can also be displayed using a graph. Complete this graph using the values of relative frequency of heads so far from the table above:

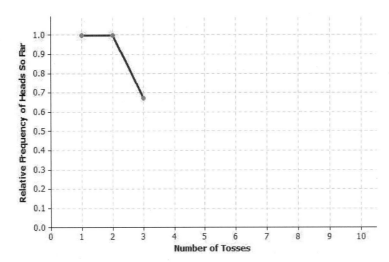

4. Beth continued tossing the coin and recording results for a total of 40 tosses. Here are the results of the next 30 tosses:

| Toss | 11 | 12 | 13 | 14 | 15 | 16 | 17 | 18 | 19 | 20 |
|---|---|---|---|---|---|---|---|---|---|---|
| Result | T | H | T | H | T | H | H | T | H | T |

| Toss | 21 | 22 | 23 | 24 | 25 | 26 | 27 | 28 | 29 | 30 |
|---|---|---|---|---|---|---|---|---|---|---|
| Result | H | T | T | H | T | T | T | T | H | T |

| Toss | 31 | 32 | 33 | 34 | 35 | 36 | 37 | 38 | 39 | 40 |
|---|---|---|---|---|---|---|---|---|---|---|
| Result | H | T | H | T | H | T | H | H | T | T |

As the number of tosses increases, the relative frequency of heads changes. Complete the following table for the 40 coin tosses:

| Number of tosses | Total number of heads so far | Relative frequency of heads so far (to the nearest hundredth) |
|---|---|---|
| 1 | | |
| 5 | | |
| 10 | | |
| 15 | | |
| 20 | | |
| 25 | | |
| 30 | | |
| 35 | | |
| 40 | | |

5. Complete the graph below using the relative frequency of heads so far from the table above for total number of tosses of 1, 5, 10, 15, 20, 25, 30, 35, and 40:

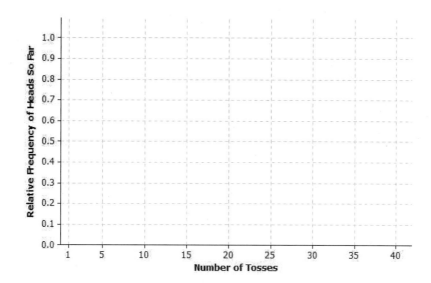

6. What do you notice about the changes in the relative frequency of number of heads so far as the number of tosses increases?

7. If you tossed the coin 100 times, what do you think the relative frequency of heads would be? Explain your answer.

8. Based on the graph and the relative frequencies, what would you estimate the probability of getting heads to be? Explain your answer.

9.  How close is your estimate in Exercise 8 to the theoretical probability of 0.5?  Would the estimate of this probability have been as good if Beth had only tossed the coin a few times instead of 40?

The value you gave in Exercise 8 is an estimate of the theoretical probability and is called an experimental or estimated probability.

### Example 2:  More Pennies!

Beth received nine more pennies.  She securely taped them together to form a small stack.  The top penny of her stack showed heads, and the bottom penny showed tails.  If Beth tosses the stack, what outcomes could she observe?

### Exercises 10–17

10.  Beth wanted to determine the probability of getting heads when she tosses the stack.  Do you think this probability is the same as the probability of getting heads with just one coin?  Explain your answer.

11.  Make a sturdy stack of 10 pennies in which one end of the stack has a penny showing heads and the other end tails.  Make sure the pennies are taped securely, or you may have a mess when you toss the stack.  Toss the stack to observe possible outcomes.  What is the sample space for tossing a stack of 10 pennies taped together?  Do you think the probability of each outcome of the sample space is equal?  Explain your answer.

12.  Record the results of 10 tosses.  Complete the following table of the relative frequencies of heads for your 10 tosses:

| Toss | 1 | 2 | 3 | 4 | 5 | 6 | 7 | 8 | 9 | 10 |
|---|---|---|---|---|---|---|---|---|---|---|
| Result | | | | | | | | | | |
| Relative frequency of heads so far | | | | | | | | | | |

13.  Based on the value of the relative frequencies of heads so far, what would you estimate the probability of getting heads to be?

14.  Toss the stack of 10 pennies another 20 times.  Complete the following table:

| Toss | 11 | 12 | 13 | 14 | 15 | 16 | 17 | 18 | 19 | 20 |
|---|---|---|---|---|---|---|---|---|---|---|
| Result | | | | | | | | | | |

| Toss | 21 | 22 | 23 | 24 | 25 | 26 | 27 | 28 | 29 | 30 |
|---|---|---|---|---|---|---|---|---|---|---|
| Result | | | | | | | | | | |

15. Summarize the relative frequencies of heads so far by completing the following table:

| Number of tosses | Total number of heads so far | Relative frequency of heads so far (to the nearest hundredth) |
|---|---|---|
| 1 | | |
| 5 | | |
| 10 | | |
| 15 | | |
| 20 | | |
| 25 | | |
| 30 | | |

16. Based on the relative frequencies for the 30 tosses, what is your estimate of the probability of getting heads? Can you compare this estimate to a theoretical probability like you did in the first example? Explain your answer.

17. Create another stack of pennies. Consider creating a stack using 5 pennies, 15 pennies, or 20 pennies taped together in the same way you taped the pennies to form a stack of 10 pennies. Again, make sure the pennies are taped securely, or you might have a mess!

Toss the stack you made 30 times. Record the outcome for each toss:

| Toss | 1 | 2 | 3 | 4 | 5 | 6 | 7 | 8 | 9 | 10 |
|---|---|---|---|---|---|---|---|---|---|---|
| Result | | | | | | | | | | |

| Toss | 11 | 12 | 13 | 14 | 15 | 16 | 17 | 18 | 19 | 20 |
|---|---|---|---|---|---|---|---|---|---|---|
| Result | | | | | | | | | | |

| Toss | 21 | 22 | 23 | 24 | 25 | 26 | 27 | 28 | 29 | 30 |
|---|---|---|---|---|---|---|---|---|---|---|
| Result | | | | | | | | | | |

**Problem Set**

1.  If you created a stack of 15 pennies taped together, do you think the probability of getting a head on a toss of the stack would be different than for a stack of 10 pennies?  Explain your answer.

2.  If you created a stack of 20 pennies taped together, what do you think the probability of getting a head on a toss of the stack would be?  Explain your answer.

3.  Based on your work in this lesson, complete the following table of the relative frequencies of heads for the stack you created:

| Number of tosses | Total number of heads so far | Relative frequency of heads so far (to the nearest hundredth) |
|---|---|---|
| 1 | | |
| 5 | | |
| 10 | | |
| 15 | | |
| 20 | | |
| 25 | | |
| 30 | | |

4.  What is your estimate of the probability that your stack of pennies will land heads up when tossed?  Explain your answer.

5.  Is there a theoretical probability you could use to compare to the estimated probability?  Explain you answer.

# Lesson 9:  Comparing Estimated Probabilities to Probabilities Predicted by a Model

## Game Show—Picking Blue!

Imagine, for a moment, the following situation:  You and your classmates are contestants on a quiz show called *Picking Blue!*  There are two bags in front of you, Bag A and Bag B.  Each bag contains red and blue chips.  You are told that one of the bags has exactly the same number of blue chips as red chips.  But you are told nothing about the number of blue and red chips in the second bag.

Each student in your class will be asked to select either Bag A or Bag B.  Starting with Bag A, a chip is randomly selected from the bag.  If a blue chip is drawn, all of the students in your class who selected Bag A win a Blue Token.  The chip is put back in the bag.  After mixing up the chips in the bag, another chip is randomly selected from the bag.  If the chip is blue, the students who picked Bag A win another Blue Token.  After the chip is placed back into the bag, the process continues until a red chip is picked. When a red chip is picked, the game moves to Bag B.  A chip from the Bag B is then randomly selected.  If it is blue, all of the students who selected Bag B win a Blue Token.  But if the chip is red, the game is over.  Just like for Bag A, if the chip is blue, the process repeats until a red chip is picked from the bag.  When the game is over, the students with the greatest number of Blue Tokens are considered the winning team.

Without any information about the bags, you would probably select a bag simply by guessing.  But surprisingly, the show's producers are going to allow you to do some research before you select a bag.  For the next 20 minutes, you can pull a chip from either one of the two bags, look at the chip, and then put the chip back in the bag.  You can repeat this process as many times as you want within the 20 minutes.  At the end of 20 minutes, you must make your final decision and select which of the bags you want to use in the game.

## Getting Started

Assume that the producers of the show do not want give away a lot of their Blue Tokens.  As a result, if one bag has the same number of red and blue chips, do you think the other bag would have more, or fewer, blue chips than red chips?  Explain your answer.

## Planning the Research

Your teacher will provide you with two bags labeled A and B.  You have 20 minutes to experiment with pulling chips one at a time from the bags.  After you examine a chip, you must put it back in the bag.  Remember, no fair peeking in the bags as that will disqualify you from the game.  You can pick chips from just one bag, or you can pick chips from one bag and then the other bag.

Use the results from 20 minutes of research to determine which bag you will choose for the game.

Provide a description outlining how you will carry out your research:

## Carrying Out the Research

Share your plan with your teacher.  Your teacher will verify whether your plan is within the rules of the quiz show.  Approving your plan does not mean, however, that your teacher is indicating that your research method offers the most accurate way to determine which bag to select.  If your teacher approves your research, carry out your plan as outlined.  Record the results from your research, as directed by your teacher.

## Playing the Game

After the research has been conducted, the competition begins.  First, your teacher will shake up Bag A.  A chip is selected.  If the chip is blue, all students who selected Bag A win an imaginary Blue Token.  The chip is put back in the bag, and the process continues. When a red chip is picked from Bag A, students selecting Bag A have completed the competition.  Your teacher will now shake up Bag B.  A chip is selected.  If it is blue, all students who selected Bag B win an imaginary Blue Token.  The process continues until a red chip is picked.  At that point, the game is over.

How many Blue Tokens did you win?

## Examining Your Results

At the end of the game, your teacher will open the bags and reveal how many blue and red chips were in each bag. Answer the following questions. (*See: Closing Questions.*) After you have answered these questions, discuss them with your class.

1.  Before you played the game, what were you trying to learn about the bags from your research?

2.  What did you expect to happen when you pulled chips from the bag with the same number of blue and red chips? Did the bag that you thought had the same number of blue and red chips yield the results you expected?

3.  How confident were you in predicting which bag had the same number of blue and red chips? Explain.

4.  What bag did you select to use in the competition and why?

5.  If you were the show's producers, how would make up the second bag? (Remember, one bag has the same number of red and blue chips.)

6.  If you picked a chip from Bag B 100 times and found that you picked each color exactly 50 times, would you know for sure that that bag was the one with equal numbers of each color?

---

**Lesson Summary:**

- The long-run relative frequencies can be used as estimated probabilities of events.
- Collecting data on a game or chance experiment is one way to estimate the probability of an outcome.
- The more data collected on the outcomes from a game or chance experiment, the closer the estimates of the probabilities are likely to be the actual probabilities.

---

Problem Set

Jerry and Michael played a game similar to *Picking Blue*. The following results are from their research using the same two bags:

Jerry's research:

|  | Number of Red chips picked | Number of Blue chips picked |
|---|---|---|
| Bag A | 2 | 8 |
| Bag B | 3 | 7 |

Michael's research:

|  | Number of Red chips picked | Number of Blue chips picked |
|---|---|---|
| Bag A | 28 | 12 |
| Bag B | 22 | 18 |

1. If all you knew about the bags were the results of Jerry's research, which bag would you select for the game? Explain your answer.

2. If all you knew about the bags were the results of Michael's research, which bag would you select for the game? Explain your answer.

3. Does Jerry's research or Michael's research give you a better indication of the make-up of the blue and red chips in each bag? Explain why you selected this research.

4. Assume there are 12 chips in each bag. Use either Jerry's or Michael's research to estimate the number of red and blue chips in each bag. Then explain how you made your estimates.

   **Bag A**

   Number of red chips:

   Number of blue chips:

   **Bag B**

   Number of red chips:

   Number of blue chips:

5. In a different game of *Picking Blue!*, two bags each contain red, blue, green, and yellow chips. One bag contains the same number of red, blue, green, and yellow chips. In the second bag, half the chips are blue. Describe a plan for determining which bag has more blue chips than any of the other colors.

# Lesson 10:  Using Simulation to Estimate a Probability

Classwork

In previous lessons, you estimated probabilities of events by collecting data empirically or by establishing a theoretical probability model.  There are real problems for which those methods may be difficult or not practical to use.  Simulation is a procedure that will allow you to answer questions about real problems by running experiments that closely resemble the real situation.

It is often important to know the probabilities of real-life events that may not have known theoretical probabilities.  Scientists, engineers, and mathematicians design simulations to answer questions that involve topics such as diseases, water flow, climate changes, or functions of an engine.  Results from the simulations are used to estimate probabilities that help researchers understand problems and provide possible solutions to these problems.

### Example 1:  Families

How likely is it that a family with three children has all boys or all girls?

Let's assume that a child is equally likely to be a boy or a girl.  Instead of observing the result of actual births, a toss of a fair coin could be used to simulate a birth.  If the toss results in heads (H), then we could say a boy was born; if the toss results in tails (T), then we could say a girl was born.  If the coin is fair (i.e., heads and tails are equally likely), then getting a boy or a girl is equally likely.

### Exercises 1–2

Suppose that a family has three children.  To simulate the genders of the three children, the coin or number cube or a card would need to be used three times, once for each child.  For example, three tosses of the coin resulted in HHT, representing a family with two boys and one girl.  Note that HTH and THH also represent two boys and one girl.

1.  Suppose a prime number (P) result of a rolled number cube simulates a boy birth, and a non-prime (N) simulates a girl birth.  Using such a number cube, list the outcomes that would simulate a boy birth, and those that simulate a girl birth.  Are the boy and girl birth outcomes equally likely?

2.  Suppose that one card is drawn from a regular deck of cards, a red card (R) simulates a boy birth and a black card (B) simulates a girl birth.  Describe how a family of three children could be simulated.

## Example 2

Simulation provides an estimate for the probability that a family of three children would have three boys or three girls by performing three tosses of a fair coin many times. Each sequence of three tosses is called a trial. If a trial results in either HHH or TTT, then the trial represents all boys or all girls, which is the event that we are interested in. These trials would be called a "success." If a trial results in any other order of H's and T's, then it is called a "failure."

The estimate for the probability that a family has either three boys or three girls based on the simulation is the number of successes divided by the number of trials. Suppose 100 trials are performed, and that in those 100 trials, 28 resulted in either HHH or TTT. Then the estimated probability that a family of three children has either three boys or three girls would be $\frac{28}{100} = 0.28$.

## Exercises 3–5

3.  Find an estimate of the probability that a family with three children will have exactly one girl using the following outcomes of 50 trials of tossing a fair coin three times per trial. Use H to represent a boy birth, and T to represent a girl birth.

| HHT | HTH | HHH | TTH | THT | THT | HTT | HHH | TTH | HHH |
| HHT | TTT | HHT | TTH | HHH | HTH | THH | TTT | THT | THT |
| THT | HHH | THH | HTT | HTH | TTT | HTT | HHH | TTH | THT |
| THH | HHT | TTT | TTH | HTT | THH | HTT | HTH | TTT | HHH |
| HTH | HTH | THT | TTH | TTT | HHT | HHT | THT | TTT | HTT |

4.  Perform a simulation of 50 trials by rolling a fair number cube in order to find an estimate of the probability that a family with three children will have exactly one girl.

    a.  Specify what outcomes of one roll of a fair number cube will represent a boy, and what outcomes will represent a girl.

b.    Simulate 50 trials, keeping in mind that one trial requires three rolls of the number cube.  List the results of your 50 trials.

c.    Calculate the estimated probability.

5.    Calculate the theoretical probability that a family with three children will have exactly one girl.

   a.    List the possible outcomes for a family with three children.  For example, one possible outcome is BBB (all three children are boys).

   b.    Assume that having a boy and having a girl are equally likely.  Calculate the theoretical probability that a family with three children will have exactly one girl.

   c.    Compare it to the estimated probabilities found in parts (a) and (b) above.

**Example 3:  Basketball Player**

Suppose that, on average, a basketball player makes about three out of every four foul shots.  In other words, she has a 75% chance of making each foul shot she takes.  Since a coin toss produces equally likely outcomes, it could not be used in a simulation for this problem.

Instead, a number cube could be used by specifying that the numbers 1, 2, or 3 represent a hit, the number 4 represents a miss, and the numbers 5 and 6 would be ignored.  Based on the following 50 trials of rolling a fair number cube, find an estimate of the probability that she makes five or six of the six foul shots she takes.

| | | | | |
|---|---|---|---|---|
| 441323 | 342124 | 442123 | 422313 | 441243 |
| 124144 | 333434 | 243122 | 232323 | 224341 |
| 121411 | 321341 | 111422 | 114232 | 414411 |
| 344221 | 222442 | 343123 | 122111 | 322131 |
| 131224 | 213344 | 321241 | 311214 | 241131 |
| 143143 | 243224 | 323443 | 324243 | 214322 |
| 214411 | 423221 | 311423 | 142141 | 411312 |
| 343214 | 123131 | 242124 | 141132 | 343122 |
| 121142 | 321442 | 121423 | 443431 | 214433 |
| 331113 | 311313 | 211411 | 433434 | 323314 |

Lesson Summary

In previous lessons, you estimated probabilities by collecting data and found theoretical probabilities by creating a model. In this lesson you used simulation to estimate probabilities in real problems and in situations for which empirical or theoretical procedures are not easily calculated.

Simulation is a method that uses an artificial process (like tossing a coin or rolling a number cube) to represent the outcomes of a real process that provides information about the probability of events. In several cases, simulations are needed to both understand the process as well as provide estimated probabilities.

Problem Set

1.  A mouse is placed at the start of the maze shown below. If it reaches station B, it is given a reward. At each point where the mouse has to decide which direction to go, assume that it is equally likely to go in either direction. At each decision point 1, 2, 3, it must decide whether to go left (L) or right (R). It cannot go backwards.

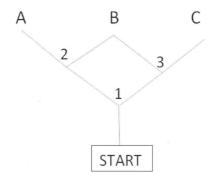

a.  Create a theoretical model of probabilities for the mouse to arrive at terminal points A, B, and C.

    i.   List the possible paths of a sample space for the paths the mouse can take. For example, if the mouse goes left at decision point 1, and then right at decision point 2, then the path would be denoted LR.

    ii.  Are the paths in your sample space equally likely? Explain.

    iii. What are the theoretical probabilities that a mouse reaches terminal points A, B, and C? Explain.

b.  Based on the following set of simulated paths, estimate the probabilities that the mouse arrives at points A, B, and C.

| | | | | | | | | | |
|----|----|----|----|----|----|----|----|----|----|
| RR | RR | RL | LL | LR | RL | LR | LL | LR | RR |
| LR | RL | LR | RR | RL | LR | RR | LL | RL | RL |
| LL | LR | LR | LL | RR | RR | RL | LL | RR | LR |
| RR | LR | RR | LR | LR | LL | LR | RL | RL | LL |

c.  How do the simulated probabilities in part (b) compare to the theoretical probabilities of part (a)?

2.  Suppose that a dartboard is made up of the $8 \times 8$ grid of squares shown below. Also, suppose that when a dart is thrown, it is equally likely to land on any one of the 64 squares. A point is won if the dart lands on one of the 16 black squares. Zero points are earned if the dart lands in a white square.

a.  For one throw of a dart, what is the probability of winning a point? Note that a point is won if the dart lands on a black square.

b.  Lin wants to use a number cube to simulate the result of one dart. She suggests that 1 on the number cube could represent a win. Getting 2, 3, or 4 could represent no point scored. She says that she would ignore getting a 5 or 6. Is Lin's suggestion for a simulation appropriate? Explain why you would use it *or,* if not, how you would change it.

c.  Suppose a game consists of throwing a dart three times. A trial consists of three rolls of the number cube. Based on Lin's suggestion in part (b) and the following simulated rolls, estimate the probability of scoring two points in three darts.

| | | | | |
|---|---|---|---|---|
| 324 | 332 | 411 | 322 | 124 |
| 224 | 221 | 241 | 111 | 223 |
| 321 | 332 | 112 | 433 | 412 |
| 443 | 322 | 424 | 412 | 433 |
| 144 | 322 | 421 | 414 | 111 |
| 242 | 244 | 222 | 331 | 224 |
| 113 | 223 | 333 | 414 | 212 |
| 431 | 233 | 314 | 212 | 241 |
| 421 | 222 | 222 | 112 | 113 |
| 212 | 413 | 341 | 442 | 324 |

d.  The theoretical probability model for winning 0, 1, 2, and 3 points in three throws of the dart as described in this problem is

i.  winning 0 points has a probability of 0.42;

ii.  winning 1 point has a probability of 0.42;

iii.  winning 2 points has a probability of 0.14;

iv.  winning 3 points has a probability of 0.02.

Use the simulated rolls in part (c) to build a model of winning 0, 1, 2, and 3 points, and compare it to the theoretical model.

# Lesson 11:  Using Simulation to Estimate a Probability

### Example 1:  Simulation

In the last lesson, we used coins, number cubes, and cards to carry out simulations.  Another option is putting identical pieces of paper or colored disks into a container, mixing them thoroughly, and then choosing one.

For example, if a basketball player typically makes five out of eight foul shots, then a colored disk could be used to simulate a foul shot.  A green disk could represent a made shot, and a red disk could represent a miss.  You could put five green and three red disks in a container, mix them, and then choose one to represent a foul shot.  If the color of the disk is green, then the shot is made.  If the color of the disk is red, then the shot is missed.  This procedure simulates one foul shot.

### Exercises 1–2

1.  Using colored disks, describe how one at-bat could be simulated for a baseball player who has a batting average of 0.300.  Note that a batting average of 0.300 means the player gets a hit (on average) three times out of every ten times at bat.  Be sure to state clearly what a color represents.

2.  Using colored disks, describe how one at-bat could be simulated for a player who has a batting average of 0.273.  Note that a batting average of 0.273 means that on average, the player gets 273 hits out of 1000 at-bats.

## Example 2:  Using Random Number Tables

Why is using colored disks not practical for the situation described in Exercise 2?  Another way to carry out a simulation is to use a random-number table, or a random-number generator.  In a random-number table, the digits 0, 1, 2, 3, 4, 5, 6, 7, 8, and 9 occur equally often in the long run.  Pages and pages of random numbers can be found online.

For example, here are three lines of random numbers.  The space after every five digits is only for ease of reading. Ignore the spaces when using the table.

25256  65205  72597  00562  12683  90674  78923  96568  32177  33855

76635  92290  88864  72794  14333  79019  05943  77510  74051  87238

07895  86481  94036  12749  24005  80718  13144  66934  54730  77140

To use the random-number table to simulate an at-bat for the 0.273 hitter in Exercise 2, you could use a three-digit number to represent one at bat.  The three-digit numbers from 000–272 could represent a hit, and the three-digit numbers from 273–999 could represent a non-hit.  Using the random numbers above, and starting at the beginning of the first line, the first three-digit random number is 252, which is between 000 and 272, so that simulated at-bat is a hit. The next three-digit random number is 566, which is a non-hit.

## Exercise 3

3.  Continuing on the first line of the random numbers above, what would the hit/non-hit outcomes be for the next six at-bats?  Be sure to state the random number and whether it simulates a hit or non-hit.

## Example 3:  Baseball Player

A batter typically gets to bat four times in a ballgame.  Consider the 0.273 hitter of the previous example.  Use the following steps (and the random numbers shown above) to estimate that player's probability of getting at least three hits (three or four) in four times at-bat.

a.  Describe what one trial is for this problem.

b.  Describe when a trial is called a success and when it is called a failure.

c. Simulate 12 trials. (Continue to work as a class, or let students work with a partner.)

d. Use the results of the simulation to estimate the probability that a 0.273 hitter gets three or four hits in four times at-bat. Compare your estimate with other groups.

## Example 4: Birth Month

In a group of more than 12 people, is it likely that at least two people, maybe more, will have the same birth-month? Why? Try it in your class.

Now suppose that the same question is asked for a group of only seven people. Are you likely to find some groups of seven people in which there is a match, but other groups in which all seven people have different birth-months? In the following exercise, you will estimate the probability that at least two people in a group of seven, were born in the same month.

## Exercises 4–7

4. What might be a good way to generate outcomes for the birth-month problem—using coins, number cubes, cards, spinners, colored disks, or random numbers?

5. How would you simulate one trial of seven birth-months?

6.  How is a success determined for your simulation?

7.  How is the simulated estimate determined for the probability that a least two in a group of seven people were born in the same month?

### Lesson Summary

In the previous lesson, you carried out simulations to estimate a probability. In this lesson, you had to provide parts of a simulation design. You also learned how random numbers could be used to carry out a simulation.

To design a simulation:

- Identify the possible outcomes and decide how to simulate them, using coins, number cubes, cards, spinners, colored disks, or random numbers.

- Specify what a trial for the simulation will look like and what a success and a failure would mean.

- Make sure you carry out enough trials to ensure that the estimated probability gets closer to the actual probability as you do more trials. There is no need for a specific number of trials at this time; however, you want to make sure to carry out enough trials so that the relative frequencies level off.

### Problem Set

1.  A model airplane has two engines. It can fly if one engine fails but is in serious trouble if both engines fail. The engines function independently of one another. On any given flight, the probability of a failure is 0.10 for each engine. Design a simulation to estimate the probability that the airplane will be in serious trouble the next time it goes up.

    a.  How would you simulate the status of an engine?

    b.  What constitutes a trial for this simulation?

    c.  What constitutes a success for this simulation?

    d.  Carry out 50 trials of your simulation, list your results, and calculate an estimate of the probability that the airplane will be in serious trouble the next time it goes up.

2.  In an effort to increase sales, a cereal manufacturer created a really neat toy that has six parts to it. One part is put into each box of cereal. Which part is in a box is not known until the box is opened. You can play with the toy without having all six parts, but it is better to have the complete set. If you are really lucky, you might only need to buy six boxes to get a complete set, but if you are very unlucky, you might need to buy many, many boxes before obtaining all six parts.

    a.  How would you represent the outcome of purchasing a box of cereal, keeping in mind that there are six different parts? There is one part in each box.

    b.  What constitutes a trial in this problem?

    c.  What constitutes a success in a trial in this problem?

    d.  Carry out 15 trials, list your results, and compute an estimate of the probability that it takes the purchase of 10 or more boxes to get all six parts.

Lesson 11:     Using Simulation to Estimate a Probability

3.  Suppose that a type A blood donor is needed for a certain surgery.  Carry out a simulation to answer the following question:  If 40% of donors have type A blood, what is an estimate of the probability that it will take at least four donors to find one with type A blood?

    a.  How would you simulate a blood donor having or not having type A?

    b.  What constitutes a trial for this simulation?

    c.  What constitutes a success for this simulation?

    d.  Carry out 15 trials, list your results, and compute an estimate for the probability that it takes at least four donors to find one with type A blood.

# Lesson 12: Using Probability to Make Decisions

### Example 1: Number Cube

Your teacher gives you a number cube with numbers 1–6 on its faces. You have never seen that particular cube before. You are asked to state a theoretical probability model for rolling it once. A probability model consists of the list of possible outcomes (the sample space) and the theoretical probabilities associated with each of the outcomes. You say that the probability model might assign a probability of $\frac{1}{6}$ to each of the possible outcomes, but because you have never seen this particular cube before, you would like to roll it a few times. (Maybe it's a trick cube.) Suppose your teacher allows you to roll it 500 times and you get the following results:

| Outcome | 1 | 2 | 3 | 4 | 5 | 6 |
|---------|----|----|----|----|----|----|
| Frequency | 77 | 92 | 75 | 90 | 76 | 90 |

### Exercises 1–2

1. If the "equally likely" model were correct, about how many of each outcome would you expect to see if the cube is rolled 500 times?

2. Based on the data from the 500 rolls, how often were odd numbers observed? How often were even numbers observed?

### Example 2: Probability Model

Two black balls and two white balls are put in a small cup whose bottom allows the four balls to fit snugly. After shaking the cup well, two patterns of colors are possible as shown. The pattern on the left shows the similar colors are opposite each other, and the pattern on the right shows the similar colors are next to or adjacent to each other.

Philippe is asked to specify a probability model for the chance experiment of shaking the cup and observing the pattern. He thinks that because there are two outcomes—like heads and tails on a coin—that the outcomes should be equally likely. Sylvia isn't so sure that the "equally likely" model is correct, so she would like to collect some data before deciding on a model.

### Exercise 3

Collect data for Sylvia. Carry out the experiment of shaking a cup that contains four balls, two black and two white, observing, and recording whether the pattern is opposite or adjacent. Repeat this process 20 times. Then, combine the data with that collected by your classmates.

3.  Do your results agree with Philippe's equally likely model, or do they indicate that Sylvia had the right idea? Explain.

**Exercises 4–5**

There are three popular brands of mixed nuts. Your teacher loves cashews, and in his experience of having purchased these brands, he suggests that not all brands have the same percentage of cashews. One has around 20% cashews, one has 25%, and one has 35%.

Your teacher has bags labeled A, B, and C representing the three brands. The bags contain red beads representing cashews and brown beads representing other types of nuts. One bag contains 20% red beads, another 25% red beads, and the third has 35% red beads. You are to determine which bag contains which percentage of cashews. You cannot just open the bags and count the beads.

4. Work as a class to design a simulation. You need to agree on what an outcome is, what a trial is, what a success is, and how to calculate the estimated probability of getting a cashew. Base your estimate on 50 trials.

5. Your teacher will give your group one of the bags labeled A, B, or C. Using your plan from part (a), collect your data. Do you think you have the 20%, 25%, or 35% cashews bag? Explain.

**Exercises 6–8**

Suppose you have two bags, A and B, in which there are an equal number of slips of paper. Positive numbers are written on the slips. The numbers are not known, but they are whole numbers between 1 and 75, inclusive. The same number may occur on more than one slip of paper in a bag.

These bags are used to play a game. In this game, you choose one of the bags and then choose one slip from that bag. If you choose bag A, and the number you choose from it is a prime number, then you win. If you choose bag B, and the number you choose from it is a power of 2, you win. Which bag should you choose?

6. Emma suggests that it doesn't matter which bag you choose because you don't know anything about what numbers are inside the bags. So she thinks that you are equally likely to win with either bag. Do you agree with her? Explain.

7. Aamir suggests that he would like to collect some data from both bags before making a decision about whether or not the model is equally likely. Help Aamir by drawing 50 slips from each bag, being sure to replace each one before choosing again. Each time you draw a slip, record whether it would have been a winner or not. Using the results, what is your estimate for the probability of drawing a prime number from bag A and drawing a power of 2 from bag B?

8. If you were to play this game, which bag would you choose? Explain why you would pick this bag.

---

**Lesson Summary**

This lesson involved knowing the probabilities for outcomes of a sample space. You used these probabilities to determine whether or not the simulation supported a given theoretical probability. The simulated probabilities, or estimated probabilities, suggested a workable process for understanding the probabilities. Only 50 trials were conducted in some examples; however, as stated in several other lessons, the more trials that are observed from a simulation, the better.

---

**Problem Set**

1. Some M&M's are "defective." For example, a defective M&M may have its "m" missing, or it may be cracked, broken, or oddly shaped. Is the probability of getting a defective M&M higher for peanut M&M's than for plain M&M's?

   Gloriann suggests the probability of getting a defective plain M&M is the same as the probability of getting a defective peanut M&M. Suzanne doesn't think this is correct because a peanut M&M is bigger than a plain M&M, and, therefore, has a greater opportunity to be damaged.

   a. Simulate inspecting a plain M&M by rolling two number cubes. Let a sum of 7 or 11 represent a defective plain M&M, and the other possible rolls represent a plain M&M that is not defective. Do 50 trials and compute an estimate of the probability that a plain M&M is defective. Record the 50 outcomes you observed. Explain your process.

   b. Simulate inspecting a peanut M&M by selecting a card from a well-shuffled deck of cards. Let a one-eyed face card and clubs represent a defective peanut M&M, and the other cards represent a peanut M&M that is not defective. Be sure to replace the chosen card after each trial and to shuffle the deck well before choosing the next card. Note that the one-eyed face cards are the King of Diamonds, Jack of Hearts, and Jack of Spades. Do 20 trials and compute an estimate of the probability that a peanut M&M is defective. Record the list of 20 cards that you observed. Explain your process.

   c. For this problem, suppose that the two simulations provide accurate estimates of the probability of a defective M&M for plain and peanut M&M's. Compare your two probability estimates and decide whether Gloriann's belief that the defective probability is the same for both types of M&M's is reasonable. Explain your reasoning.

---

2. One at a time, mice are placed at the start of the maze shown below. There are four terminal stations at A, B, C, and D. At each point where a mouse has to decide in which direction to go, assume that it is equally likely for it to choose any of the possible directions. A mouse cannot go backwards.

In the following simulated trials, L stands for Left, R for Right, and S for Straight. Estimate the probability that a mouse finds station C where the food is. No food is at A, B, or D. The following data were collected on 50 simulated paths that the mice took.

LR RL RL LL LS LS RL RR RR RL

RL LR LR RR LR LR LL LS RL LR

RR LS RL RR RL LR LR LL LS RR

RL RL RL RR RR RR LR LL LL RR

RR LS RR LR RR RR LL RR LS LS

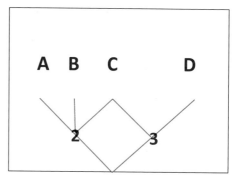

a. What paths constitute a success and what paths constitute a failure?

b. Use the data to estimate the probability that a mouse finds food. Show your calculation.

c. Paige suggests that it is equally likely that a mouse gets to any of the four terminal stations. What does your simulation suggest about whether her equally likely model is believable? If it is not believable, what do your data suggest is a more believable model?

d. Does your simulation support the following theoretical probability model? Explain.
   i.   The probability a mouse finds terminal point A is 0.167.
   ii.  The probability a mouse finds terminal point B is 0.167.
   iii. The probability a mouse finds terminal point C is 0.417.
   iv.  The probability a mouse finds terminal point D is 0.250.

# Lesson 13: Populations, Samples, and Generalizing from a Sample to a Population

## Classwork

In this lesson, you will learn about collecting data from a sample that is selected from a population. You will also learn about summary values for both a population and a sample and think about what can be learned about the population by looking at a sample from that population.

### Exercises 1–4: Collecting Data

1. Describe what you would do if you had to collect data to investigate the following statistical questions using either a sample statistic or a population characteristic. Explain your reasoning in each case.

   a. How might you collect data to answer the question, "Does the soup taste good?"

   b. How might you collect data to answer the question, "How many movies do students in your class see in a month?"

   c. How might you collect data to answer the question, "What is the median price of a home in our town?"

   d. How might you collect data to answer the question, "How many pets do people own in my neighborhood?"

   e. How might you collect data to answer the question, "What is the typical number of absences in math classes at your school on a given day?"

f.  How might you collect data to answer the question, "What is the typical lifetime of a particular brand flashlight battery?"

g.  How might you collect data to answer the question, "What percentage of girls and of boys in your school have a curfew?"

h.  How might you collect data to answer the question, "What is the most common blood type of students in my class?"

A **population** is the entire set of objects (people, animals, plants, etc.) from which data might be collected. A **sample** is a subset of the population. Numerical summary values calculated using data from an entire population are called **population characteristics**. Numerical summary values calculated using data from a sample are called **statistics**.

2.  For which of the scenarios in Exercise 1 did you describe collecting data from a population, and which from a sample?

3.  Think about collecting data in the scenarios above. Give at least two reasons you might want to collect data from a sample rather than from the entire population.

4.  Make up a result you might get in response to the situations in Exercise 1, and identify whether the result would be based on a population characteristic or a sample statistic.

    a.  Does the soup taste good?

    b.  How many movies do your classmates see in a month?

c.   What is the median price of a home in our town?

d.   How many pets do people in my neighborhood own?

e.   What is the typical number of absences in math classes at your school on a given day?"

f.   What is the typical lifetime of a particular brand of flashlight batteries?

g.   What percentage of girls and of boys in your school that have a curfew?

h.   What is the most common blood type of my classmates?

## Exercise 5:  Population or Sample?

5.   Indicate whether the following statements are summarizing data collected to answer a statistical question from a population or from a sample.  Identify references in the statement as population characteristics or sample statistics.

a.   54% of the responders to a poll at a university indicated that wealth needed to be distributed more evenly among people.

b.   Are students in the Bay Shore school district proficient on the state assessments in mathematics?  After all the tests taken by the students in the Bay Shore schools were evaluated, over 52% of those students were at or above proficient on the state assessment.

c. Does talking on mobile phones while driving distract people? Researchers measured the reaction times of 38 study participants as they talked on mobile phones and found that the average level of distraction from their driving was rated 2.25 out of 5.

d. Did most people living in New York in 2010 have at least a high school education? Based on the data collected from all New York residents in 2010 by the United States Census Bureau, 84.6% of people living in New York had at least a high school education.

e. Were there more deaths than births in the United States between July 2011 and July 2012? Data from a health service agency indicated that there were 2% more deaths than births in the U.S. during that timeframe.

f. What is the fifth best-selling book in the United States? Based on the sales of books in the United States, the fifth best-selling book was Oh, the Places You'll Go! by Dr. Seuss.

### Exercises 6–8: A Census

6. When data are collected from an entire population, it is called a census. The United States takes a census of its population every ten years, with the last one in 2010. Go to http://ri.essortment.com/unitedstatesce_rlta.htm to find the history of the U.S. census.

   a. Identify three things that you found to be interesting.

   b. Why is the census important in the United States?

7. Go to the site: www.census.gov/2010census/popmap/ipmtext.php?fl=36
   Select the state of New York.

   a. How many people were living in New York for the 2010 census?

b. Estimate the ratio of those 65 and older to those under 18 years old. Why is this important to think about?

c. Is the ratio a population characteristic or a statistic? Explain your thinking.

8. The American Community Survey (ACS) takes samples from a small percentage of the US population in years between the Censuses. (www.census.gov/acs/www/about_the_survey/american_community_survey/)

   a. What is the difference between the way the ACS collects information about the U.S. population and the way the U.S. Census Bureau collects information?

   b. In 2011, the ACS sampled workers living in New York about commuting to work each day. Why do you think these data are important for the state to know?

   c. Suppose that from a sample of 200,000 New York workers, 32,400 reported traveling more than an hour to work each day. From this information, statisticians determined that between 16% and 16.4% of the workers in the state traveled more than an hour to work every day in 2011. If there were 8,437,512 workers in the entire population, about how many traveled more than an hour to work each day?

   d. Reasoning from a sample to the population is called making an inference about a population characteristic. Identify the statistic involved in making the inference in part (c).

   e. The data about traveling time to work suggest that across the United States typically between 79.8 and 80% of commuters travel alone, 10 to 10.2% carpool, and 4.9 to 5.1% use public transportation. Survey your classmates to find out how a worker in their family gets to work. How do the results compare to the national data? What might explain any differences?

### Lesson Summary

The focus of this lesson was on collecting information either from a **population**, which is the entire set of elements in the group of interest, or a subset of the population, called a **sample**. One example of data being collected from a population is the US Census, which collects data from every person in the United States every ten years. The Census Bureau also samples the population to study things that affect the economy and living conditions of people in the U.S. in more detail. When data from a population are used to calculate a numerical summary, the value is called a **population characteristic**. When data from a sample are used to calculate a numerical summary, the value is called a **sample statistic**. Sample statistics can be used to learn about population characteristics.

## Problem Set

1.  The lunch program at Blake Middle School is being revised to align with the new nutritional standards that reduce calories and increase servings of fruit and vegetables. The administration decided to do a census of all students at Blake Middle School by giving a survey to all students about the school lunches.

    http://frac.org/federal-foodnutrition-programs/school-breakfast-program/school-meal-nutrition-standards

    a.  Name some questions that you would include in the survey. Explain why you think those questions would be important to ask.

    b.  Read through the paragraph below that describes some of the survey results. Then, identify the population characteristics and the sample statistics.

    > About $\frac{3}{4}$ of the students surveyed eat the school lunch regularly. The median number of days per month that students at Blake Middle School ate a school lunch was 18 days. 36% of students responded that their favorite fruit is bananas. The survey results for Tanya's 7$^{th}$ grade homeroom showed that the median number of days per month that her classmates ate lunch at school was 22 and only 20% liked bananas. The fiesta salad was approved by 78% of the group of students who tried it, but when it was put on the lunch menu, only 40% of the students liked it. Of the seventh graders as a whole, 73% liked spicy jicama strips, but only 2 out 5 of all the middle school students liked them.

2.  For each of the following questions: (1) describe how you would collect data to answer the question, and (2) describe whether it would result in a sample statistic or a population characteristic.

    a.  Where should the eighth grade class go for their class trip?

    b.  What is the average number of pets per family for families that live in your town?

    c.  If people tried a new diet, what percentage would have an improvement in cholesterol reading?

    d.  What is the average grade point of students who got accepted to a particular state university?

    e.  What is a typical number of home runs hit in a particular season for major league baseball players?

3. Identify a question that would lead to collecting data from the given set as a population, and one where the data could be a sample from a larger population.

   a. All students in your school

   b. Your state

4. Suppose that researchers sampled attendees of a certain movie and found that the mean age was 17-years old. Based on this observation, which of the following would be most likely.

   a. The mean ages of all of the people who went to see the movie was 17-years old.

   b. About a fourth of the people who went to see the movie were older than 51.

   c. The mean age of all people who went to see the movie would probably be in an interval around 17, say maybe between 15 and 19.

   d. The median age of those who attended the movie was 17-years old as well.

5. The headlines proclaimed: "Education Impacts Work-Life Earnings Five Times More Than Other Demographic Factors, Census Bureau Reports." According to a U.S. Census Bureau study, education levels had more effect on earnings over a 40-year span in the workforce than any other demographic factor. www.census.gov/newsroom/releases/archives/education/cb11-153.html

   a. The article stated that the estimated impact on annual earnings between a professional degree and an 8th grade education was roughly five times the impact of gender, which was $13,000. What would the difference in annual earnings be with a professional degree and with an eighth grade education?

   b. Explain whether you think the data are from a population or a sample, and identify either the population characteristic or the sample statistic.

## History of the U.S. Census

Do you realize that our census came from the Constitution of the United States of America? It was nowhere near being the very first census ever done, though. The word "census" is Latin, and it means to "tax." Archeologists have found ancient records from the Egyptians dating as far back as 3000 B.C.

In the year of 1787, the United States became the first nation to make a census mandatory in its constitution. Article One, Section Two of this historic document directs that "Representatives and direct taxes shall be apportioned among the several states ... according to their respective numbers ..." It then goes on to describe how the "numbers" or "people" of the United States would be counted and when.

Therefore, the first census was started in the year of 1790. The members of Congress gave the responsibility of visiting every house and every establishment and filling out the paperwork to the federal marshals. It took a total of eighteen months, but the tally was finally in on March of 1792. The results were given to President Washington. The first census consisted of only a few simple questions about the number of people in the household and their ages. (The census was primarily used to determine how many young men were available for wars.)

Every ten years thereafter, a census has been performed. In 1810, however, Congress decided that, while the population needed to be counted, other information needed to be gathered, also. Thus, the first census in the field of manufacturing began. Other censuses [are also conducted] on: agriculture, construction, mining, housing, local governments, commerce, transportation, and business.

Over the years, the census changed and evolved many times over. Finally, in 1950, the UNIVAC (which stood for Universal Automatic Computer) was used to tabulate the results of that year's census. It would no longer be done manually. Then, in 1960, censuses were sent via the United States Postal Service. The population was asked to complete the censuses and then wait for a visit from a census official.

Nowadays, as with the 2000 census, questionnaires were sent out to the population. The filled-out censuses are supposed to be returned via the mail to the government Census Bureau. There are two versions of the census, a short form and then a longer, more detailed form. If the census is not filled out and sent back by the required date, a census worker pays a visit to each and every household who failed to do so. The worker then verbally asks the head of the household the questions and fills in the appropriate answers on a printed form.

Being an important part of our Constitution, the census is a useful way for the Government to find out, not only the number of the population, but also other useful information.

eSsortment http://ri.essortment.com/unitedstatesce_rlta.htm

# Lesson 14:  Selecting a Sample

## Classwork

As you learned in Lesson 13, sampling is a central concept in statistics.  Examining every element in a population is usually impossible.  So research and articles in the media typically refer to a "sample" from a population.  In this lesson, you will begin to think about how to choose a sample.

### Exercises 1–2:  What is Random?

1.  Write down a sequence of heads/tails you think would typically occur if you tossed a coin 20 times.  Compare your sequence to the ones written by some of your classmates.  How are they alike?  How are they different?

2.  Working with a partner, toss a coin 20 times and write down the sequence of heads and tails you get.
    a.  Compare your results with your classmates.

    b.  How are your results from actually tossing the coin different from the sequences you and your classmates wrote down?

    c.  Toni claimed she could make up a set of numbers that would be random.  What would you say to her?

**Exercises 3–11: Length of Words in the Poem *Casey at the Bat***

3.  Suppose you wanted to learn about the lengths of the words in the poem *Casey at the Bat*. You plan to select a sample of eight words from the poem and use these words to answer the following statistical question: On average, how long is a word in the poem? What is the population of interest here?

4.  Look at the poem, *Casey at the Bat* by Ernest Thayer, and select eight words you think are representative of words in the poem. Record the number of letters in each word you selected. Find the mean number of letters in the words you chose.

5.  A random sample is a sample in which every possible sample of the same size has an equal chance of being chosen. Do you think the set of words you wrote down was random? Why or why not?

6.  Working with a partner, follow your teacher's instruction for randomly choosing eight words. Begin with the title of the poem and count a hyphenated word as one word.

    a.  Record the eight words you randomly selected and find the mean number of letters in those words.

    b.  Compare the mean of your random sample to the mean you found in Exercise 4. Explain how you found the mean for each sample.

7. As a class, compare the means from Exercise 4 and the means from Exercise 6. Your teacher will provide a chart to compare the means. Record your mean from Exercise 4 and your mean for Exercise 6 on this chart.

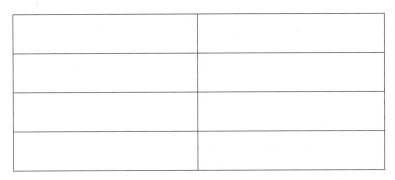

8. Do you think means from Exercise 4 or the means from Exercise 6 are more representative of the mean of all of the words in the poem? Explain your choice.

9. The actual mean of the words in the poem *Casey at the Bat* is 4.2 letters. Based on the fact that the population mean is 4.2 letters, are the means from Exercise 4 or means from Exercise 6 a better representation the mean of the population. Explain your answer.

10. How did population mean of 4.2 letters compare to the mean of your random sample from Exercise 6 and to the mean you found in Exercise 4?

11. Summarize how you would estimate the mean number of letters in the words of another poem based on what you learned in the above exercises.

## Lesson Summary

When choosing a sample, you want the sample to be representative of a population. When you try to select a sample just by yourself, you do not usually do very well, like the words you chose from the poem to find the mean number of letters. One way to help ensure that a sample is representative of the population is to take a random sample, a sample in which every element of the population has an equal chance of being selected. You can take a random sample from a population by numbering the elements in the population, putting the numbers in a bag and shaking the bag to mix the numbers. Then draw numbers out of a bag and use the elements that correspond to the numbers you draw in your sample, as you did to get a sample of the words in the poem.

## Problem Set

1.  Would any of the following provide a random sample of letters used in text of the book *Harry Potter and the Sorcerer's Stone* by J.K. Rowling? Explain your reasoning.

    a.  Use the first letter of every word of a randomly chosen paragraph

    b.  Number all of the letters in the words in a paragraph of the book, cut out the numbers, and put them in a bag. Then, choose a random set of numbers from the bag to identify which letters you will use.

    c.  Have a family member or friend write down a list of their favorite words and count the number of times each of the letters occurs.

2.  Indicate whether the following are random samples from the given population and explain why or why not.

    a.  Population: All students in school; sample includes every fifth student in the hall outside of class.

    b.  Population: Students in your class; sample consists of students that have the letter "s" in their last name.

    c.  Population: Students in your class; sample selected by putting their names in a hat and drawing the sample from the hat.

    d.  Population: People in your neighborhood; sample includes those outside in the neighborhood at 6:00 p.m.

    e.  Population: Everyone in a room; sample selected by having everyone tosses a coin and those with heads are the sample.

3.   Consider the two sample distributions of the number of letters in randomly selected words shown below:

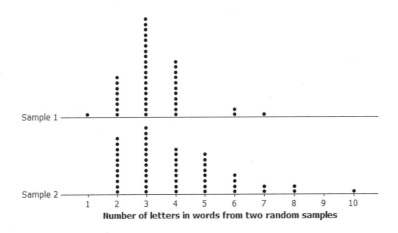

a.   Describe each distribution using statistical terms as much as possible.

b.   Do you think the two samples came from the same poem?  Why or why not?

4.   What questions about samples and populations might you want to ask if you saw the following headlines in a newspaper?

a.   "Peach Pop is the top flavor according to 8 out of 10 people."

b.   "Candidate X looks like a winner! 10 out of 12 people indicate they will vote for Candidate X."

c.   "Students Overworked.  Over half of 400 people surveyed think students spend too many hours on homework."

d.   "Action/adventure was selected as the favorite movie type by an overwhelming 75% of those surveyed."

# Lesson 15:  Random Sampling

In this lesson, you will investigate taking random samples and how random samples from the same population vary.

**Exercises 1–5:  Sampling Pennies**

1.  Do you think different random samples from the same population will be fairly similar?  Explain your reasoning.

2.  The plot below shows the number of years since being minted (the penny age) for 150 pennies that JJ had collected over the past year.  Describe the shape, center, and spread of the distribution.

### Dot Plot of Population of Penny Ages

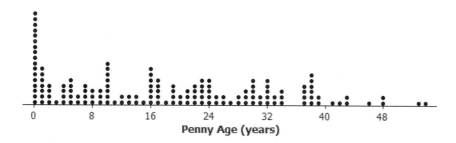

3. Place ten dots on the number line that you think might be the distribution of a sample of 10 pennies from the jar.

4. Select a random sample of 10 pennies and make a dot plot of the ages. Describe the distribution of the penny ages in your sample. How does it compare to the population distribution?

5. Compare your sample distribution to the sample distributions on the board.

   a. What do you observe?

   b. How does your sample distribution compare to those on the board?

**Exercises 6–9: Grocery Prices and Rounding**

6. Look over some of the grocery prices for this activity. Consider the following statistical question, "Do the store the store owners price the merchandise with cents that are closer to a higher dollar value or a lower dollar value?" Describe a plan that might answer that question that does not involve working with all 100 items.

7.  Do the store the store owners price the merchandise with cents that are closer to a higher dollar value or a lower dollar value?  To investigate this question in one situation, you will look at some grocery prices in weekly flyers and advertising for local grocery stores.

    a.  How would you round $3.49 and $4.99 to the nearest dollar?

    b.  If the advertised price was three for $4.35, how much would you expect to pay for one item?

    c.  Do you think more grocery prices will round up or round down?  Explain your thinking.

8.  Follow your teacher's instructions to cut out the items and their prices from the weekly flyers, and put them in a bag.  Select a random sample of 25 items without replacement, and record the items and their prices in the table below.

| | rounded |
|---|---|
| | |
| | |
| | |
| | |
| | |
| | |
| | |
| | |
| | |
| | |
| | |
| | |
| | |

Example of chart suggested:

| Student | Number of times prices were rounded to the higher value. | Number of times the prices were rounded to the lower value. | Percent of prices rounded up. |
|---------|---------|---------|---------|
| Bettina | 20 | 5 | 80% |

9.  Round each of the prices in your sample to the nearest dollar and count the number of times you rounded up and the number of times you rounded down.

    a.  Given the results of your sample, how would you answer the question:  Are grocery prices in the weekly ads at the local grocery closer to a higher dollar value or a lower dollar value?

    b.  Share your results with classmates who used the same flyer or ads.  Looking at the results of several different samples, how would you answer the question in part (a)?

    c.  Identify the population, sample and sample statistic \used to answer the statistical question.

    d.  Bettina says that over half of all the prices in the grocery store will round up.  What would you say to her?

### Lesson Summary

In this lesson, you took random samples in two different scenarios. In the first scenario, you physically reached into a jar and drew a random sample from a population of pennies. In the second scenario, you drew items from a bag and recorded the prices. In both activities, you investigated how random samples of the same size from the same population varied. Even with sample sizes of 10, the sample distributions of pennies were somewhat similar to the population (the distribution of penny ages was skewed right, and the samples all had 0 as an element). In both cases, the samples tended to have similar characteristics. For example, the samples of prices from the same store had about the same percent of prices that rounded to the higher dollar value.

### Problem Set

1.  Look at the distribution of years since the pennies were minted from Example 1. Which of the following box plots seem like they might not have come from a random sample from that distribution? Explain your thinking.

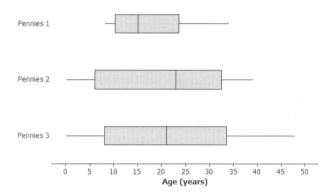

**Box Plots of Three Random Samples of Penny Ages**

2.  Given the following sample of scores on a physical fitness test, from which of the following populations might the sample have been chosen? Explain your reasoning.

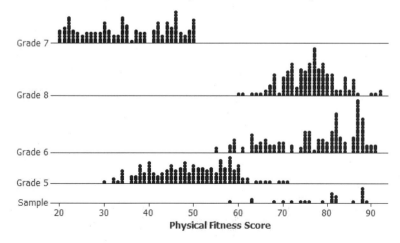

**Dot Plots of Four Populations and One Sample**

3.   Consider the distribution below:

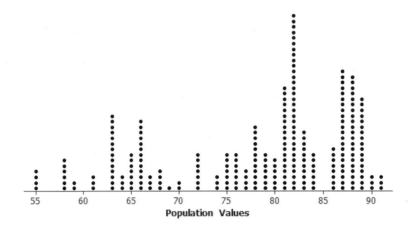

Population Values

a.   What would you expect the distribution of a random sample of size 10 from this population to look like?

b.   Random samples of different sizes that were selected from the population in part (a) are displayed below. How did your answer to part (a) compare to these samples of size 10?

**Dot Plots of Five Samples of Different Sizes**

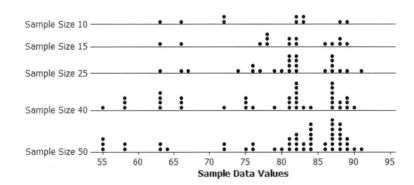

Sample Data Values

c.   Why is it reasonable to think that these samples could have come from the above population?

d.   What do you observe about the sample distributions as the sample size increases?

4.   Based on your random sample of prices from Exercise 2, answer the following questions:

a.   It looks like a lot of the prices end in 9. Do your sample results support that claim? Why or why not?

b.   What is the typical price of the items in your sample? Explain how you found the price and why you chose that method.

5. The sample distributions of prices for three different random samples of 25 items from a grocery store are shown below.

   a. How do the distributions compare?

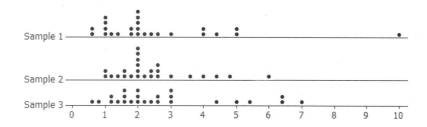

Dot Plots of Three Samples

   b. Thomas says that if he counts the items in his cart at that grocery store and multiplies by $2, he will have a pretty good estimate of how much he will have to pay. What do you think of his strategy?

# Lesson 16: Methods for Selecting a Random Sample

## Classwork

In this lesson, you will obtain random numbers to select a random sample. You will also design a plan for selecting a random sample to answer a statistical question about a population.

### Example 1: Sampling Children's Books

What is the longest book you have ever read? *The Hobbit* has 95,022 words and *The Cat in the Hat* has 830 words. Popular books vary in the number of words they have—not just the number of *different* words, but the total number of words. The table below shows the total number of words in some of those books. The histogram displays the total number of words in 150 best-selling children's books with fewer than 100,000 words.

| Book | Words | Book | Words | Book | Words |
|------|-------|------|-------|------|-------|
| Black Beauty | 59,635 | Charlie and the Chocolate Factory | 30,644 | The Hobbit | 95,022 |
| Catcher in the Rye | 73,404 | Old Yeller | 35,968 | Judy Moody was in a Mood | 11,049 |
| The Adventures of Tom Sawyer | 69,066 | Cat in the Hat | 830 | Treasure Island | 66,950 |
| The Secret Garden | 80,398 | Green Eggs and Ham | 702 | Magic Tree House Lions at Lunchtime | 5,313 |
| Mouse and the Motorcycle | 22,416 | Little Bear | 1,630 | Philosopher's Stone | 77,325 |
| The Wind in the Willows | 58,424 | Red Badge of Courage | 47,180 | Chamber of Secrets | 84,799 |
| My Father's Dragon | 7,682 | Anne Frank: The Diary of a Young Girl | 82,762 | Junie B. Jones and the Stupid Smelly Bus | 6,570 |
| Frog and Toad All Year | 1,727 | Midnight for Charlie Bone | 65,006 | White Mountains | 44,763 |
| Book of Three | 46,926 | The Lion, The Witch and the Wardrobe | 36,363 | Double Fudge | 38,860 |

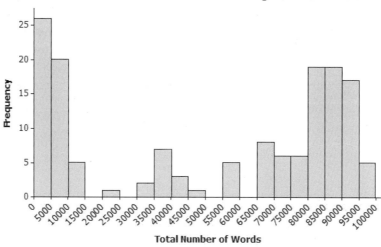

**Total Number of Words for 150 Best-Selling Children's Books**

## Exercises 1–2

1.  From the table, choose two books with which you are familiar and describe their locations in the data distribution shown in the histogram.

2.  Put dots on the number line below that you think would represent a random sample of size 10 from the number of words distribution above.

### Example 2:  Using Random Numbers to Select a Sample

The histogram indicates the differences in the number of words in the collection of 150 books.  How many words are typical for a best-selling children's book?  Answering this question would involve collecting data, and there would be variability in that data.  This makes the question a statistical question.  Think about the 150 books used to create the histogram above as a population.  How would you go about collecting data to determine the typical number of words for the books in this population?

How would you choose a random sample from the collection of 150 books discussed in this lesson?

The data for the number of words in the 150 best-selling children's books are listed below.  Select a random sample of the number of words for 10 books.

| | | | | | | | | | | |
|---|---|---|---|---|---|---|---|---|---|---|
| Books 1–10 | 59,635 | 82,762 | 92,410 | 75,340 | 8,234 | 59,705 | 92,409 | 75,338 | 8,230 | 82,768 |
| Books 11–20 | 73,404 | 65,006 | 88,250 | 2,100 | 81,450 | 72,404 | 88,252 | 2,099 | 81,451 | 65,011 |
| Books 21–30 | 69,066 | 36,363 | 75,000 | 3,000 | 80,798 | 69,165 | 75,012 | 3,010 | 80,790 | 36,361 |
| Books 31–40 | 80,398 | 95,022 | 71,200 | 3,250 | 81,450 | 80,402 | 71,198 | 3,252 | 81,455 | 95,032 |
| Books 41–50 | 22,416 | 11,049 | 81,400 | 3,100 | 83,475 | 22,476 | 81,388 | 3,101 | 83,472 | 11,047 |
| Books 51–60 | 58,424 | 66,950 | 92,400 | 2,750 | 9,000 | 58,481 | 92,405 | 2,748 | 9,002 | 66,954 |
| Books 61–70 | 7,682 | 5,313 | 83,000 | 87,000 | 89,170 | 7,675 | 83,021 | 87,008 | 89,167 | 5,311 |
| Books 71–80 | 1,727 | 77,325 | 89,010 | 862 | 88,365 | 1,702 | 89,015 | 860 | 88,368 | 77,328 |
| Books 81–90 | 46,926 | 84,799 | 88,045 | 927 | 89,790 | 46,986 | 88,042 | 926 | 89,766 | 84,796 |
| Books 91–100 | 30,644 | 6,570 | 90,000 | 8,410 | 91,010 | 30,692 | 90,009 | 8,408 | 91,015 | 6,574 |
| Books 101–110 | 35,968 | 44,763 | 89,210 | 510 | 9,247 | 35,940 | 89,213 | 512 | 9,249 | 44,766 |
| Books 111–120 | 830 | 8,700 | 92,040 | 7,891 | 83,150 | 838 | 92,037 | 7,889 | 83,149 | 8,705 |
| Books 121–130 | 702 | 92,410 | 94,505 | 38,860 | 81,110 | 712 | 94,503 | 87,797 | 81,111 | 92,412 |
| Books 131–140 | 1,630 | 88,250 | 97,000 | 7,549 | 8,245 | 1,632 | 97,002 | 7,547 | 8,243 | 88,254 |
| Books 141–150 | 47,180 | 75,000 | 89,241 | 81,234 | 8,735 | 47,192 | 89,239 | 81,238 | 8,739 | 75,010 |

## Exercises 3–6

3.  Follow your teacher's instructions to generate a set of 10 random numbers. Find the total number of words corresponding to each book identified by your random numbers.

4.  Choose two more different random samples of size 10 from the data and make a dot plot of each of the three samples.

```
    |-----|-----|-----|-----|-----|-----|-----|-----|-----|-----|
    0    2000       4000       6000       8000      10000
```

5.  If your teacher randomly chooses 10 books from your summer vacation reading list, would you be likely to get many books with a lot of words? Explain your thinking using statistical terms.

6.  If you were to compare your samples to your classmates' samples, do you think your answer to Exercise 5 would change? Why or why not?

## Exercises 7–9:  A Statistical Study of Balance and Grade

7.  Is the following question a statistical question, "Do sixth graders or seventh graders tend to have better balance?" Why or why not?

8.  Berthio's class decided to measure balance by finding out how long people can stand on one foot.

    a.  How would you rephrase the question above to create a statistical question using this definition of balance? Explain your reasoning.

    b.  What should the class think about to be consistent in how they collect the data if they actually have people stand on one foot and measure the time?

9.  Work with your class to devise a plan to select a random sample of sixth graders and a random sample of seventh graders to measure their balance using Berthio's method. Then, write a paragraph describing how you will collect data to determine whether there is a difference in how long sixth graders and seventh graders can stand on one leg. Your plan should answer the following questions:

    ▪ What is the population? How will samples be selected from the population? And, why is it important that they be random samples?

    ▪ How would you conduct the activity?

    ▪ What sample statistics will you calculate, and how you would display and analyze the data?

    ▪ What you would accept as evidence that there actually is a difference in how long sixth graders can stand on one leg compared to seventh graders.

## Lesson Summary

In this lesson, you collected data on total number of words by selecting a random sample of children's books. You also observed that several different samples of the same size had some characteristics in common with each other and with the population. In the second activity, you designed a statistical study. First, you considered a statistical question. Then, you went through the statistical process beginning with the question, and then thinking about how to choose a random sample, how students would take part, what data you would collect to answer the question, and how you would display and analyze the data.

## Problem Set

1.  The suggestions below for how to choose a random sample of students at your school were made and vetoed. Explain why you think each was vetoed.

    a.  Use every fifth person you see in the hallway before class starts.

    b.  Use all of the students taking math the same time as your class meets.

    c.  Have students who come to school early do the activity before school starts.

    d.  Have everyone in the class find two friends to be in the sample.

2.  A teacher decided to collect homework from a random sample of her students, rather than grading every paper every day.

    a.  Describe how she might choose a random sample of five students from her class of 35 students.

    b.  Suppose every day for 75 days throughout an entire semester she chooses a random sample of five students. Do you think some students will never get selected? Why or why not?

3.  Think back to earlier lessons in which you chose a random sample. Describe how you could have used a random-number generator to select a random sample in each case.

    a.  A random sample of the words in the poem "Casey at the Bat"

    b.  A random sample of the grocery prices on a weekly flyer.

4.  Sofia decided to use a different plan for selecting a random sample of books from the population of 150 top-selling children's books from Example 2. She generated ten random numbers between 1 and 100,000 to stand for the possible number of pages in any of the books. Then she found the books that had the number of pages specified in the sample. What would you say to Sofia?

5.  Find an example from a newspaper, magazine or another source that used a sample. Describe the population, the sample, the sample statistic, how you think the sample might have been chosen, and whether or not you think the sample was random.

# Lesson 17: Sampling Variability

### Example 1: Estimating a Population Mean

The owners of a gym have been keeping track of how long each person spends at the gym. Eight hundred of these times (in minutes) are shown in the population tables located at the end of the lesson. These 800 times will form the population that you will investigate in this lesson.

Look at the values in the population. Can you find the longest time spent in the gym in the population? Can you find the shortest?

On average, roughly how long do you think people spend at the gym? In other words, by just looking at the numbers in the two tables, make an estimate of the population mean.

You could find the population mean by typing all 800 numbers into a calculator or a computer, adding them up, and dividing by 800. This would be extremely time-consuming, and usually it is not possible to measure every value in a population.

Instead of doing a calculation using every value in the population, we will use a random sample to find the mean of the sample. The sample mean will then be used as an estimate of the population mean.

### Example 2: Selecting a Sample Using a Table of Random Digits

The table of random digits provided with this lesson will be used to select items from a population to produce a random sample from the population. The list of digits are determined by a computer program that simulates a random selection of the digits 0, 1, 2, 3, 4, 5, 6, 7, 8, or 9. Imagine that each of these digits is written on a slip of paper and placed in a bag. After thoroughly mixing the bag, one slip is drawn and its digit is recorded in this list of random digits. The slip is then returned to the bag and another slip is selected. The digit on this slip is recorded, and then returned to the bag. The process is repeated over and over. The resulting list of digits is called a random-number table.

How could you use a table of random digits to take a random sample?

Step 1: Place the table of random digits in front of you. Without looking at the page, place the eraser end of your pencil somewhere on the table. Start using the table of random digits at the number closest to where your eraser touched the paper. This digit and the following two specify which observation from the population tables will be the first observation in your sample.

For example, suppose the eraser end of your pencil lands on the twelfth number in row 3 of the random digit table. This number is 5 and the two following numbers are 1 and 4. This means that the first observation in your sample is observation number 514 from the population. Find observation number 514 in the population table. Do this by going to Row 51 and moving across to the column heading "4." This observation is 53, so the first observation in your sample is 53.

If the number from the random-number table is any number 800 or greater, you will ignore this number and use the next three digits in the table.

Step 2:  Continue using the table of random digits from the point you reached, and select the other four observations in your sample like you did above.

For example, continuing on from the position in the example given in Step 1:

- The next number from the random-digit table is 716, and observation 716 is 63
- The next number from the random-digit table is 565, and observation 565 is 31.
- The next number from the random-digit table is 911, and there is no observation 911.  So we ignore these three digits.
- The next number from the random-digit table is 928, and there is no observation 928.  So we ignore these three digits.
- The next number from the random-digit table is 303, and observation 303 is 70.
- The next number from the random-digit table is 677, and observation 677 is 42.

## Exercises 1–4

Initially, you will select just five values from the population to form your sample.  This is a very small sample size, but it is a good place to start to understand the ideas of this lesson.

The values found in Example 2 are used to illustrate the answers to the following questions, but each student should select their own random sample of five.

1.  Use the table of random number to select five values from the population of times.  What are the five observations in your sample?

2.  For the sample that you selected, calculate the sample mean.

3.  You selected a random sample and calculated the sample mean in order to estimate the population mean.  Do you think that the mean of these five observations is exactly correct for the population mean?  Could the population mean be greater than the number you calculated?  Could the population mean be less than the number you calculated?

4.  In practice, you only take one sample in order to estimate a population characteristic. But, for the purposes of this lesson, suppose you were to take another random sample from the same population of times at the gym. Could the new sample mean be closer to the population mean than the mean of these five observations? Could it be further from the population mean?

**Exercises 5–7**

As a class, you will now investigate sampling variability by taking several samples from the same population. Each sample will have a different sample mean. This variation provides an example of sampling variability.

5.  Place the table of random digits in front of you and, without looking at the page, place the eraser end of your pencil somewhere on the table of random numbers. Start using the table of random digits at the number closest to where your eraser touches the paper. This digit and the following two specify which observation from the population tables will be the first observation in your sample. Write this three-digit number and the corresponding data value from the population in the space below.

6.  Continue moving to the right in the table of random digits from the place ended in Exercise 5. Use three digits at a time. Each set of three digits specifies which observation in the population is the next number in your sample. Continue until you have four more observations, and write these four values in the space below.

7.  Calculate the mean of the five values that form your sample. Round your answer to the nearest tenth. Show your work and your sample mean in the space below.

**Exercises 8–11**

You will now use the sample means from Exercise 7 from the entire class to make a dot plot.

8.  Write the sample means for everyone in the class in the space below.

9.  Use all the sample means to make a dot plot using the axis given below.  (Remember, if you have repeated or close values, stack the dots one above the other.)

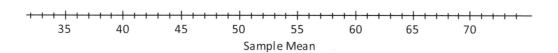

10. What do you see in the dot plot that demonstrates sampling variability?

11. Remember that in practice you only take one sample.  (In this lesson, many samples were taken in order to demonstrate the concept of sampling variability.)  Suppose that a statistician plans to take a random sample of size 5 from the population of times spent at the gym and that he or she will use the sample mean as an estimate of the population mean.  Approximately how far can the statistician expect the sample mean to be from the population mean?

**Population**

|     | 0  | 1  | 2  | 3  | 4  | 5  | 6  | 7  | 8  | 9  |
|-----|----|----|----|----|----|----|----|----|----|----|
| 00  | 45 | 58 | 49 | 78 | 59 | 36 | 52 | 39 | 70 | 51 |
| 01  | 50 | 45 | 45 | 66 | 71 | 55 | 65 | 33 | 60 | 51 |
| 02  | 53 | 83 | 40 | 51 | 83 | 57 | 75 | 38 | 43 | 77 |
| 03  | 49 | 49 | 81 | 57 | 42 | 36 | 22 | 66 | 68 | 52 |
| 04  | 60 | 67 | 43 | 60 | 55 | 63 | 56 | 44 | 50 | 58 |
| 05  | 64 | 41 | 67 | 73 | 55 | 69 | 63 | 46 | 50 | 65 |
| 06  | 54 | 58 | 53 | 55 | 51 | 74 | 53 | 55 | 64 | 16 |
| 07  | 28 | 48 | 62 | 24 | 82 | 51 | 64 | 45 | 41 | 47 |
| 08  | 70 | 50 | 38 | 16 | 39 | 83 | 62 | 50 | 37 | 58 |
| 09  | 79 | 62 | 45 | 48 | 42 | 51 | 67 | 68 | 56 | 78 |
| 10  | 61 | 56 | 71 | 55 | 57 | 77 | 48 | 65 | 61 | 62 |
| 11  | 65 | 40 | 56 | 47 | 44 | 51 | 38 | 68 | 64 | 40 |
| 12  | 53 | 22 | 73 | 62 | 82 | 78 | 84 | 50 | 43 | 43 |
| 13  | 81 | 42 | 72 | 49 | 55 | 65 | 41 | 92 | 50 | 60 |
| 14  | 56 | 44 | 40 | 70 | 52 | 47 | 30 | 9  | 58 | 53 |
| 15  | 84 | 64 | 64 | 34 | 37 | 69 | 57 | 75 | 62 | 67 |
| 16  | 45 | 58 | 49 | 78 | 59 | 36 | 52 | 39 | 70 | 51 |
| 17  | 50 | 45 | 45 | 66 | 71 | 55 | 65 | 33 | 60 | 51 |
| 18  | 53 | 83 | 40 | 51 | 83 | 57 | 75 | 38 | 43 | 77 |
| 19  | 49 | 49 | 81 | 57 | 42 | 36 | 22 | 66 | 68 | 52 |
| 20  | 60 | 67 | 43 | 60 | 55 | 63 | 56 | 44 | 50 | 58 |
| 21  | 64 | 41 | 67 | 73 | 55 | 69 | 63 | 46 | 50 | 65 |
| 22  | 54 | 58 | 53 | 55 | 51 | 74 | 53 | 55 | 64 | 16 |
| 23  | 20 | 48 | 62 | 24 | 82 | 51 | 64 | 45 | 41 | 47 |
| 24  | 70 | 50 | 38 | 16 | 39 | 83 | 62 | 50 | 37 | 58 |
| 25  | 79 | 62 | 45 | 48 | 42 | 51 | 67 | 68 | 56 | 78 |
| 26  | 61 | 56 | 71 | 55 | 57 | 77 | 48 | 65 | 61 | 62 |
| 27  | 65 | 40 | 56 | 47 | 44 | 51 | 38 | 68 | 64 | 40 |
| 28  | 53 | 22 | 73 | 62 | 82 | 78 | 84 | 50 | 43 | 43 |
| 29  | 81 | 42 | 72 | 49 | 55 | 65 | 41 | 92 | 50 | 60 |
| 30  | 56 | 44 | 40 | 70 | 52 | 47 | 30 | 9  | 58 | 53 |
| 31  | 84 | 64 | 64 | 34 | 37 | 69 | 57 | 75 | 62 | 67 |
| 32  | 45 | 58 | 49 | 78 | 59 | 36 | 52 | 39 | 70 | 51 |
| 33  | 50 | 45 | 45 | 66 | 71 | 55 | 65 | 33 | 60 | 51 |
| 34  | 53 | 83 | 40 | 51 | 83 | 57 | 75 | 38 | 43 | 77 |
| 35  | 49 | 49 | 81 | 57 | 42 | 36 | 22 | 66 | 68 | 52 |
| 36  | 60 | 67 | 43 | 60 | 55 | 63 | 56 | 44 | 50 | 58 |
| 37  | 64 | 41 | 67 | 73 | 55 | 69 | 63 | 46 | 50 | 65 |
| 38  | 54 | 58 | 53 | 55 | 51 | 74 | 53 | 55 | 64 | 16 |
| 39  | 28 | 48 | 62 | 24 | 82 | 51 | 64 | 45 | 41 | 47 |

**Population, continued**

|    | 0 | 1 | 2 | 3 | 4 | 5 | 6 | 7 | 8 | 9 |
|----|----|----|----|----|----|----|----|----|----|----|
| **40** | 53 | 70 | 59 | 62 | 33 | 31 | 74 | 44 | 46 | 68 |
| **41** | 37 | 51 | 84 | 47 | 46 | 33 | 53 | 54 | 70 | 74 |
| **42** | 35 | 45 | 48 | 45 | 56 | 60 | 66 | 60 | 65 | 57 |
| **43** | 42 | 81 | 67 | 64 | 60 | 79 | 46 | 48 | 67 | 56 |
| **44** | 41 | 21 | 41 | 58 | 48 | 38 | 50 | 53 | 73 | 38 |
| **45** | 35 | 28 | 43 | 43 | 55 | 39 | 75 | 45 | 68 | 36 |
| **46** | 64 | 31 | 31 | 40 | 84 | 79 | 47 | 63 | 48 | 46 |
| **47** | 34 | 36 | 54 | 61 | 33 | 16 | 50 | 60 | 52 | 55 |
| **48** | 53 | 52 | 48 | 47 | 77 | 37 | 66 | 51 | 61 | 64 |
| **49** | 40 | 44 | 45 | 22 | 36 | 64 | 50 | 49 | 64 | 39 |
| **50** | 45 | 69 | 67 | 33 | 55 | 61 | 62 | 38 | 51 | 43 |
| **51** | 55 | 39 | 46 | 56 | 53 | 50 | 44 | 42 | 40 | 60 |
| **52** | 11 | 36 | 56 | 69 | 72 | 73 | 71 | 48 | 58 | 52 |
| **53** | 81 | 47 | 36 | 54 | 81 | 59 | 50 | 42 | 80 | 69 |
| **54** | 40 | 43 | 30 | 54 | 61 | 13 | 73 | 65 | 52 | 40 |
| **55** | 71 | 78 | 71 | 61 | 54 | 79 | 63 | 47 | 49 | 73 |
| **56** | 53 | 70 | 59 | 62 | 33 | 31 | 74 | 44 | 46 | 68 |
| **57** | 37 | 51 | 84 | 47 | 46 | 33 | 53 | 54 | 70 | 74 |
| **58** | 35 | 45 | 48 | 45 | 56 | 60 | 66 | 60 | 65 | 57 |
| **59** | 42 | 81 | 67 | 64 | 60 | 79 | 46 | 48 | 67 | 56 |
| **60** | 41 | 21 | 41 | 58 | 48 | 38 | 50 | 53 | 73 | 38 |
| **61** | 35 | 28 | 43 | 43 | 55 | 39 | 75 | 45 | 68 | 36 |
| **62** | 64 | 31 | 31 | 40 | 84 | 79 | 47 | 63 | 48 | 46 |
| **63** | 34 | 36 | 54 | 61 | 33 | 16 | 50 | 60 | 52 | 55 |
| **64** | 53 | 52 | 48 | 47 | 77 | 37 | 66 | 51 | 61 | 64 |
| **65** | 40 | 44 | 45 | 22 | 36 | 64 | 50 | 49 | 64 | 39 |
| **66** | 45 | 69 | 67 | 33 | 55 | 61 | 62 | 38 | 51 | 43 |
| **67** | 55 | 39 | 46 | 56 | 53 | 50 | 44 | 42 | 40 | 60 |
| **68** | 11 | 36 | 56 | 69 | 72 | 73 | 71 | 48 | 58 | 52 |
| **69** | 81 | 47 | 36 | 54 | 81 | 59 | 50 | 42 | 80 | 69 |
| **70** | 40 | 43 | 30 | 54 | 61 | 13 | 73 | 65 | 52 | 40 |
| **71** | 71 | 78 | 71 | 61 | 54 | 79 | 63 | 47 | 49 | 73 |
| **72** | 53 | 70 | 59 | 62 | 33 | 31 | 74 | 44 | 46 | 68 |
| **73** | 37 | 51 | 84 | 47 | 46 | 33 | 53 | 54 | 70 | 74 |
| **74** | 35 | 45 | 48 | 45 | 56 | 60 | 66 | 60 | 65 | 57 |
| **75** | 42 | 81 | 67 | 64 | 60 | 79 | 46 | 48 | 67 | 56 |
| **76** | 41 | 21 | 41 | 58 | 48 | 38 | 50 | 53 | 73 | 38 |
| **77** | 35 | 28 | 43 | 43 | 55 | 39 | 75 | 45 | 68 | 36 |
| **78** | 64 | 31 | 31 | 40 | 84 | 79 | 47 | 63 | 48 | 46 |
| **79** | 34 | 36 | 54 | 61 | 33 | 16 | 50 | 60 | 52 | 55 |

## Table of Random Digits

| Row | | | | | | | | | | | | | | | | | | | | |
|---|---|---|---|---|---|---|---|---|---|---|---|---|---|---|---|---|---|---|---|---|
| 1 | 6 | 6 | 7 | 2 | 8 | 0 | 0 | 8 | 4 | 0 | 0 | 4 | 6 | 0 | 3 | 2 | 2 | 4 | 6 | 8 |
| 2 | 8 | 0 | 3 | 1 | 1 | 1 | 1 | 2 | 7 | 0 | 1 | 9 | 1 | 2 | 7 | 1 | 3 | 3 | 5 | 3 |
| 3 | 5 | 3 | 5 | 7 | 3 | 6 | 3 | 1 | 7 | 2 | 5 | 5 | 1 | 4 | 7 | 1 | 6 | 5 | 6 | 5 |
| 4 | 9 | 1 | 1 | 9 | 2 | 8 | 3 | 0 | 3 | 6 | 7 | 7 | 4 | 7 | 5 | 9 | 8 | 1 | 8 | 3 |
| 5 | 9 | 0 | 2 | 9 | 9 | 7 | 4 | 6 | 3 | 6 | 6 | 3 | 7 | 4 | 2 | 7 | 0 | 0 | 1 | 9 |
| 6 | 8 | 1 | 4 | 6 | 4 | 6 | 8 | 2 | 8 | 9 | 5 | 5 | 2 | 9 | 6 | 2 | 5 | 3 | 0 | 3 |
| 7 | 4 | 1 | 1 | 9 | 7 | 0 | 7 | 2 | 9 | 0 | 9 | 7 | 0 | 4 | 6 | 2 | 3 | 1 | 0 | 9 |
| 8 | 9 | 9 | 2 | 7 | 1 | 3 | 2 | 9 | 0 | 3 | 9 | 0 | 7 | 5 | 6 | 7 | 1 | 7 | 8 | 7 |
| 9 | 3 | 4 | 2 | 2 | 9 | 1 | 9 | 0 | 7 | 8 | 1 | 6 | 2 | 5 | 3 | 9 | 0 | 9 | 1 | 0 |
| 10 | 2 | 7 | 3 | 9 | 5 | 9 | 9 | 3 | 2 | 9 | 3 | 9 | 1 | 9 | 0 | 5 | 5 | 1 | 4 | 2 |
| 11 | 0 | 2 | 5 | 4 | 0 | 8 | 1 | 7 | 0 | 7 | 1 | 3 | 0 | 4 | 3 | 0 | 6 | 4 | 4 | 4 |
| 12 | 8 | 6 | 0 | 5 | 4 | 8 | 8 | 2 | 7 | 7 | 0 | 1 | 0 | 1 | 7 | 1 | 3 | 5 | 3 | 4 |
| 13 | 4 | 2 | 6 | 4 | 5 | 2 | 4 | 2 | 6 | 1 | 7 | 5 | 6 | 6 | 4 | 0 | 8 | 4 | 1 | 2 |
| 14 | 4 | 4 | 9 | 8 | 7 | 3 | 4 | 3 | 8 | 2 | 9 | 1 | 5 | 3 | 5 | 9 | 8 | 9 | 2 | 9 |
| 15 | 6 | 4 | 8 | 0 | 0 | 0 | 4 | 2 | 3 | 8 | 1 | 8 | 4 | 0 | 9 | 5 | 0 | 9 | 0 | 4 |
| 16 | 3 | 2 | 3 | 8 | 4 | 8 | 8 | 6 | 2 | 9 | 1 | 0 | 1 | 9 | 9 | 3 | 0 | 7 | 3 | 5 |
| 17 | 6 | 6 | 7 | 2 | 8 | 0 | 0 | 8 | 4 | 0 | 0 | 4 | 6 | 0 | 3 | 2 | 2 | 4 | 6 | 8 |
| 18 | 8 | 0 | 3 | 1 | 1 | 1 | 1 | 2 | 7 | 0 | 1 | 9 | 1 | 2 | 7 | 1 | 3 | 3 | 5 | 3 |
| 19 | 5 | 3 | 5 | 7 | 3 | 6 | 3 | 1 | 7 | 2 | 5 | 5 | 1 | 4 | 7 | 1 | 6 | 5 | 6 | 5 |
| 20 | 9 | 1 | 1 | 9 | 2 | 8 | 3 | 0 | 3 | 6 | 7 | 7 | 4 | 7 | 5 | 9 | 8 | 1 | 8 | 3 |
| 21 | 9 | 0 | 2 | 9 | 9 | 7 | 4 | 6 | 3 | 6 | 6 | 3 | 7 | 4 | 2 | 7 | 0 | 0 | 1 | 9 |
| 22 | 8 | 1 | 4 | 6 | 4 | 6 | 8 | 2 | 8 | 9 | 5 | 5 | 2 | 9 | 6 | 2 | 5 | 3 | 0 | 3 |
| 23 | 4 | 1 | 1 | 9 | 7 | 0 | 7 | 2 | 9 | 0 | 9 | 7 | 0 | 4 | 6 | 2 | 3 | 1 | 0 | 9 |
| 24 | 9 | 9 | 2 | 7 | 1 | 3 | 2 | 9 | 0 | 3 | 9 | 0 | 7 | 5 | 6 | 7 | 1 | 7 | 8 | 7 |
| 25 | 3 | 4 | 2 | 2 | 9 | 1 | 9 | 0 | 7 | 8 | 1 | 6 | 2 | 5 | 3 | 9 | 0 | 9 | 1 | 0 |
| 26 | 2 | 7 | 3 | 9 | 5 | 9 | 9 | 3 | 2 | 9 | 3 | 9 | 1 | 9 | 0 | 5 | 5 | 1 | 4 | 2 |
| 27 | 0 | 2 | 5 | 4 | 0 | 8 | 1 | 7 | 0 | 7 | 1 | 3 | 0 | 4 | 3 | 0 | 6 | 4 | 4 | 4 |
| 28 | 8 | 6 | 0 | 5 | 4 | 8 | 8 | 2 | 7 | 7 | 0 | 1 | 0 | 1 | 7 | 1 | 3 | 5 | 3 | 4 |
| 29 | 4 | 2 | 6 | 4 | 5 | 2 | 4 | 2 | 6 | 1 | 7 | 5 | 6 | 6 | 4 | 0 | 8 | 4 | 1 | 2 |
| 30 | 4 | 4 | 9 | 8 | 7 | 3 | 4 | 3 | 8 | 2 | 9 | 1 | 5 | 3 | 5 | 9 | 8 | 9 | 2 | 9 |
| 31 | 6 | 4 | 8 | 0 | 0 | 0 | 4 | 2 | 3 | 8 | 1 | 8 | 4 | 0 | 9 | 5 | 0 | 9 | 0 | 4 |
| 32 | 3 | 2 | 3 | 8 | 4 | 8 | 8 | 6 | 2 | 9 | 1 | 0 | 1 | 9 | 9 | 3 | 0 | 7 | 3 | 5 |
| 33 | 6 | 6 | 7 | 2 | 8 | 0 | 0 | 8 | 4 | 0 | 0 | 4 | 6 | 0 | 3 | 2 | 2 | 4 | 6 | 8 |
| 34 | 8 | 0 | 3 | 1 | 1 | 1 | 1 | 2 | 7 | 0 | 1 | 9 | 1 | 2 | 7 | 1 | 3 | 3 | 5 | 3 |
| 35 | 5 | 3 | 5 | 7 | 3 | 6 | 3 | 1 | 7 | 2 | 5 | 5 | 1 | 4 | 7 | 1 | 6 | 5 | 6 | 5 |
| 36 | 9 | 1 | 1 | 9 | 2 | 8 | 3 | 0 | 3 | 6 | 7 | 7 | 4 | 7 | 5 | 9 | 8 | 1 | 8 | 3 |
| 37 | 9 | 0 | 2 | 9 | 9 | 7 | 4 | 6 | 3 | 6 | 6 | 3 | 7 | 4 | 2 | 7 | 0 | 0 | 1 | 9 |
| 38 | 8 | 1 | 4 | 6 | 4 | 6 | 8 | 2 | 8 | 9 | 5 | 5 | 2 | 9 | 6 | 2 | 5 | 3 | 0 | 3 |
| 39 | 4 | 1 | 1 | 9 | 7 | 0 | 7 | 2 | 9 | 0 | 9 | 7 | 0 | 4 | 6 | 2 | 3 | 1 | 0 | 9 |
| 40 | 9 | 9 | 2 | 7 | 1 | 3 | 2 | 9 | 0 | 3 | 9 | 0 | 7 | 5 | 6 | 7 | 1 | 7 | 8 | 7 |

**Lesson Summary**

A population characteristic is estimated by taking a random sample from the population and calculating the value of a statistic for the sample. For example, a population mean is estimated by selecting a random sample from the population and calculating the sample mean.

The value of the sample statistic (e.g., the sample mean) will vary based on the random sample that is selected. This variation from sample to sample in the values of the sample statistic is called **sampling variability**.

## Problem Set

1. Yousef intends to buy a car. He wishes to estimate the mean fuel efficiency (in miles per gallon) of all cars available at this time. Yousef selects a random sample of 10 cars and looks up their fuel efficiencies on the Internet. The results are shown below.

   22  25  29  23  31  29  28  22  23  27

   a. Yousef will estimate the mean fuel efficiency of all cars by calculating the mean for his sample. Calculate the sample mean, and record your answer below. (Be sure to show your work.)

   b. In practice, you only take one sample to estimate a population characteristic. However, if Yousef were to take another random sample of 10 cars from the same population, would he likely get the same value for the sample mean?

   c. What if Yousef were to take *many* random samples of 10 cars? Would the entire sample means be the same?

   d. Using this example, explain what sampling variability means.

2. Think about the mean number of siblings (brothers and sisters) for all students at your school.

   a. Approximately what do you think is the value of the mean number of siblings for the population of all students at your school?

   b. How could you find a better estimate of this population mean?

   c. Suppose that you have now selected a random sample of students from your school. You have asked all of the students in your sample how many siblings they have. How will you calculate the sample mean?

   d. If you had taken a different sample, would the sample mean have taken the same value?

   e. There are many different samples of students that you could have selected. These samples produce many different possible sample means. What is the phrase used for this concept?

   f. Does the phrase you gave in part (e) apply only to sample means?

# Lesson 18: Sampling Variability and the Effect of Sample Size

### Example 1: Sampling Variability

The previous lesson investigated the statistical question, "What is the typical time spent at the gym?" by selecting random samples from the population of 800 gym members. Two different dot plots of sample means calculated from random samples from the population are displayed below. The first dot plot represents the means of 20 samples with each sample having 5 data points. The second dot plot represents the means of 20 samples with each sample having 15 data points.

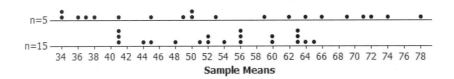

Based on the first dot plot, Jill answered the statistical question by indicating the mean time people spent at the gym was between 34 and 78 minutes. She decided that a time approximately in the middle of that interval would be her estimate of the mean time the 800 people spent at the gym. She estimated 52 minutes. Scott answered the question using the second dot plot. He indicated that the mean time people spent at the gym was between 41 and 65 minutes. He also selected a time of 52 minutes to answer the question.

- Describe the differences in the two dot plots.

- Which dot plot do you feel more confident in using to answer the statistical question? Explain your answer.

- In general, do you want sampling variability to be large or small? Explain.

**Exercises 1–3**

In the previous lesson, you saw a population of 800 times spent at the gym. You will now select a random sample of size 15 from that population. You will then calculate the sample mean.

1.  Start by selecting a three-digit number from the table of random digits. Place the random-digit table in front of you. Without looking at the page, place the eraser end of your pencil somewhere on the table of random digits. Start using the table of random digits at the digit closest to your eraser. This digit and the following two specify which observation from the population will be the first observation in your sample. Write the value of this observation in the space below. (Discard any three-digit number that is 800 or larger, and use the next three digits from the random-digit table.)

2.  Continue moving to the right in the table of random digits from the point that you reached in part (a). Each three-digit number specifies a value to be selected from the population. Continue in this way until you have selected 14 more values from the population. This will make 15 values altogether. Write the values of all 15 observations in the space below.

3.  Calculate the mean of your 15 sample values. Write the value of your sample mean below. Round your answer to the nearest tenth. (Be sure to show your work.)

**Exercises 4–6**

You will now use the sample means from Exercise 3 for the entire class to make a dot plot.

4.  Write the sample means for everyone in the class in the space below.

5.  Use all the sample means to make a dot plot using the axis given below.  (Remember, if you have repeated values or values close to each other, stack the dots one above the other.)

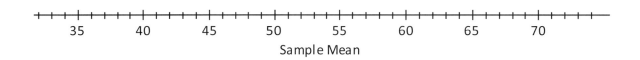

Sample Mean

6.  In the previous lesson, you drew a dot plot of sample means for samples of size 5.  How does the dot plot above (of sample means for samples of size 15) compare to the dot plot of sample means for samples of size 5?  For which sample size (5 or 15) does the sample mean have the greater sampling variability?

This exercise illustrates the notion that the greater the sample size, the smaller the sampling variability of the sample mean.

**Exercises 7–8**

7.  Remember that in practice you only take one sample.  Suppose that a statistician plans to take a random sample of size 15 from the population of times spent at the gym and will use the sample mean as an estimate of the population mean.  Based on the dot plot of sample means that your class collected from the population, approximately how far can the statistician expect the sample mean to be from the population mean? (The actual population mean is 53.9 minutes.)

8.  How would your answer in Exercise 7 compare to the equivalent mean of the distances for a sample of size 5?

**Exercises 9–11**

Suppose everyone in your class selected a random sample of size 25 from the population of times spent at the gym.

9.  What do you think the dot plot of the class's sample means would look like?  Make a sketch using the axis below.

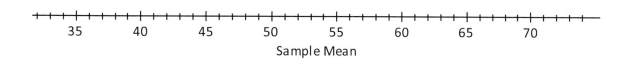

Sample Mean

10. Suppose that a statistician plans to estimate the population mean using a sample of size 25.  According to your sketch, approximately how far can the statistician expect the sample mean to be from the population mean?

11. Suppose you have a choice of using a sample of size 5, 15, or 25.  Which of the three makes the sampling variability of the sample mean the smallest?  Why would you choose the sample size that makes the sampling variability of the sample mean as small as possible?

Lesson Summary

Suppose that you plan to take a random sample from a population. You will use the value of a sample statistic to estimate the value of a population characteristic. You want the sampling variability of the sample statistic to be small so that the sample statistic is likely to be close to the value of the population characteristic.

When you increase the sample size, the sampling variability of the sample mean is decreased.

## Problem Set

1. The owner of a new coffee shop is keeping track of how much each customer spends (in dollars). One hundred of these amounts are shown in the table below. These amounts will form the *population* for this question.

|   | 0 | 1 | 2 | 3 | 4 | 5 | 6 | 7 | 8 | 9 |
|---|---|---|---|---|---|---|---|---|---|---|
| 0 | 6.18 | 4.67 | 4.01 | 4.06 | 3.28 | 4.47 | 4.86 | 4.91 | 3.96 | 6.18 |
| 1 | 4.98 | 5.42 | 5.65 | 2.97 | 2.92 | 7.09 | 2.78 | 4.20 | 5.02 | 4.98 |
| 2 | 3.12 | 1.89 | 4.19 | 5.12 | 4.38 | 5.34 | 4.22 | 4.27 | 5.25 | 3.12 |
| 3 | 3.90 | 4.47 | 4.07 | 4.80 | 6.28 | 5.79 | 6.07 | 7.64 | 6.33 | 3.90 |
| 4 | 5.55 | 4.99 | 3.77 | 3.63 | 5.21 | 3.85 | 7.43 | 4.72 | 6.53 | 5.55 |
| 5 | 4.55 | 5.38 | 5.83 | 4.10 | 4.42 | 5.63 | 5.57 | 5.32 | 5.32 | 4.55 |
| 6 | 4.56 | 7.67 | 6.39 | 4.05 | 4.51 | 5.16 | 5.29 | 6.34 | 3.68 | 4.56 |
| 7 | 5.86 | 4.75 | 4.94 | 3.92 | 4.84 | 4.95 | 4.50 | 4.56 | 7.05 | 5.86 |
| 8 | 5.00 | 5.47 | 5.00 | 5.70 | 5.71 | 6.19 | 4.41 | 4.29 | 4.34 | 5.00 |
| 9 | 5.12 | 5.58 | 6.16 | 6.39 | 5.93 | 3.72 | 5.92 | 4.82 | 6.19 | 5.12 |

a. Place the table of random digits in front of you. Select a starting point without looking at the page. Then, taking two digits at a time, select a random sample of size 5 from the population above. Write the 5 values in the space below. (For example, suppose you start at the third digit of row four of the random digit table. Taking two digits gives you 19. In the population above, go to the row labeled "1" and move across to the column labeled 9. This observation is 4.98, and that will be the first observation in your sample. Then continue in the random-digit table from the point you reached.)

Calculate the mean for your sample, showing your work. Round your answer to the nearest thousandth.

b. Using the same approach as in part (a), select a random sample of size 20 from the population.

Calculate the mean for your sample of size 20. Round your answer to the nearest thousandth.

c. Which of your sample means is likely to be the better estimate of the population mean? Explain your answer in terms of sampling variability.

2.  Two dot plots are shown below.  One of the dot plots shows the values of some sample means from random samples of size 5 from the population given in Problem 1.  The other dot plot shows the values of some sample means from random samples of size 20 from the population given in Problem 1.

**Dot Plot A**

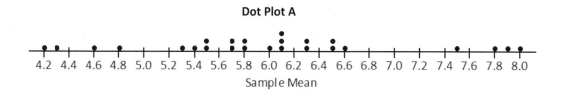

Sample Mean

**Dot Plot B**

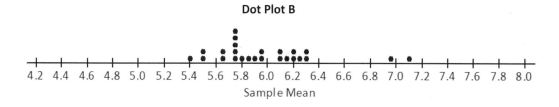

Sample Mean

Which dot plot is for sample means from samples of size 5, and which dot plot is for sample means from samples of size 20?  Explain your reasoning.

The sample means from samples of size 5 are shown in Dot Plot _____.

The sample means from samples of size 20 are shown in Dot Plot _____.

3.  You are going to use a random sample to estimate the mean travel time for getting to school for all the students in your grade. You will select a random sample of students from your grade.  Explain why you would like the sampling variability of the sample mean to be *small*.

# Lesson 19: Understanding Variability in Estimates

## Classwork

In a previous lesson, you selected several random samples from a population. You recorded values of a numerical variable. You then calculated the mean for each sample, saw that there was variability in the sample means, and created a distribution of sample means to better see the sampling variability. You then considered larger samples and saw that the variability in the distribution decreased when the sample size increases. In this lesson, you will use a similar process to investigate variability in sample proportions.

### Example 1: Sample Proportion

Your teacher will give your group a bag that contains colored cubes, some of which are red. With your classmates, you are going to build a distribution of sample proportions.

### Exercises 1–6

1. Each person in your group should randomly select a sample of 10 cubes from the bag. Record the data for your sample in the table below.

| Cube | Outcome (Color) |
|------|-----------------|
| 1 | |
| 2 | |
| 3 | |
| 4 | |
| 5 | |
| 6 | |
| 7 | |
| 8 | |
| 9 | |
| 10 | |

2. What is the proportion of red cubes in your sample of 10?

   This value is called the sample proportion. The sample proportion is found by dividing the number of "successes" (in this example, the number of red cubes) by the total number of observations in the sample.

3.  Write your sample proportion on a post-it note and place it on the number line that your teacher has drawn on the board.  Place your note above the value on the number line that corresponds to your sample proportion.

    The graph of all the students' sample proportions is called a sampling distribution of the sample proportions.

4.  Describe the shape of the distribution.

5.  Describe the variability in the sample proportions.

6.  Based on the distribution, answer the following:

    a.  What do you think is the population proportion?

    b.  How confident are you of your estimate?

## Example 2:  Sampling Variability

What do you think would happen to the sampling distribution if everyone in class took a random sample of 30 cubes from the bag?  To help answer this question you will repeat the random sampling you did in Exercise 1, except now you will draw a random sample of 30 cubes instead of 10.

## Exercises 7–15

What do you think would happen to the sampling distribution if everyone in class took a random sample of 30 cubes from the bag?  To help answer this question you will repeat the random sampling you did in Exercise 1, except now you will draw a random sample of 30 cubes instead of 10.

7.  Take a random sample of 30 cubes from the bag.  Carefully record the outcome of each draw.

| Cube | Outcome (Color) | Cube | Outcome (Color) |
|------|-----------------|------|-----------------|
| 1    |                 | 16   |                 |
| 2    |                 | 17   |                 |
| 3    |                 | 18   |                 |
| 4    |                 | 19   |                 |
| 5    |                 | 20   |                 |
| 6    |                 | 21   |                 |
| 7    |                 | 22   |                 |
| 8    |                 | 23   |                 |
| 9    |                 | 24   |                 |
| 10   |                 | 25   |                 |
| 11   |                 | 26   |                 |
| 12   |                 | 27   |                 |
| 13   |                 | 28   |                 |
| 14   |                 | 29   |                 |
| 15   |                 | 30   |                 |

8.  What is the proportion of red cubes in your sample of 30?

9.  Write your sample proportion on a post-it note and place the note on the number line that your teacher has drawn on the board.  Place your note above the value on the number line that corresponds to your sample proportion.

10. Describe the shape of the distribution.

11. Describe the variability in the sample proportions.

12. Based on the distribution, answer the following:

    a. What do you think is the population proportion?

    b. How confident are you of your estimate?

    c. If you were taking a random sample of 30 cubes and determined the proportion that was red, do you think your sample proportion will be within 0.05 of the population proportion? Explain.

13. Compare the sampling distribution based on samples of size 10 to the sampling distribution based on samples of size 30.

14. As the sample size increased from 10 to 30 describe what happened to the sampling variability of the sample proportions.

15. What do you think would happen to the variability of the sample proportions if the sample size for each sample was 50 instead of 30? Explain.

## Problem Set

1.  A class of seventh graders wanted to find the proportion of M&M's that are red.  Each seventh grader took a random sample of 20 M&M's from a very large container of M&M's.  Following is the proportion of red M&M's each student found.

    | 0.15 | 0    | 0.1 | 0.1 | 0.05 | 0.1  | 0.2  | 0.05 | 0.1 |
    |------|------|-----|-----|------|------|------|------|-----|
    | 0.1  | 0.15 | 0.2 | 0   | 0.1  | 0.15 | 0.15 | 0.1  | 0.2 |
    | 0.3  | 0.1  | 0.1 | 0.2 | 0.1  | 0.15 | 0.1  | 0.05 | 0.3 |

    a.  Construct a dot plot of the sample proportions.

    b.  Describe the shape of the distribution.

    c.  Describe the variability of the distribution.

    d.  Suppose the seventh grade students had taken random samples of size 50.  Describe how the sampling distribution would change from the one you constructed in part (a).

2.  A group of seventh graders wanted to estimate the proportion of middle school students who suffer from allergies.  One group of seventh graders each took a random sample of 10 middle school students, and another group of seventh graders each took a random sample of 40 middle school students.  Below are two sampling distributions of the sample proportions of middle school students who said that they suffer from allergies.  Which dot plot is based on random samples of size 40?  How can you tell?

**Dot Plot A:**

**Dot Plot of Sample Proportion**

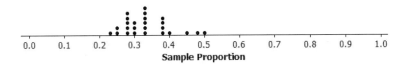

**Dot Plot B:**

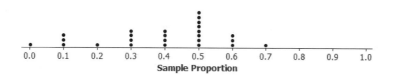

Dot Plot of Sample Proportion

3.  The nurse in your school district would like to study the proportion of middle school students who usually get at least eight hours of sleep on school nights. Suppose each student in your class plans on taking a random sample of 20 middle school students from your district, and each calculates a sample proportion of students who said that they usually get at least eight hours of sleep on school nights.

    a.  Do you expect everyone in your class to get the same value for their sample proportion? Explain.

    b.  Suppose each student in class increased the sample size from 20 to 40. Describe how you could reduce the sampling variability.

# Lesson 20: Estimating a Population Proportion

## Classwork

In a previous lesson, each student in your class selected a random sample from a population and calculated the sample proportion. It was observed that there was sampling variability in the sample proportions, and as the sample size increased, the variability decreased. In this lesson, you will investigate how sample proportions can be used to estimate population proportions.

### Example 1:  Mean of Sample Proportions

A class of 30 seventh graders wanted to estimate the proportion of middle school students who were vegetarians. Each seventh grader took a random sample of 20 middle-school students. Students were asked the question, "Are you a vegetarian?" One sample of 20 students had three students who said that they were vegetarians. For this sample, the sample proportion is $\frac{3}{20}$ or 0.15. Following are the proportions of vegetarians the seventh graders found in 30 samples. Each sample was of size 20 students. The proportions are rounded to the nearest hundredth.

| | | | | | | | | | |
|------|------|------|------|------|------|------|------|------|------|
| 0.15 | 0.10 | 0.15 | 0.00 | 0.10 | 0.15 | 0.10 | 0.10 | 0.05 | 0.20 |
| 0.25 | 0.15 | 0.25 | 0.25 | 0.30 | 0.20 | 0.10 | 0.20 | 0.05 | 0.10 |
| 0.10 | 0.30 | 0.15 | 0.05 | 0.25 | 0.15 | 0.20 | 0.10 | 0.20 | 0.15 |

### Exercises 1–9

1.  The first student reported a sample proportion of 0.15. Interpret this value in terms of the summary of the problem in the example.

2.  Another student reported a sample proportion of 0. Did this student do something wrong when selecting the sample of middle school students?

3.  Assume you were part of this seventh grade class and you got a sample proportion of 0.20 from a random sample of middle school students.  Based on this sample proportion, what is your estimate for the proportion of all middle school students who are vegetarians?

4.  Construct a dot plot of the 30 sample proportions.

5.  Describe the shape of the distribution.

6.  Using the 30 class results listed above, what is your estimate for the proportion of all middle school students who are vegetarians?  Explain how you made this estimate.

7.  Calculate the mean of the 30 sample proportions.  How close is this value to the estimate you made in Exercise 6?

8.  The proportion of all middle school students who are vegetarians is 0.15.  This is the actual proportion for the entire population of middle school students used to select the samples.  How the mean of the 30 sample proportions compares with the actual population proportion depends on the students' samples.

9.  Do the sample proportions in the dot plot tend to cluster around the value of the population proportion?  Are any of the sample proportions far away from 0.15?  List the proportions that are far away from 0.15.

## Example 2:  Estimating Population Proportion

Two hundred middle school students at Roosevelt Middle School responded to several survey questions.  A printed copy of the responses the students gave to various questions is provided with this lesson.

The data are organized in columns and are summarized by the following table:

| Column Heading | Description |
| --- | --- |
| ID | Numbers from 1 to 200 |
| Travel to School | Method used to get to school<br>Walk, car, rail, bus, bike, skateboard, boat, other |
| Favorite Season | Summer, fall, winter, spring |
| Allergies | Yes or no |
| Favorite School Subject | Art, English, languages, social studies, history, geography, music, science, computers, math, PE, other |
| Favorite Music | Classical, Country, heavy metal, jazz, pop, punk rock, rap, reggae, R&B, rock and roll, techno, gospel, other |
| What superpower would you like? | Invisibility, super strength, telepathy, fly, freeze time |

The last column in the data file is based on the question:  Which of the following superpowers would you most like to have?  The choices were:  invisibility, super-strength, telepathy, fly, or freeze time.

The class wants to determine the proportion of Roosevelt Middle School students who answered freeze time to the last question.  You will use a sample of the Roosevelt Middle School population to estimate the proportion of the students who answered freeze time to the last question.

**Exercises 10–17**

A random sample of 20 student responses is needed. You are provided the random number table you used in a previous lesson. A printed list of the 200 Roosevelt Middle School students is also provided. In small groups, complete the following exercise:

10. Select a random sample of 20 student responses from the data file. Explain how you selected the random sample.

11. In the table below list the 20 responses for your sample.

|    | Response |
|----|----------|
| 1  |          |
| 2  |          |
| 3  |          |
| 4  |          |
| 5  |          |
| 6  |          |
| 7  |          |
| 8  |          |
| 9  |          |
| 10 |          |
| 11 |          |
| 12 |          |
| 13 |          |
| 14 |          |
| 15 |          |
| 16 |          |
| 17 |          |
| 18 |          |
| 19 |          |
| 20 |          |

12. Estimate the population proportion of students who responded "freeze time" by calculating the sample proportion of the 20 sampled students who responded "freeze time" to the question.

13. Combine your sample proportion with other students' sample proportions and create a dot plot of the distribution of the sample proportions of students who responded "freeze time" to the question.

14. By looking at the dot plot, what is the value of the proportion of the 200 Roosevelt Middle School students who responded "freeze time" to the question?

15. Usually you will estimate the proportion of Roosevelt Middle School students using just a single sample proportion. How different was your sample proportion from your estimate based on the dot plot of many samples?

16. Circle your sample proportion on the dot plot. How does your sample proportion compare with the mean of all the sample proportions?

17. Calculate the mean of all of the sample proportions. Locate the mean of the sample proportions in your dot plot; mark this position with an "$X$." How does the mean of the sample proportions compare with your sample proportion?

Lesson Summary

The sample proportion from a random sample can be used to estimate a population proportion. The sample proportion will not be exactly equal to the population proportion, but values of the sample proportion from random samples tend to cluster around the actual value of the population proportion.

Problem Set

1.  A class of 30 seventh graders wanted to estimate the proportion of middle school students who played a musical instrument. Each seventh grader took a random sample of 25 middle school students and asked each student whether or not they played a musical instrument. Following are the sample proportions the seventh graders found in 30 samples.

    | 0.80 | 0.64 | 0.72 | 0.60 | 0.60 | 0.72 | 0.76 | 0.68 | 0.72 | 0.68 |
    |------|------|------|------|------|------|------|------|------|------|
    | 0.72 | 0.68 | 0.68 | 0.76 | 0.84 | 0.60 | 0.80 | 0.72 | 0.76 | 0.80 |
    | 0.76 | 0.60 | 0.80 | 0.84 | 0.68 | 0.68 | 0.70 | 0.68 | 0.64 | 0.72 |

    a.  The first student reported a sample proportion of 0.8. What does this value mean in terms of this scenario?

    b.  Construct a dot plot of the 30 sample proportions.

    c.  Describe the shape of the distribution.

    d.  Describe the variability of the distribution.

    e.  Using the 30 class sample proportions listed above, what is your estimate for the proportion of all middle school students who played a musical instrument? Explain how you made this estimate.

2.  Select another variable or column from the data file that is of interest. Take a random sample of 30 students from the list and record the response to your variable of interest of each of the 30 students.

    a.  Based on your random sample what is your estimate for the proportion of all middle school students?

    b.  If you selected a second random sample of 30, would you get the same sample proportion for the second random sample that you got for the first random sample? Explain why or why not.

**Table of Random Digits**

| Row | | | | | | | | | | | | | | | | | | | |
|---|---|---|---|---|---|---|---|---|---|---|---|---|---|---|---|---|---|---|---|
| 1 | 6 | 6 | 7 | 2 | 8 | 0 | 0 | 8 | 4 | 0 | 0 | 4 | 6 | 0 | 3 | 2 | 2 | 4 | 6 | 8 |
| 2 | 8 | 0 | 3 | 1 | 1 | 1 | 1 | 2 | 7 | 0 | 1 | 9 | 1 | 2 | 7 | 1 | 3 | 3 | 5 | 3 |
| 3 | 5 | 3 | 5 | 7 | 3 | 6 | 3 | 1 | 7 | 2 | 5 | 5 | 1 | 4 | 7 | 1 | 6 | 5 | 6 | 5 |
| 4 | 9 | 1 | 1 | 9 | 2 | 8 | 3 | 0 | 3 | 6 | 7 | 7 | 4 | 7 | 5 | 9 | 8 | 1 | 8 | 3 |
| 5 | 9 | 0 | 2 | 9 | 9 | 7 | 4 | 6 | 3 | 6 | 6 | 3 | 7 | 4 | 2 | 7 | 0 | 0 | 1 | 9 |
| 6 | 8 | 1 | 4 | 6 | 4 | 6 | 8 | 2 | 8 | 9 | 5 | 5 | 2 | 9 | 6 | 2 | 5 | 3 | 0 | 3 |
| 7 | 4 | 1 | 1 | 9 | 7 | 0 | 7 | 2 | 9 | 0 | 9 | 7 | 0 | 4 | 6 | 2 | 3 | 1 | 0 | 9 |
| 8 | 9 | 9 | 2 | 7 | 1 | 3 | 2 | 9 | 0 | 3 | 9 | 0 | 7 | 5 | 6 | 7 | 1 | 7 | 8 | 7 |
| 9 | 3 | 4 | 2 | 2 | 9 | 1 | 9 | 0 | 7 | 8 | 1 | 6 | 2 | 5 | 3 | 9 | 0 | 9 | 1 | 0 |
| 10 | 2 | 7 | 3 | 9 | 5 | 9 | 9 | 3 | 2 | 9 | 3 | 9 | 1 | 9 | 0 | 5 | 5 | 1 | 4 | 2 |
| 11 | 0 | 2 | 5 | 4 | 0 | 8 | 1 | 7 | 0 | 7 | 1 | 3 | 0 | 4 | 3 | 0 | 6 | 4 | 4 | 4 |
| 12 | 8 | 6 | 0 | 5 | 4 | 8 | 8 | 2 | 7 | 7 | 0 | 1 | 0 | 1 | 7 | 1 | 3 | 5 | 3 | 4 |
| 13 | 4 | 2 | 6 | 4 | 5 | 2 | 4 | 2 | 6 | 1 | 7 | 5 | 6 | 6 | 4 | 0 | 8 | 4 | 1 | 2 |
| 14 | 4 | 4 | 9 | 8 | 7 | 3 | 4 | 3 | 8 | 2 | 9 | 1 | 5 | 3 | 5 | 9 | 8 | 9 | 2 | 9 |
| 15 | 6 | 4 | 8 | 0 | 0 | 0 | 4 | 2 | 3 | 8 | 1 | 8 | 4 | 0 | 9 | 5 | 0 | 9 | 0 | 4 |
| 16 | 3 | 2 | 3 | 8 | 4 | 8 | 8 | 6 | 2 | 9 | 1 | 0 | 1 | 9 | 9 | 3 | 0 | 7 | 3 | 5 |
| 17 | 6 | 6 | 7 | 2 | 8 | 0 | 0 | 8 | 4 | 0 | 0 | 4 | 6 | 0 | 3 | 2 | 2 | 4 | 6 | 8 |
| 18 | 8 | 0 | 3 | 1 | 1 | 1 | 1 | 2 | 7 | 0 | 1 | 9 | 1 | 2 | 7 | 1 | 3 | 3 | 5 | 3 |
| 19 | 5 | 3 | 5 | 7 | 3 | 6 | 3 | 1 | 7 | 2 | 5 | 5 | 1 | 4 | 7 | 1 | 6 | 5 | 6 | 5 |
| 20 | 9 | 1 | 1 | 9 | 2 | 8 | 3 | 0 | 3 | 6 | 7 | 7 | 4 | 7 | 5 | 9 | 8 | 1 | 8 | 3 |
| 21 | 9 | 0 | 2 | 9 | 9 | 7 | 4 | 6 | 3 | 6 | 6 | 3 | 7 | 4 | 2 | 7 | 0 | 0 | 1 | 9 |
| 22 | 8 | 1 | 4 | 6 | 4 | 6 | 8 | 2 | 8 | 9 | 5 | 5 | 2 | 9 | 6 | 2 | 5 | 3 | 0 | 3 |
| 23 | 4 | 1 | 1 | 9 | 7 | 0 | 7 | 2 | 9 | 0 | 9 | 7 | 0 | 4 | 6 | 2 | 3 | 1 | 0 | 9 |
| 24 | 9 | 9 | 2 | 7 | 1 | 3 | 2 | 9 | 0 | 3 | 9 | 0 | 7 | 5 | 6 | 7 | 1 | 7 | 8 | 7 |
| 25 | 3 | 4 | 2 | 2 | 9 | 1 | 9 | 0 | 7 | 8 | 1 | 6 | 2 | 5 | 3 | 9 | 0 | 9 | 1 | 0 |
| 26 | 2 | 7 | 3 | 9 | 5 | 9 | 9 | 3 | 2 | 9 | 3 | 9 | 1 | 9 | 0 | 5 | 5 | 1 | 4 | 2 |
| 27 | 0 | 2 | 5 | 4 | 0 | 8 | 1 | 7 | 0 | 7 | 1 | 3 | 0 | 4 | 3 | 0 | 6 | 4 | 4 | 4 |
| 28 | 8 | 6 | 0 | 5 | 4 | 8 | 8 | 2 | 7 | 7 | 0 | 1 | 0 | 1 | 7 | 1 | 3 | 5 | 3 | 4 |
| 29 | 4 | 2 | 6 | 4 | 5 | 2 | 4 | 2 | 6 | 1 | 7 | 5 | 6 | 6 | 4 | 0 | 8 | 4 | 1 | 2 |
| 30 | 4 | 4 | 9 | 8 | 7 | 3 | 4 | 3 | 8 | 2 | 9 | 1 | 5 | 3 | 5 | 9 | 8 | 9 | 2 | 9 |
| 31 | 6 | 4 | 8 | 0 | 0 | 0 | 4 | 2 | 3 | 8 | 1 | 8 | 4 | 0 | 9 | 5 | 0 | 9 | 0 | 4 |
| 32 | 3 | 2 | 3 | 8 | 4 | 8 | 8 | 6 | 2 | 9 | 1 | 0 | 1 | 9 | 9 | 3 | 0 | 7 | 3 | 5 |
| 33 | 6 | 6 | 7 | 2 | 8 | 0 | 0 | 8 | 4 | 0 | 0 | 4 | 6 | 0 | 3 | 2 | 2 | 4 | 6 | 8 |
| 34 | 8 | 0 | 3 | 1 | 1 | 1 | 1 | 2 | 7 | 0 | 1 | 9 | 1 | 2 | 7 | 1 | 3 | 3 | 5 | 3 |
| 35 | 5 | 3 | 5 | 7 | 3 | 6 | 3 | 1 | 7 | 2 | 5 | 5 | 1 | 4 | 7 | 1 | 6 | 5 | 6 | 5 |
| 36 | 9 | 1 | 1 | 9 | 2 | 8 | 3 | 0 | 3 | 6 | 7 | 7 | 4 | 7 | 5 | 9 | 8 | 1 | 8 | 3 |
| 37 | 9 | 0 | 2 | 9 | 9 | 7 | 4 | 6 | 3 | 6 | 6 | 3 | 7 | 4 | 2 | 7 | 0 | 0 | 1 | 9 |
| 38 | 8 | 1 | 4 | 6 | 4 | 6 | 8 | 2 | 8 | 9 | 5 | 5 | 2 | 9 | 6 | 2 | 5 | 3 | 0 | 3 |
| 39 | 4 | 1 | 1 | 9 | 7 | 0 | 7 | 2 | 9 | 0 | 9 | 7 | 0 | 4 | 6 | 2 | 3 | 1 | 0 | 9 |
| 40 | 9 | 9 | 2 | 7 | 1 | 3 | 2 | 9 | 0 | 3 | 9 | 0 | 7 | 5 | 6 | 7 | 1 | 7 | 8 | 7 |

| ID | Travel to School | Favorite Season | Allergies | Favorite School Subject | Favorite Music | Superpower |
|----|------------------|-----------------|-----------|-------------------------|----------------|------------|
| 1 | Car | Spring | Yes | English | Pop | Freeze time |
| 2 | Car | Summer | Yes | Music | Pop | Telepathy |
| 3 | Car | Summer | No | Science | Pop | Fly |
| 4 | Walk | Fall | No | Computers and technology | Pop | Invisibility |
| 5 | Car | Summer | No | Art | Country | Telepathy |
| 6 | Car | Summer | No | Physical education | Rap/Hip hop | Freeze time |
| 7 | Car | Spring | No | Physical education | Pop | Telepathy |
| 8 | Car | Winter | No | Art | Other | Fly |
| 9 | Car | Summer | No | Physical education | Pop | Fly |
| 10 | Car | Spring | No | Mathematics and statistics | Pop | Telepathy |
| 11 | Car | Summer | Yes | History | Rap/Hip hop | Invisibility |
| 12 | Car | Spring | No | Art | Rap/Hip hop | Freeze time |
| 13 | Bus | Winter | No | Computers and technology | Rap/Hip hop | Fly |
| 14 | Car | Winter | Yes | Social studies | Rap/Hip hop | Fly |
| 15 | Car | Summer | No | Art | Pop | Freeze time |
| 16 | Car | Fall | No | Mathematics and statistics | Pop | Fly |
| 17 | Bus | Winter | No | Science | Rap/Hip hop | Freeze time |
| 18 | Car | Spring | Yes | Art | Pop | Telepathy |
| 19 | Car | Fall | Yes | Science | Pop | Telepathy |
| 20 | Car | Summer | Yes | Physical education | Rap/Hip hop | Invisibility |
| 21 | Car | Spring | Yes | Science | Pop | Invisibility |
| 22 | Car | Winter | Yes | Mathematics and statistics | Country | Invisibility |
| 23 | Car | Summer | Yes | Art | Pop | Invisibility |
| 24 | Bus | Winter | Yes | Other | Pop | Telepathy |
| 25 | Bus | Summer | Yes | Science | Other | Fly |
| 26 | Car | Summer | No | Science | Pop | Fly |
| 27 | Car | Summer | Yes | Music | Pop | Telepathy |
| 28 | Car | Summer | No | Physical education | Country | Super strength |
| 29 | Car | Fall | Yes | Mathematics and statistics | Country | Telepathy |
| 30 | Car | Summer | Yes | Physical education | Rap/Hip hop | Telepathy |
| 31 | Boat | Winter | No | Computers and technology | Gospel | Invisibility |
| 32 | Car | Spring | No | Physical education | Pop | Fly |
| 33 | Car | Spring | No | Physical education | Pop | Fly |
| 34 | Car | Summer | No | Mathematics and statistics | Classical | Fly |
| 35 | Car | Fall | Yes | Science | Jazz | Telepathy |
| 36 | Car | Spring | No | Science | Rap/Hip hop | Telepathy |
| 37 | Car | Summer | No | Music | Country | Telepathy |
| 38 | Bus | Winter | No | Mathematics and statistics | Pop | Fly |
| 39 | Car | Spring | No | Art | Classical | Freeze time |
| 40 | Car | Winter | Yes | Art | Pop | Fly |
| 41 | Walk | Summer | Yes | Physical education | Rap/Hip hop | Fly |
| 42 | Bus | Winter | Yes | Physical education | Gospel | Invisibility |

Lesson 20:     Estimating a Population Proportion

| 43 | Bus | Summer | No | Art | Other | Invisibility |
|----|-----|--------|-----|-----|-------|--------------|
| 44 | Car | Summer | Yes | Computers and technology | Other | Freeze time |
| 45 | Car | Fall | Yes | Science | Pop | Fly |
| 46 | Car | Summer | Yes | Music | Rap/Hip hop | Fly |
| 47 | Car | Spring | No | Science | Rap/Hip hop | Invisibility |
| 48 | Bus | Spring | No | Music | Pop | Telepathy |
| 49 | Car | Summer | Yes | Social studies | Techno/Electronic | Telepathy |
| 50 | Car | Summer | Yes | Physical education | Pop | Telepathy |
| 51 | Car | Spring | Yes | Other | Other | Telepathy |
| 52 | Car | Summer | No | Art | Pop | Fly |
| 53 | Car | Summer | Yes | Other | Pop | Telepathy |
| 54 | Car | Summer | Yes | Physical education | Rap/Hip hop | Invisibility |
| 55 | Bus | Summer | Yes | Physical education | Other | Super strength |
| 56 | Car | Summer | No | Science | Rap/Hip hop | Invisibility |
| 57 | Car | Winter | No | Languages | Rap/Hip hop | Super strength |
| 58 | Car | Fall | Yes | English | Pop | Fly |
| 59 | Car | Winter | No | Science | Pop | Telepathy |
| 60 | Car | Summer | No | Art | Pop | Invisibility |
| 61 | Car | Summer | Yes | Other | Pop | Freeze time |
| 62 | Bus | Spring | No | Science | Pop | Fly |
| 63 | Car | Winter | Yes | Mathematics and statistics | Other | Freeze time |
| 64 | Car | Summer | No | Social studies | Classical | Fly |
| 65 | Car | Winter | Yes | Science | Pop | Telepathy |
| 66 | Car | Winter | No | Science | Rock and roll | Fly |
| 67 | Car | Summer | No | Mathematics and statistics | Rap/Hip hop | Super strength |
| 68 | Car | Fall | No | Music | Rock and roll | Super strength |
| 69 | Car | Spring | No | Other | Other | Invisibility |
| 70 | Car | Summer | Yes | Mathematics and statistics | Rap/Hip hop | Telepathy |
| 71 | Car | Winter | No | Art | Other | Fly |
| 72 | Car | Spring | Yes | Mathematics and statistics | Pop | Telepathy |
| 73 | Car | Winter | Yes | Computers and technology | Techno/Electronic | Telepathy |
| 74 | Walk | Winter | No | Physical education | Techno/Electronic | Fly |
| 75 | Walk | Summer | No | History | Rock and roll | Fly |
| 76 | Skateboard/Scooter/Rollerblade | Winter | Yes | Computers and technology | Techno/Electronic | Freeze time |
| 77 | Car | Spring | Yes | Science | Other | Telepathy |
| 78 | Car | Summer | No | Music | Rap/Hip hop | Invisibility |
| 79 | Car | Summer | No | Social studies | Pop | Invisibility |
| 80 | Car | Summer | No | Other | Rap/Hip hop | Telepathy |
| 81 | | Spring | Yes | History | Rap/Hip hop | Invisibility |
| 82 | Car | Summer | No | Art | Pop | Invisibility |

| 83 | Walk | Spring | No | Languages | Jazz | Super strength |
| 84 | Car | Fall | No | History | Jazz | Invisibility |
| 85 | Car | Summer | No | Physical education | Rap/Hip hop | Freeze time |
| 86 | Car | Spring | No | Mathematics and statistics | Pop | Freeze time |
| 87 | Bus | Spring | Yes | Art | Pop | Telepathy |
| 88 | Car | Winter | No | Mathematics and statistics | Other | Invisibility |
| 89 | Car | Summer | Yes | Physical education | Country | Telepathy |
| 90 | Bus | Summer | No | Computers and technology | Other | Fly |
| 91 | Car | Winter | No | History | Pop | Telepathy |
| 92 | Walk | Winter | No | Science | Classical | Telepathy |
| 93 | Bicycle | Summer | No | Physical education | Pop | Invisibility |
| 94 | Car | Summer | No | English | Pop | Telepathy |
| 95 | Car | Summer | Yes | Physical education | Pop | Fly |
| 96 | Car | Winter | No | Science | Other | Freeze time |
| 97 | Car | Winter | No | Other | Rap/Hip hop | Super strength |
| 98 | Car | Summer | Yes | Physical education | Rap/Hip hop | Freeze time |
| 99 | Car | Spring | No | Music | Classical | Telepathy |
| 100 | Car | Spring | Yes | Science | Gospel | Telepathy |
| 101 | Car | Summer | Yes | History | Pop | Super strength |
| 102 | Car | Winter | Yes | English | Country | Freeze time |
| 103 | Car | Spring | No | Computers and technology | Other | Telepathy |
| 104 | Car | Winter | No | History | Other | Invisibility |
| 105 | Car | Fall | No | Music | Pop | Telepathy |
| 106 | Car | Fall | No | Science | Pop | Telepathy |
| 107 | Car | Winter | No | Art | Heavy metal | Fly |
| 108 | Car | Spring | Yes | Science | Rock and roll | Fly |
| 109 | Car | Fall | Yes | Music | Other | Fly |
| 110 | Car | Summer | Yes | Social studies | Techno/ Electronic | Telepathy |
| 111 | Car | Spring | No | Physical education | Pop | Fly |
| 112 | Car | Summer | No | Physical education | Pop | Fly |
| 113 | Car | Summer | Yes | Social studies | Pop | Freeze time |
| 114 | Car | Summer | Yes | Computers and technology | Gospel | Freeze time |
| 115 | Car | Winter | Yes | Other | Rap/Hip hop | Telepathy |
| 116 | Car | Summer | Yes | Science | Country | Telepathy |
| 117 | Car | Fall | | Music | Country | Fly |
| 118 | Walk | Summer | No | History | Pop | Telepathy |
| 119 | Car | Spring | Yes | Art | Pop | Freeze time |
| 120 | Car | Fall | Yes | Physical education | Rap/Hip hop | Fly |
| 121 | Car | Spring | No | Music | Rock and roll | Telepathy |
| 122 | Car | Fall | No | Art | Pop | Invisibility |
| 123 | Car | Summer | Yes | Physical education | Rap/Hip hop | Fly |
| 124 | | Summer | No | Computers and technology | Pop | Telepathy |
| 125 | Car | Fall | No | Art | Pop | Fly |

| 126 | Bicycle | Spring | No | Science | Pop | Invisibility |
| 127 | Car | Summer | No | Social studies | Gospel | Fly |
| 128 | Bicycle | Winter | No | Social studies | Rap/Hip hop | Fly |
| 129 | Car | Summer | Yes | Mathematics and statistics | Pop | Invisibility |
| 130 | Car | Fall | Yes | Mathematics and statistics | Country | Telepathy |
| 131 | Car | Winter | Yes | Music | Gospel | Super strength |
| 132 | Rail (Train/Tram /Subway) | Fall | Yes | Art | Other | Fly |
| 133 | Walk | Summer | No | Social studies | Pop | Invisibility |
| 134 | Car | Summer | Yes | Music | Pop | Freeze time |
| 135 | Car | Winter | No | Mathematics and statistics | Pop | Telepathy |
| 136 | Car | Fall | Yes | Music | Pop | Telepathy |
| 137 | Car | Summer | Yes | Computers and technology | Other | Freeze time |
| 138 | Car | Summer | Yes | Physical education | Pop | Telepathy |
| 139 | Car | Summer | Yes | Social studies | Other | Telepathy |
| 140 | Car | Spring | Yes | Physical education | Other | Freeze time |
| 141 | Car | Fall | Yes | Science | Country | Telepathy |
| 142 | Car | Spring | Yes | Science | Pop | Invisibility |
| 143 | Car | Summer | No | Other | Rap/Hip hop | Freeze time |
| 144 | Car | Summer | No | Other | Other | Fly |
| 145 | Car | Summer | No | Languages | Pop | Freeze time |
| 146 | Car | Summer | Yes | Physical education | Pop | Telepathy |
| 147 | Bus | Winter | No | History | Country | Invisibility |
| 148 | Car | Spring | No | Computers and technology | Other | Telepathy |
| 149 | Bus | Winter | Yes | Science | Pop | Invisibility |
| 150 | Car | Summer | No | Social studies | Rap/Hip hop | Invisibility |
| 151 | Car | Summer | No | Physical education | Pop | Invisibility |
| 152 | Car | Summer | Yes | Physical education | Pop | Super strength |
| 153 | Car | Summer | No | Mathematics and statistics | Pop | Fly |
| 154 | Car | Summer | No | Art | Rap/Hip hop | Freeze time |
| 155 | Car | Winter | Yes | Other | Classical | Freeze time |
| 156 | Car | Summer | Yes | Computers and technology | Other | Telepathy |
| 157 | Car | Spring | No | Other | Pop | Freeze time |
| 158 | Car | Winter | Yes | Music | Country | Fly |
| 159 | Car | Winter | No | History | Jazz | Invisibility |
| 160 | Car | Spring | Yes | History | Pop | Fly |
| 161 | Car | Winter | Yes | Mathematics and statistics | Other | Telepathy |
| 162 | Car | Fall | No | Science | Country | Invisibility |
| 163 | Car | Winter | No | Science | Other | Fly |
| 164 | Car | Summer | No | Science | Pop | Fly |
| 165 | Skateboard /Scooter/Ro llerblade | Spring | Yes | Social studies | Other | Freeze time |
| 166 | Car | Winter | Yes | Art | Rap/Hip hop | Fly |

| 167 | Car | Summer | Yes | Other | Pop | Freeze time |
|-----|-----|--------|-----|-------|-----|-------------|
| 168 | Car | Summer | No | English | Pop | Telepathy |
| 169 | Car | Summer | No | Other | Pop | Invisibility |
| 170 | Car | Summer | Yes | Physical education | Techno/ Electronic | Freeze time |
| 171 | Car | Summer | No | Art | Pop | Telepathy |
| 172 | Car | Summer | No | Physical education | Rap/Hip hop | Freeze time |
| 173 | Car | Winter | Yes | Mathematics and statistics | Other | Invisibility |
| 174 | Bus | Summer | Yes | Music | Pop | Freeze time |
| 175 | Car | Winter | No | Art | Pop | Fly |
| 176 | Car | Fall | No | Science | Rap/Hip hop | Fly |
| 177 | Car | Winter | Yes | Social studies | Pop | Telepathy |
| 178 | Car | Fall | No | Art | Other | Fly |
| 179 | Bus | Spring | No | Physical education | Country | Fly |
| 180 | Car | Winter | No | Music | Other | Telepathy |
| 181 | Bus | Summer | No | Computers and technology | Rap/Hip hop | Freeze time |
| 182 | Car | Summer | Yes | Physical education | Rap/Hip hop | Invisibility |
| 183 | Car | Summer | Yes | Music | Other | Telepathy |
| 184 | Car | Spring | No | Science | Rap/Hip hop | Invisibility |
| 185 | Rail (Train/Tram /Subway) | Summer | No | Physical education | Other | Freeze time |
| 186 | Car | Summer | Yes | Mathematics and statistics | Rap/Hip hop | Fly |
| 187 | Bus | Winter | Yes | Mathematics and statistics | Other | Super strength |
| 188 | Car | Summer | No | Mathematics and statistics | Other | Freeze time |
| 189 | Rail (Train/Tram /Subway) | Fall | Yes | Music | Jazz | Fly |
| 190 | Car | Summer | Yes | Science | Pop | Super strength |
| 191 | Car | Summer | Yes | Science | Techno/ Electronic | Freeze time |
| 192 | Car | Spring | Yes | Physical education | Rap/Hip hop | Freeze time |
| 193 | Car | Summer | Yes | Physical education | Rap/Hip hop | Freeze time |
| 194 | Car | Winter | No | Physical education | Rap/Hip hop | Telepathy |
| 195 | Car | Winter | No | Music | Jazz | Freeze time |
| 196 | Walk | Summer | Yes | History | Country | Freeze time |
| 197 | Car | Spring | No | History | Rap/Hip hop | Freeze time |
| 198 | Car | Fall | Yes | Other | Pop | Freeze time |
| 199 | Car | Spring | Yes | Science | Other | Freeze time |
| 200 | Bicycle | Winter | Yes | Other | Rap/Hip hop | Freeze time |

# Lesson 21: Why Worry about Sampling Variability?

There are three bags, Bag $A$, Bag $B$, and Bag $C$, with 100 numbers in each bag. You and your classmates will investigate the population mean (the mean of all 100 numbers) in each bag. Each set of numbers has the same range. However, the population means of each set may or may not be the same. We will see who can uncover the mystery of the bags!

## Exercises 1–5

1. To begin your investigation, start by selecting a random sample of ten numbers from Bag $A$. Remember to mix the numbers in the bag first. Then, select one number from the bag. Do not put it back into the bag. Write the number in the chart below. Continue selecting one number at a time until you have selected ten numbers. Mix up the numbers in the bag between each selection.

| Selection | 1 | 2 | 3 | 4 | 5 | 6 | 7 | 8 | 9 | 10 |
|---|---|---|---|---|---|---|---|---|---|---|
| Bag $A$ | | | | | | | | | | |

a. Create a dot plot of your sample of ten numbers. Use a dot to represent each number in the sample.

b. Do you think the mean of all the numbers in Bag $A$ might be 10? Why or why not?

c. Based on the dot plot, what would you estimate the mean of the numbers in Bag $A$ to be? How did you make your estimate?

d. Do you think your sample mean will be close to the population mean? Why or why not?

e. Is your sample mean the same as your neighbors' sample means? Why or why not?

2. Repeat the process by selecting a random sample of ten numbers from Bag $B$.

| Selection | 1 | 2 | 3 | 4 | 5 | 6 | 7 | 8 | 9 | 10 |
|-----------|---|---|---|---|---|---|---|---|---|----|
| Bag $B$ | | | | | | | | | | |

a. Create a dot plot of your sample of ten numbers. Use a dot to represent each of the numbers in the sample.

b. Based on your dot plot, do you think the mean of the numbers in Bag $B$ is the same or different than the mean of the numbers in Bag $A$? Explain your thinking.

3. Repeat the process once more by selecting a random sample of ten numbers from Bag $C$.

| Selection | 1 | 2 | 3 | 4 | 5 | 6 | 7 | 8 | 9 | 10 |
|-----------|---|---|---|---|---|---|---|---|---|----|
| Bag $C$ | | | | | | | | | | |

a. Create a dot plot of your sample of ten numbers. Use a dot to represent each of the numbers in the sample.

b. Based on your dot plot, do you think the mean of the numbers in Bag $C$ is the same or different than the mean of the numbers in Bag $A$? Explain your thinking.

4. Are your dot plots of the three bags the same as the dot plots of other students in your class? Why or why not?

5. Calculate the mean of the numbers for each of the samples from Bag $A$, Bag, $B$, and Bag $C$.

| | Mean of the sample of numbers |
|---|---|
| Bag $A$ | |
| Bag $B$ | |
| Bag $C$ | |

   a. Are the sample means you calculated the same as the sample means of other members of your class? Why or why not?

   b. How do your sample means for Bag $A$ and for Bag $B$ compare?

   c. Calculate the difference of sample mean for Bag $A$ minus sample mean for Bag $B$ ($Mean_A - Mean_B$). Based on this difference, can you be sure which bag has the larger population mean? Why or why not?

## Exercises 6–10

6. Based on the class dot plots of the sample means, do you think the mean of the numbers in Bag $A$ and the mean of the numbers in Bag $B$ are different? Do you think the mean of the numbers in Bag $A$ and the mean of the numbers in Bag $C$ are different? Explain your answers.

7.  Based on the difference between sample mean of Bag $A$ and the sample mean of Bag $B$, $(Mean_A - Mean_B)$, that you calculated in Exercise 5, do you think that the two populations (Bags $A$ and $B$) have different means or do you think that the two population means might be the same?

8.  Based on this difference, can you be sure which bag has the larger population mean?  Why or why not?

9.  Is your difference in sample means the same as your neighbors' differences?  Why or why not?

10. Plot your difference of the means, $(Mean_A - Mean_B)$, on a class dot plot.  Describe the distribution of differences plotted on the graph.  Remember to discuss center and spread.

**Exercises 11–13**

11. Why are the differences in the sample means of Bag $A$ and Bag $B$ not always 0?

12. Does the class dot plot contain differences that were relatively far away from 0?  If yes, why do you think this happened?

13. Suppose you will take a sample from a new bag.  How big would the difference in the sample mean for Bag $A$ and the sample mean for the new bag ($Mean_A - Mean_{new}$) have to be before you would be convinced that the population mean for the new bag is different than the population mean of Bag $A$?  Use the class dot plot of the differences in sample means for Bags $A$ and $B$ (which have equal population means) to help you answer this question.

The differences in the class dot plot occur because of sampling variability—the chance variability from one sample to another.  In Exercise 13, you were asked about how great the difference in sample means would need to be before you have convincing evidence that one population mean is larger than another population mean.  A "*meaningful*" difference between two sample means is one that is unlikely to have occurred by chance if the population means are equal.  In other words, the difference is one that is greater than would have been expected just due to sampling variability.

**Exercises 14–17**

14. Calculate the sample mean of Bag $A$ minus the sample mean of Bag $C$, ($Mean_A - Mean_C$).

15. Plot your difference ($Mean_A - Mean_C$) on a class dot plot.

16. How do the centers of the class dot plots for ($Mean_A - Mean_B$) and ($Mean_A - Mean_C$) compare?

17. Each bag has a population mean that is either 10.5 or 14.5.  State what you think the population mean is for each bag.  Explain your choice for each bag.

Lesson Summary

- Remember to think about sampling variability—the chance variability from sample to sample.
- Beware of making decisions based just on the fact that two sample means are not equal.
- Consider the distribution of the difference in sample means when making a decision.

Problem Set

Below are three dot plots. Each dot plot represents the differences in sample means for random samples selected from two populations (Bag $A$ and Bag $B$). For each distribution, the differences were found by subtracting the sample means of Bag $B$ from the sample means of Bag $A$ (sample mean $A$ − sample mean $B$).

1.  Does the graph below indicate that the population mean of $A$ is larger than the population mean of $B$? Why or why not?

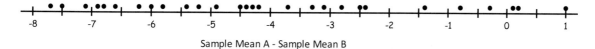

Sample Mean A - Sample Mean B

2.  Use the graph above to estimate the difference in the population means $(A − B)$.

3.  Does the graph below indicate that the population mean of $A$ is larger than the population mean of $B$? Why or why not?

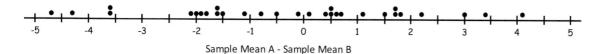

Sample Mean A - Sample Mean B

4.  Does the graph below indicate that the population mean of $A$ is larger than the population mean of $B$? Why or why not?

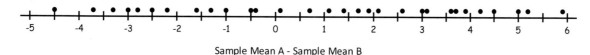

Sample Mean A - Sample Mean B

5.  In the above graph, how many differences are greater than 0? How many differences are less than 0? What might this tell you?

6.  In Problem 4, the population mean for Bag $A$ is really larger than the population mean for Bag $B$. Why is it possible to still get so many negative differences in the graph?

# Lesson 22:  Using Sample Data to Decide if Two Population Means are Different

## Classwork

In previous lessons, you worked with one population.  Many statistical questions involve comparing two populations.
For example:

- On average, do boys and girls differ on quantitative reasoning?
- Do students learn basic arithmetic skills better with or without calculators?
- Which of two medications is more effective in treating migraine headaches?
- Does one type of car get better mileage per gallon of gasoline than another type?
- Does one type of fabric decay in landfills faster than another type?
- Do people with diabetes heal more slowly than people who do not have diabetes?

In this lesson, you will begin to explore how big of a difference there needs to be in sample means in order for the
difference to be considered meaningful.  The next lesson will extend that understanding to making informal inferences
about population differences.

### Example 1

Tamika's mathematics project is to see whether boys or girls are faster in solving a KenKen-type puzzle.  She creates a
puzzle and records the following times that it took to solve the puzzle (in seconds) for a random sample of 10 boys from
her school and a random sample of 11 girls from her school:

|       |    |    |    |    |    |    |    |    |    |    |    | Mean | MAD  |
|-------|----|----|----|----|----|----|----|----|----|----|----|------|------|
| Boys  | 39 | 38 | 27 | 36 | 40 | 27 | 43 | 36 | 34 | 33 |    | 35.3 | 4.04 |
| Girls | 41 | 41 | 33 | 42 | 47 | 38 | 41 | 36 | 36 | 32 | 46 | 39.4 | 3.96 |

### Exercises 1–3

1.  On the same scale, draw dot plots for the boys' data and for the girls' data.  Comment on the amount of overlap
    between the two dot plots.  How are the dot plots the same, and how are they different?

2. Compare the variability in the two data sets using the MAD (mean absolute deviation). Is the variability in each sample about the same? Interpret the MAD in the context of the problem.

3. In the previous lesson, you learned that a difference between two sample means is considered to be meaningful if the difference is more than what you would expect to see just based on sampling variability. The difference in the sample means of the boys' times and the girls' times is 4.1 seconds (39.4 seconds−35.3 seconds). This difference is approximately 1 MAD.

   a. If 4 sec. is used to approximate the values of 1 MAD (mean absolute deviation) for both boys and for girls, what is the interval of times that are within 1 MAD of the sample mean for boys?

   b. Of the 10 sample means for boys, how many of them are within that interval?

   c. Of the 11 sample means for girls, how many of them are within the interval you calculated in part (a)?

   d. Based on the dot plots, do you think that the difference between the two sample means is a meaningful difference? That is, are you convinced that the mean time for all girls at the school (not just this sample of girls) is different from the mean time for all boys at the school? Explain your choice based on the dot plots.

## Example 2

How good are you at estimating a minute? Work in pairs. Flip a coin to determine which person in the pair will go first. One of you puts your head down and raises your hand. When your partner says "start," keep your head down and your hand raised. When you think a minute is up, put your hand down. Your partner will record how much time has passed. Note that the room needs to be quiet. Switch roles except this time you talk with your partner during the period when the person with his or her head down is indicating when they think a minute is up. Note that the room will not be quiet.

| Group | Estimate for a minute | | | | | | | | | | | | |
|---|---|---|---|---|---|---|---|---|---|---|---|---|---|
| Quiet | | | | | | | | | | | | | |
| Talking | | | | | | | | | | | | | |

## Exercises 4–7

Use your class data to complete the following.

4. Calculate the mean minute time for each group. Then find the difference between the "quiet" mean and the "talking" mean.

5. On the same scale, draw dot plots of the two data distributions and discuss the similarities and differences in the two distributions.

6.  Calculate the mean absolute deviation (MAD) for each data set.  Based on the MADs, compare the variability in each sample.  Is the variability about the same?  Interpret the MADs in the context of the problem.

7.  Based on your calculations, is the difference in mean time estimates meaningful?  Part of your reasoning should involve the number of MADs that separate the two sample means.  Note that if the MADs differ, use the larger one in determining how many MADs separate the two means.

**Lesson Summary**

Variability is a natural occurrence in data distributions. Two data distributions can be compared by describing how far apart their sample means are. The amount of separation can be measured in terms of how many MADs separate the means. (Note that if the two sample MADs differ, the larger of the two is used to make this calculation.)

**Problem Set**

1. A school is trying to decide which reading program to purchase.

   a. How many MADs separate the mean reading comprehension score for a standard program ($mean = 67.8$, $MAD = 4.6, n = 24$) and an activity-based program ($mean = 70.3, MAD = 4.5, n = 27$)?

   b. What recommendation would you make based on this result?

2. Does a football filled with helium go farther than one filled with air? Two identical footballs were used, one filled with helium, and one filled with air to the same pressure. Matt was chosen from your team to do the kicking. You did not tell Matt which ball he was kicking. The data (yards) follow.

   | Air | 25 | 23 | 28 | 29 | 27 | 32 | 24 | 26 | 22 | 27 | 31 | 24 | 33 | 26 | 24 | 28 | 30 |
   |--------|----|----|----|----|----|----|----|----|----|----|----|----|----|----|----|----|----|
   | Helium | 24 | 19 | 25 | 25 | 22 | 24 | 28 | 31 | 22 | 26 | 24 | 23 | 22 | 21 | 21 | 23 | 25 |

   |        | Mean | MAD |
   |--------|------|-----|
   | Air    |      |     |
   | Helium |      |     |

   a. Calculate the difference between the sample mean distance for the football filled with air, and for the one filled with helium.

   b. On the same scale, draw dot plots of the two distributions and discuss the variability in each distribution.

   c. Calculate the mean absolute deviations (MAD) for each distribution. Based on the MADs, compare the variability in each distribution. Is the variability about the same? Interpret the MADs in the context of the problem.

   d. Based on your calculations, is the difference in mean distance meaningful? Part of your reasoning should involve the number of MADs that separate the sample means. Note that if the MADs differ, use the larger one in determining how many MADs separate the two means.

3.  Suppose that your classmates were debating about whether going to college is really worth it. Based on the following data of annual salaries (rounded to the nearest thousands of dollars) for college graduates and high school graduates with no college experience, does it appear that going to college is indeed worth the effort? The data are from people in their second year of employment.

| College Grad | 41 | 67 | 53 | 48 | 45 | 60 | 59 | 55 | 52 | 52 | 50 | 59 | 44 | 49 | 52 |
|---|---|---|---|---|---|---|---|---|---|---|---|---|---|---|---|
| High School Grad | 23 | 33 | 36 | 29 | 25 | 43 | 42 | 38 | 27 | 25 | 33 | 41 | 29 | 33 | 35 |

a.  Calculate the difference between the sample mean salary for college graduates and for high school graduates.

b.  On the same scale, draw dot plots of the two distributions and discuss the variability in each distribution.

c.  Calculate the mean absolute deviations (MAD) for each distribution. Based on the MADs, compare the variability in each distribution. Is the variability about the same? Interpret the MADs in the context of the problem.

d.  Based on your calculations, is going to college worth the effort? Part of your reasoning should involve the number of MADs that separate the sample means.

# Lesson 23: Using Sample Data to Decide if Two Population Means are Different

## Classwork

In the previous lesson, you described how far apart the means of two data sets are in terms of the MAD (mean absolute deviation), a measure of variability. In this lesson, you will extend that idea to informally determine when two sample means computed from random samples are far enough apart from each other so to imply that the population means also differ in a "meaningful" way. Recall that a "meaningful" difference between two means is a difference that is greater than would have been expected just due to sampling variability.

### Example 1: Texting

With texting becoming so popular, Linda wanted to determine if middle school students memorize *real* words more or less easily than *fake* words. For example, real words are "food," "car," "study," "swim;" whereas fake words are "stk," "fonw," "cqur," "ttnsp." She randomly selected 28 students from all middle school students in her district and gave half of them a list of 20 real words and the other half a list of 20 fake words.

### Exercises 1–3

1. How do you think Linda might have randomly selected 28 students from all middle school students in her district?

2. Why do you think Linda selected the students for her study randomly? Explain.

3.  She gave the selected students one minute to memorize their list after which they were to turn the list over and after two minutes write down all the words that they could remember.  Afterwards, they calculated the number of correct "words" that they were able to write down.  Do you think a penalty should be given for an incorrect "word" written down?  Explain your reasoning.

**Exercises 4–7**

Suppose the data (number of correct words recalled) she collected were:

For students given the real words list:  $8, 11, 12, 8, 4, 7, 9, 12, 12, 9, 14, 11, 5, 10$.

For students given the fake words list:  $3, 5, 4, 4, 4, 7, 11, 9, 7, 7, 1, 3, 3, 7$.

4.  On the same scale, draw dot plots for the two data sets.

5.  From looking at the dot plots, write a few sentences comparing the distribution of the number of correctly recalled real words with the distribution of number of correctly recalled fake words.  In particular, comment on which type of word, if either, that students recall better.  Explain.

6.    Linda made the following calculations for the two data sets:

|                     | Mean | MAD  |
|---------------------|------|------|
| Real words recalled | 9.43 | 2.29 |
| Fake words recalled | 5.36 | 2.27 |

In the previous lesson, you calculated the number of MADs that separated two sample means. You used the larger MAD to make this calculation if the two MADs were not the same. How many MADs separate the mean number of real words recalled and the mean number of fake words recalled for the students in the study?

7.    In the last lesson, our work suggested that if the number of MADs that separate the two sample means is 2 or more, then it is reasonable to conclude that not only do the means differ in the samples, but that the means differ in the populations as well. If the number of MADs is less than 2, then you can conclude that the difference in the sample means might just be sampling variability and that there may not be a meaningful difference in the population means. Using these criteria, what can Linda conclude about the difference in population means based on the sample data that she collected? Be sure to express your conclusion in the context of this problem.

Example 2

Ken, an eighth grade student, was interested in doing a statistics study involving sixth grade and eleventh grade students in his school district. He conducted a survey on four numerical variables and two categorical variables (grade level and gender). His Excel population database for the 265 sixth graders and 175 eleventh graders in his district has the following description:

| Column | Name     | Description                                                                                                                                                                               |
|--------|----------|-------------------------------------------------------------------------------------------------------------------------------------------------------------------------------------------|
| 1      | ID       | ID numbers are from 1 through 440<br>    1 – 128 Sixth grade females<br>129 – 265 Sixth grade males<br>266 – 363 Eleventh grade females<br>364 – 440 Eleventh grade males |
| 2      | Texting  | Number of minutes per day text (whole number)                                                                                                                                             |
| 3      | ReacTime | Time in seconds to respond to a computer screen stimulus (two decimal places)                                                                                                             |
| 4      | Homework | Total number of hours per week spend on doing homework (one decimal place)                                                                                                                |
| 5      | Sleep    | Number of hours per night sleep (one decimal place)                                                                                                                                       |

**Exercise 8**

8.  Ken decides to base his study on a random sample of 20 sixth graders and a random sample of 20 eleventh graders. The sixth graders have IDs 1–265 and the eleventh graders are numbered 266–440. Advise him on how to randomly sample 20 sixth graders and 20 eleventh graders from his data file.

**Exercises 9–13**

Suppose that from a random-number generator:

The random ID numbers for his 20 sixth graders are:
231 15 19 206 86 183 233 253 142 36 195 139 75 210 56 40 66 114 127 9

The random ID numbers for his 20 eleventh graders are:
391 319 343 426 307 360 289 328 390 350 279 283 302 287 269 332 414 267 428

9.  For each set, find the homework hours data from the population database that corresponds to these randomly selected ID numbers.

10. On the same scale, draw dot plots for the two sample data sets.

11. From looking at the dot plots, list some observations comparing the number of hours per week that sixth graders spend on doing homework and the number of hours per week that eleventh graders spend on doing homework.

12. Calculate the mean and MAD for each of the data sets. How many MADs separate the two sample means? (Use the larger MAD to make this calculation if the sample MADs are not the same.)

| | Mean (hr.) | MAD (hr.) |
|---|---|---|
| Sixth Grade | | |
| Eleventh Grade | | |

13. Ken recalled Linda suggesting that if the number of MADs is greater than or equal to $2$ then it would be reasonable to think that the population of all sixth grade students in his district and the population of all eleventh grade students in his district have different means. What should Ken conclude based on his homework study?

**Lesson Summary**

To determine if the mean value of some numerical variable differs for two populations, take random samples from each population. It is very important that the samples be random samples. If the number of MADs that separate the two sample means is 2 or more, then it is reasonable to think that the populations have different means. Otherwise, the population means are considered to be the same.

**Problem Set**

1. Based on Ken's population database, compare the amount of sleep that sixth grade females get on average to the amount of sleep that eleventh grade females get on average.

   Find the data for 15 sixth grade females based on the following random ID numbers:
   65  1  67  101  106  87  85  95  120  4  64  74  102  31  128

   Find the data for 15 eleventh grade females based on the following random ID numbers:
   348  313  297  351  294  343  275  354  311  328  274  305  288  267  301

2. On the same scale, draw dot plots for the two sample data sets.

3. Looking at the dot plots, list some observations comparing the number of hours per week that sixth graders spend on doing homework and the number of hours per week that eleventh graders spend on doing homework.

4. Calculate the mean and MAD for each of the data sets. How many MADs separate the two sample means? (Use the larger MAD to make this calculation if the sample MADs are not the same.)

   |  | Mean (hr.) | MAD (hr.) |
   |---|---|---|
   | Sixth Grade Females |  |  |
   | Eleventh Grade Females |  |  |

5. Recall that if the number of MADs in the difference of two sample means is greater than or equal to 2, then it would be reasonable to think that the population means are different. Using this guideline, what can you say about the average number of hours of sleep per night for all sixth grade females in the population compared to all eleventh grade females in the population?

# Mathematics Curriculum

## Copy Ready Material

Name _____     Date_____

# Lesson 1:  Chance Experiments

Exit Ticket

Decide where each of the following events would be located on the scale below.  Place the letter for each event on the appropriate place on the probability scale.

## Probability Scale

| 0 | | 1/2 | | 1 |
|---|---|---|---|---|
| Impossible | Unlikely | Equally Likely to Occur or Not Occur | Likely | Certain |

The numbers from 1 to 10 are written on small pieces of paper and placed in a bag.  A piece of paper will be drawn from the bag.

    A.   A piece of paper with a 5 is drawn from the bag.

    B.   A piece of paper with an even number is drawn.

    C.   A piece of paper with a 12 is drawn.

    D.   A piece of paper with a number other than 1 is drawn.

    E.   A piece of paper with a number divisible by 5 is drawn.

Name _____    Date_____

# Lesson 2:  Estimating Probabilities by Collecting Data

Exit Ticket

In the following problems, round all of your decimal answers to 3 decimal places.  Round all of your percents to the nearest tenth of a percent.

A student randomly selected crayons from a large bag of crayons.  Below is the number of each color the student selected.  Now, suppose the student were to randomly select one crayon from the bag.

| Color | Number |
|-------|--------|
| Brown | 10 |
| Blue | 5 |
| Yellow | 3 |
| Green | 3 |
| Orange | 3 |
| Red | 6 |

1.  What is the estimate for the probability of selecting a blue crayon from the bag?  Express your answer as a fraction, decimal or percent.

2.  What is the estimate for the probability of selecting a brown crayon from the bag?

3.  What is the estimate for the probability of selecting a red crayon *or* a yellow crayon from the bag?

4.  What is the estimate for the probability of selecting a pink crayon from the bag?

5.  Which color is most likely to be selected?

6.  If there are 300 crayons in the bag, how many will be red?  Justify your answer.

Name _____    Date_____

# Lesson 3:  Chance Experiments with Equally Likely Outcomes

Exit Ticket

The numbers from 1– 10 are written on note cards and placed in a bag.  One card will be drawn from the bag at random.

1.  List the sample space for this experiment.

2.  Are the events selecting an even number and selecting an odd number equally likely?  Explain your answer.

3.  Are the events selecting a number divisible by 3 and selecting a number divisible by 5 equally likely?  Explain your answer.

Name _____ Date_____

# Lesson 4: Calculating Probabilities for Chance Experiments with Equally Likely Outcomes

Exit Ticket

An experiment consists of randomly drawing a cube from a bag containing three red and two blue cubes.

1. What is the sample space of this experiment?

2. List the probability of each outcome in the sample space.

3. Is the probability of selecting a red cube equal to the probability of selecting a blue cube? Explain.

Name _____    Date_____

# Lesson 5:  Chance Experiments with Outcomes that are Not Equally Likely

Exit Ticket

Carol is sitting on the bus on the way home from school and is thinking about the fact that she has three homework assignments to do tonight.  The table below shows her estimated probabilities of completing 0, 1, 2, or all 3 of the assignments.

| Number of Homework Assignments Completed | 0 | 1 | 2 | 3 |
|---|---|---|---|---|
| Probability | $\frac{1}{6}$ | $\frac{2}{9}$ | $\frac{5}{18}$ | $\frac{1}{3}$ |

1.  Writing your answers as fractions in lowest terms, find the probability that Carol completes
    a.  exactly one assignment.

    b.  more than one assignment.

    c.  at least one assignment.

2.  Find the probability that the number of homework assignments Carol completes is not exactly 2.

3.  Carol has a bag of colored chips.  She has 3 red chips, 10 blue chips, and 7 green chips in the bag.  Estimate the probability (as a fraction or decimal) of Carol reaching into her bag and pulling out a green chip.

Name _____     Date_____

# Lesson 6:  Using Tree Diagrams to Represent a Sample Space and to Calculate Probabilities

Exit Ticket

In a laboratory experiment, two mice will be placed in a simple maze with one decision point where a mouse can turn either left ($L$) or right ($R$).  When the first mouse arrives at the decision point, the direction it chooses is recorded.  Then, the process is repeated for the second mouse.

1.  Draw a tree diagram where the first stage represents the decision made by the first mouse, and the second stage represents the decision made by the second mouse.  Determine all four possible decision outcomes for the two mice.

2.  If the probability of turning left is 0.5, and the probability of turning right is 0.5 for each mouse, what is the probability that only one of the two mice will turn left?

3.  If the researchers add food in the simple maze such that the probability of each mouse turning left is now 0.7, what is the probability that only one of the two mice will turn left?

Name _____     Date_____

# Lesson 7:  Calculating Probabilities of Compound Events

Exit Ticket

In a laboratory experiment, three mice will be placed in a simple maze that has just one decision point where a mouse can turn either left ($L$) or right ($R$).  When the first mouse arrives at the decision point, the direction he chooses is recorded.  The same is done for the second and the third mice.

1.  Draw a tree diagram where the first stage represents the decision made by the first mouse, and the second stage represents the decision made by the second mouse, and so on.  Determine all eight possible outcomes of the decisions for the three mice.

2.  Use the tree diagram from Question 1 to help answer the following question.  If, for each mouse, the probability of turning left is 0.5 and the probability of turning right is 0.5, what is the probability that only one of the three mice will turn left?

3.  If the researchers conducting the experiment add food in the simple maze such that the probability of each mouse turning left is now 0.7, what is the probability that only one of the three mice will turn left?  To answer question, use the tree diagram from Question 1.

Name _____    Date_____

# Lesson 8:  The Difference Between Theoretical Probabilities and Estimated Probabilities

Exit Ticket

1.  Which of the following graphs would NOT represent the relative frequencies of heads when tossing 1 penny? Explain your answer.

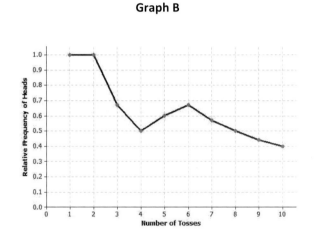

2.  Jerry indicated that after tossing a penny 30 times, the relative frequency of heads was 0.47 (to the nearest hundredth).  He indicated that after 31 times, the relative frequency of heads was 0.55.  Are Jerry's summaries correct?  Why or why not?

3.   Jerry observed 5 heads in 100 tosses of his coin.  Do you think this was a fair coin?  Why or why not?

Name _____     Date_____

# Lesson 10:  Using Simulation to Estimate a Probability

Exit Ticket

1.  Nathan is your school's star soccer player.  When he takes a shot on goal, he scores half of the time on average.  Suppose that he takes six shots in a game.  To estimate probabilities of the number of goals Nathan makes, use simulation with a number cube.  One roll of a number cube represents one shot.

    a.   Specify what outcome of a number cube you want to represent a goal scored by Nathan in one shot.

    b.   For this problem, what represents a trial of taking six shots?

    c.   Perform and list the results of ten trials of this simulation.

    d.   Identify the number of goals Nathan made in each of the ten trials you did in part (c).

    e.   Based on your ten trials, what is your estimate of the probability that Nathan scores three goals if he takes six shots in a game?

2.  Suppose that Pat scores 40% of the shots he takes in a soccer game.  If he takes six shots in a game, what would one simulated trial look like using a number cube in your simulation?

Name _____   Date_____

# Lesson 11:  Using a Simulation to Estimate a Probability

Exit Ticket

Liang wants to form a chess club.  His principal says that he can do that if Liang can find six players, including him.  How would you conduct a simulated model that estimates the probability that Liang will find at least five other players to join the club if he asks eight players who have a 70% chance of agreeing to join the club?  Suggest a simulation model for Liang by describing how you would do the following parts.

a.   Specify the device you want to use to simulate one person being asked.

b.   What outcome(s) of the device would represent the person agreeing to be a member?

c.   What constitutes a trial using your device in this problem?

d.   What constitutes a success using your device in this problem?

e.   Based on 50 trials, using the method you have suggested, how would you calculate the estimate for the probability that Liang will be able to form a chess club?

Name _____   Date_____

# Lesson 12:  Using Probability to Make Decisions

Exit Ticket

There are four pieces of bubble gum left in a quarter machine.  Two are red and two are yellow.  Chandra puts two quarters in the machine.  One piece is for her and one is for her friend, Kay.  If the two pieces are the same color, she's happy because they won't have to decide who gets what color.  Chandra claims that they are equally likely to get the same color because the colors are either the same or they are different.  Check her claim by doing a simulation.

a.  Name a device to use to simulate getting a piece of bubble gum.  Specify what outcome of the device represents a red piece and what outcome represents yellow.

b.  Define what a trial is for your simulation.

c.  Define what constitutes a success in a trial of your simulation.

d.  Perform and list 50 simulated trials.  Based on your results, is Chandra's equally likely model is correct?

Name _____     Date _____

Round all decimal answers to the nearest hundredth.

1.  Each student in a class of 38 students was asked to report how many siblings (brothers and sisters) he or she has. The data are summarized in the table below.

| Number of Siblings | 0 | 1 | 2 | 3 | 4 | 5 | 6 |
|---|---|---|---|---|---|---|---|
| Count | 8 | 13 | 12 | 3 | 1 | 0 | 1 |

a.  Based on these data, estimate the probability that a randomly selected student from this class is an only child.

b.  Based on these data, estimate the probability that a randomly selected student from this class has three or more siblings.

c.  Consider the following probability distribution for the number of siblings:

| Number of Siblings | 0 | 1 | 2 | 3 | 4 | 5 | 6 |
|---|---|---|---|---|---|---|---|
| Probability | 0.15 | 0.35 | 0.30 | 0.10 | 0.05 | 0.03 | 0.02 |

Explain how you could use simulation to estimate the probability that you will need to ask at least five students if each is an only child before you find the first one that is an only child.

2.  A cell phone company wants to predict the probability of a seventh grader in your city, City A, owning a cell phone.  Records from another city, City B, indicate that 201 of 1,000 seventh graders own a cell phone.

   a.  Assuming the probability of a seventh grader owning a cell phone is similar for the two cities, estimate the probability that a randomly selected seventh grader from City A owns a cell phone.

   b.  The company estimates the probability that a randomly selected seventh-grade male owns a cell phone is 0.25.  Does this imply that the probability that a randomly selected seventh-grade female owns a cell phone is 0.75?  Explain.

   c.  According to these data, which of the following is more likely?

      • a seventh-grade male owning a cell phone
      • a seventh grader owning a cell phone

   Explain your choice.

Suppose the cell phone company sells three different plans to its customers:

- Pay-as-you-go:  Customer is charged per minute for each call.
- Unlimited minutes:  Customer pays a flat fee per month and can make unlimited calls with no additional charges.
- Basic plan:  Customer not charged per minute unless the customer exceeds 500 minutes in the month; then, the customer is charged per minute for the extra minutes.

Consider the chance experiment of selecting a customer at random and recording which plan they purchased.

d.  What outcomes are in the sample space for this chance experiment?

e.  The company wants to assign probabilities to these three plans.  Explain what is wrong with each of the following probability assignments.

Case 1:  Probability of pay-as-you-go = 0.40, probability of unlimited minutes = 0.40, and probability of basic plan = 0.30.

Case 2:  Probability of pay-as-you-go = 0.40, probability of unlimited minutes = 0.70, and probability of basic plan = 0.10.

Now consider the chance experiment of randomly selecting a cell phone customer and recording both the cell phone plan for that customer and whether or not the customer exceeded 500 minutes last month.

f.   One possible outcome of this chance experiment is (pay-as-go, over 500).  What are the other possible outcomes in this sample space?

g.   Assuming the outcomes of this chance experiment are equally likely, what is the probability that the selected cell phone customer had a basic plan and did not exceed 500 minutes last month?

h.   Suppose the company random selects 500 of its customers and finds that 140 of these customers purchased the basic plan and did not exceed 500 minutes.  Would this cause you to question the claim that the outcomes of the chance experiment described in (g) are equally likely?  Explain why or why not.

3.  In the game of Darts, players throw darts at a circle divided into 20 wedges. In one variation of the game, the score for a throw is equal to the wedge number that the dart hits. So, if the dart hits anywhere in the 20 wedge, you earn 20 points for that throw.

a.  If you are equally likely to land in any wedge, what is the probability you will score 20 points?

b.  If you are equally likely to land in any wedge, what is the probability you will land in the "upper right" and score 20, 1, 18, 4, 13, or 6 points?

c.  Below are the results of 100 throws for one player. Does this player appear to have a tendency to land in the "upper right" more often than we would expect if the player was equally likely to land in any wedge?

| Points | 1 | 2 | 3 | 4 | 5 | 6 | 7 | 8 | 9 | 10 | 11 | 12 | 13 | 14 | 15 | 16 | 17 | 18 | 19 | 20 |
|--------|---|---|---|---|---|---|---|---|---|----|----|----|----|----|----|----|----|----|----|----|
| Count  | 7 | 9 | 2 | 6 | 6 | 3 | 5 | 2 | 4 | 7  | 2  | 6  | 4  | 6  | 5  | 7  | 4  | 6  | 5  | 4  |

Name _____     Date_____

# Lesson 13:  Populations, Samples, and Generalizing from a Sample to a Population

Exit Ticket

What is the difference between a population characteristic and a sample statistic?  Give an example to support your answer.  Clearly identify the population and sample in your example.

Name _____    Date_____

# Lesson 14:  Selecting a Sample

Exit Ticket

Write down three things you learned about taking a sample from the work we have done today.

Handout 1

\1\ Casey at the Bat

The Outlook wasn't brilliant for the Mudville nine that day:  The score stood four to two, \2\ with but one inning more to play.  And then when Cooney died at first, and Barrows did the same,  A \3\ sickly silence fell upon the patrons of the game.

A straggling few got up to go in deep despair. The \4\ rest  Clung to that hope which springs eternal in the human breast;  They thought, if only Casey could get but \5\ a whack at that -  We'd put up even money, now, with Casey at the bat.

But Flynn preceded Casey, as \6\ did also Jimmy Blake,  And the former was a lulu and the latter was a cake;  So upon that stricken \7\ multitude grim melancholy sat, For there seemed but little chance of Casey's getting to the bat.

But Flynn let drive \8\ a single, to the wonderment of all,  And Blake, the much despised, tore the cover off the ball;  And when \9\ the dust had lifted, and the men saw what had occurred,  There was Jimmy safe at second and Flynn a \10\ hugging third.

Then from five thousand throats and more there rose a lusty yell;  It rumbled through the valley, it \11\ rattled in the dell;  It knocked upon the mountain and recoiled upon the flat,  For Casey, mighty Casey, was advancing \12\ to the bat.

There was ease in Casey's manner as he stepped into his place;  There was pride in Casey's \13\ bearing and a smile on Casey's face.  And when, responding to the cheers, he lightly doffed his hat,  No stranger \14\ in the crowd could doubt 'twas Casey at the bat.

Ten thousand eyes were on him as he rubbed his \15\ hands with dirt;   Five thousand tongues applauded when he wiped them on his shirt.  Then while the writhing pitcher ground \16\ the ball into his hip,  Defiance gleamed in Casey's eye, a sneer curled Casey's lip.

And now the leather covered \17\ sphere came hurtling through the air,  And Casey stood a-watching it in haughty grandeur there.  Close by the sturdy batsman \18\ the ball unheeded sped-  "That ain't my style," said Casey. "Strike one," the umpire said.

From the benches, black with \19\ people, there went up a muffled roar,  Like the beating of the storm waves on a stern and distant shore. \20\  "Kill him! Kill the umpire!" shouted someone on the stand;  And its likely they'd a-killed him had not Casey raised \21\ his hand.

With a smile of Christian charity great Casey's visage shone;  He stilled the rising tumult; he bade the \22\ game go on;  He signaled to the pitcher, and once more the spheroid flew;  But Casey still ignored it, and \23\ the umpire  said, "Strike two."

"Fraud!" cried the maddened thousands, and echo answered fraud;  But one scornful look from Casey \24\ and the audience was awed.  They saw his face grow stern and cold, they saw his muscles strain,  And they \25\ knew that Casey wouldn't let that ball go by again.

The sneer is gone from Casey's lip, his teeth are \26\ clenched in hate;  He pounds with cruel violence his bat upon the plate.  And now the pitcher holds the ball, \27\ and now he lets it go,  And now the air is shattered by the force of Casey's blow.

Oh, somewhere \28\ in this favored land the sun is shining bright;  The band is playing somewhere, and somewhere hearts are light,  And \29\ somewhere men are laughing, and somewhere children shout;  But there is no joy in Mudville - mighty Casey has struck out.

by Ernest Lawrence Thayer

Name _____   Date_____

# Lesson 15:  Random Sampling

Exit Ticket

Identify each as true or false.  Explain your reasoning in each case.

1.  The values of a sample statistic for different random samples of the same size from the same population will be the same.

2.  Random samples from the same population will vary from sample to sample.

3.  If a random sample is chosen from a population that has a large cluster of points at the maximum, the sample is likely to have at least one element near the maximum.

**100 Grocery Items (2013 prices)**

| | | | |
|---|---|---|---|
| T-Bone steaks<br>$ 6.99 (1 lb.) | Porterhouse steaks<br>$ 7.29 (1 lb.) | Pasta sauce<br>$ 2.19 (16 oz.) | Ice cream cups<br>$ 7.29 (6 cups) |
| Hot dog buns<br>$ 0.88 (6 buns) | Baking chips<br>$ 2.99 (12 oz.) | Cheese chips<br>$ 2.09 (12 oz.) | Cookies<br>$ 1.77 (15 oz.) |
| Kidney beans<br>$ 0.77 (15 oz.) | Box of oatmeal<br>$ 1.77 (18 oz.) | Soup<br>$ 0.77 (14 oz.) | Chicken breasts<br>$ 7.77 (1.5 lb.) |
| Pancake syrup<br>$ 2.99 (28 oz.) | Cranberry juice<br>$ 2.77 (64 oz.) | Asparagus<br>$3.29 | Seedless cucumbers<br>$ 1.29 |
| Avocado<br>$ 1.30 | Sliced pineapple<br>$ 2.99 | Box of tea<br>$ 4.29 (16 tea bags) | Cream cheese<br>$ 2.77 (16 oz.) |
| Italian roll<br>$ 1.39 (1 roll) | Turkey breast<br>$ 4.99 (1 lb.) | Meatballs<br>$ 5.79 (26 oz.) | Chili<br>$ 1.35 (15 oz.) |
| Peanut butter<br>$ 1.63 (12 oz.) | Green beans<br>$ 0.99 (1 lb.) | Apples<br>$ 1.99 (1 lb.) | Mushrooms<br>$ 0.69 (8 oz.) |
| Brown sugar<br>$ 1.29 (32 oz.) | Confectioners' sugar<br>$ 1.39 (32 oz.) | Zucchini<br>$ 0.79 (1 lb.) | Yellow onions<br>$ 0.99 (1 lb.) |
| Green peppers<br>$ 0.99 (1 count) | Mozzarella cheese<br>$2.69 (8 oz.) | Frozen chicken<br>$ 6.49 (48 oz.) | Olive oil<br>$ 2.99 (17 oz.) |
| Dark chocolate<br>$ 2.99 (9 oz.) | Cocoa mix<br>$ 3.33 (1 package) | Margarine<br>$ 1.48 (16 oz.) | Mac and Cheese<br>$ 0.66 (6 oz. box) |
| Birthday cake<br>$ 9.49 (7 in.) | Crab legs<br>$ 19.99 (1 lb.) | Cooked shrimp<br>$ 12.99 (32 oz.) | Crab legs<br>$ 19.99 (1 lb.) |
| Cooked shrimp<br>$ 12.99 (32 oz.) | Ice cream<br>$ 4.49 (1 qt.) | Pork chops<br>$ 1.79 (1 lb.) | Bananas<br>$ 0.44 (1 lb.) |
| Chocolate milk<br>$ 2.99 (1 gal.) | Beef franks<br>$ 3.35 (1 lb.) | Sliced bacon<br>$ 5.49 (1 lb.) | Fish fillets<br>$ 6.29 (1 lb.) |

| | | | |
|---|---|---|---|
| Pears<br>$ 1.29 (1 lb.) | Tangerine<br>$ 3.99 (3 lb.) | Orange juice<br>$ 2.98 (59 oz.) | Cherry pie<br>$ 4.44 (8 in.) |
| Grapes<br>$ 1.28 (1 lb.) | Peaches<br>$ 1.28 (1 lb.) | Melon<br>$ 1.69 (1 melon) | Tomatoes<br>$ 1.49 (1 lb.) |
| Shredded cheese<br>$ 1.88 (12 oz.) | Soda<br>$ 0.88 (1 can) | Roast beef<br>$ 6.49 (1 lb.) | Coffee<br>$ 6.49 (1 lb.) |
| Feta cheese<br>$ 4.99 (1 lb.) | Pickles<br>1.69 (12 oz. jar) | Load of rye bread<br>$ 2.19 | Crackers<br>$ 2.69 (7.9 oz.) |
| Purified water<br>$ 3.47 (35 pk.) | BBQ sauce<br>$ 2.19 (24 oz.) | Ketchup<br>$ 2.29 (34 oz.) | Chili sauce<br>$ 1.77 (12 oz.) |
| Sugar<br>$ 1.77 (5 lb.) | Flour<br>$ 2.11 (4 lb.) | Breakfast cereal<br>$ 2.79 (9 oz.) | Cane sugar<br>$ 2.39 (4 lb.) |
| Cheese sticks<br>$ 1.25 (10 oz.) | Cheese spread<br>$ 2.49 (45 oz.) | Coffee creamer<br>$ 2.99 (12 oz.) | Candy bars<br>$ 7.77 (40 oz.) |
| Pudding mix<br>$ 0.98 (6 oz.) | Fruit drink<br>$ 1.11 (24 oz.) | Biscuit mix<br>$ 0.89 (4 oz.) | Sausages<br>$ 2.38 (13 oz.) |
| Meat balls<br>$ 1.98 (10 oz.) | Apple juice<br>$ 1.48 (64 oz.) | Ice cream sandwich<br>$1.98 (12 ct.) | Cottage cheese<br>$ 1.98 (24 oz.) |
| Frozen vegetables<br>$ 0.88 (10 oz.) | English muffins<br>$ 1.68 (6 ct.) | String cheese<br>$ 6.09 (24 oz.) | Baby greens<br>$ 2.98 (10 oz.) |
| Caramel apples<br>$ 3.11 (1 apple) | Pumpkin mix<br>$ 3.50 (1 lb.) | Chicken salad<br>$ 0.98 (2 oz.) | Whole wheat bread<br>$ 1.55 (1 loaf) |
| Tuna<br>$ 0.98 (2.5 oz.) | Nutrition bar<br>$ 2.19 (1 bar) | Potato chips<br>$ 2.39 (12 oz.) | 2% milk<br>$3.13 (gal.) |

Name _____    Date_____

# Lesson 16:  Methods for Selecting a Random Sample

Exit Ticket

1. Name two things to consider when you are planning how to select a random sample.

2. Consider a population consisting of the 200 seventh graders at a particular middle school.  Describe how you might select a random sample of 20 students from a list of the students in this population.

Name _____    Date_____

# Lesson 17:  Sampling Variability

Suppose that you want to estimate the mean time per evening students at your school spend doing homework.  You will do this using a random sample of 30 students.

1.  Suppose that you have a list of all the students at your school.  The students are numbered $1, 2, 3, \ldots$.  One way to select the random sample of students is to use the random-digit table from today's class, taking three digits at a time.  If you start at the third digit of row 9, what is the number of the first student you would include in your sample?

2.  Suppose that you have now selected your random sample and you have asked each student how long he or she spends doing homework each evening.  How will you use these results to estimate the mean time spent doing homework for *all* students?

3.  Explain what is meant by *sampling variability* in this context.

Name _____    Date_____

# Lesson 18:  Sampling Variability and the Effect of Sample Size

Exit Ticket

Suppose that you wanted to estimate the mean time per evening spent doing homework for students at your school. You decide to do this by taking a random sample of students from your school.  You will calculate the mean time spent doing homework for your sample.  You will then use your sample mean as an estimate of the population mean.

1.   The sample mean has "sampling variability."  Explain what this means.

2.   When you are using a sample statistic to estimate a population characteristic, do you want the sampling variability of the sample statistic to be large or small?  Explain why.

3.   Think about your estimate of the mean time spent doing homework for students at your school.  Given a choice of using a sample of size 20 or a sample of size 40, which should you choose?  Explain your answer.

Name _____     Date_____

# Lesson 19:  Understanding Variability when Estimating a Population Proportion

Exit Ticket

A group of seventh graders took repeated samples of size 20 from a bag of colored cubes.  The dot plot below shows the sampling distribution of the sample proportion of blue cubes in the bag.

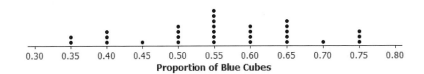

1.  Describe the shape of the distribution.

2.  Describe the variability of the distribution.

3.  Predict how the dot plot would look differently if the sample sizes had been 40 instead of 20.

Name _____   Date_____

# Lesson 20:  Estimating a Population Proportion

Exit Ticket

Thirty seventh graders each took a random sample of 10 middle school students and asked each student whether or not they like pop music.  Then they calculated the proportion of students who like pop music for each sample.  The dot plot below shows the distribution of the sample proportions.

**Dot Plot of Sample Proportions for n=10**

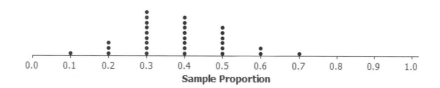

Sample Proportion

1.   There are three dots above 0.2.  What does each dot represent in terms of this scenario?

2.   Based on the dot plot, do you think the proportion of the middle school students at this school who like pop music is 0.6?  Explain why or why not.

**Table of Random Digits**

| Row | | | | | | | | | | | | | | | | | | | | |
|-----|---|---|---|---|---|---|---|---|---|---|---|---|---|---|---|---|---|---|---|---|
| 1 | 6 | 6 | 7 | 2 | 8 | 0 | 0 | 8 | 4 | 0 | 0 | 4 | 6 | 0 | 3 | 2 | 2 | 4 | 6 | 8 |
| 2 | 8 | 0 | 3 | 1 | 1 | 1 | 1 | 2 | 7 | 0 | 1 | 9 | 1 | 2 | 7 | 1 | 3 | 3 | 5 | 3 |
| 3 | 5 | 3 | 5 | 7 | 3 | 6 | 3 | 1 | 7 | 2 | 5 | 5 | 1 | 4 | 7 | 1 | 6 | 5 | 6 | 5 |
| 4 | 9 | 1 | 1 | 9 | 2 | 8 | 3 | 0 | 3 | 6 | 7 | 7 | 4 | 7 | 5 | 9 | 8 | 1 | 8 | 3 |
| 5 | 9 | 0 | 2 | 9 | 9 | 7 | 4 | 6 | 3 | 6 | 6 | 3 | 7 | 4 | 2 | 7 | 0 | 0 | 1 | 9 |
| 6 | 8 | 1 | 4 | 6 | 4 | 6 | 8 | 2 | 8 | 9 | 5 | 5 | 2 | 9 | 6 | 2 | 5 | 3 | 0 | 3 |
| 7 | 4 | 1 | 1 | 9 | 7 | 0 | 7 | 2 | 9 | 0 | 9 | 7 | 0 | 4 | 6 | 2 | 3 | 1 | 0 | 9 |
| 8 | 9 | 9 | 2 | 7 | 1 | 3 | 2 | 9 | 0 | 3 | 9 | 0 | 7 | 5 | 6 | 7 | 1 | 7 | 8 | 7 |
| 9 | 3 | 4 | 2 | 2 | 9 | 1 | 9 | 0 | 7 | 8 | 1 | 6 | 2 | 5 | 3 | 9 | 0 | 9 | 1 | 0 |
| 10 | 2 | 7 | 3 | 9 | 5 | 9 | 9 | 3 | 2 | 9 | 3 | 9 | 1 | 9 | 0 | 5 | 5 | 1 | 4 | 2 |
| 11 | 0 | 2 | 5 | 4 | 0 | 8 | 1 | 7 | 0 | 7 | 1 | 3 | 0 | 4 | 3 | 0 | 6 | 4 | 4 | 4 |
| 12 | 8 | 6 | 0 | 5 | 4 | 8 | 8 | 2 | 7 | 7 | 0 | 1 | 0 | 1 | 7 | 1 | 3 | 5 | 3 | 4 |
| 13 | 4 | 2 | 6 | 4 | 5 | 2 | 4 | 2 | 6 | 1 | 7 | 5 | 6 | 6 | 4 | 0 | 8 | 4 | 1 | 2 |
| 14 | 4 | 4 | 9 | 8 | 7 | 3 | 4 | 3 | 8 | 2 | 9 | 1 | 5 | 3 | 5 | 9 | 8 | 9 | 2 | 9 |
| 15 | 6 | 4 | 8 | 0 | 0 | 0 | 4 | 2 | 3 | 8 | 1 | 8 | 4 | 0 | 9 | 5 | 0 | 9 | 0 | 4 |
| 16 | 3 | 2 | 3 | 8 | 4 | 8 | 8 | 6 | 2 | 9 | 1 | 0 | 1 | 9 | 9 | 3 | 0 | 7 | 3 | 5 |
| 17 | 6 | 6 | 7 | 2 | 8 | 0 | 0 | 8 | 4 | 0 | 0 | 4 | 6 | 0 | 3 | 2 | 2 | 4 | 6 | 8 |
| 18 | 8 | 0 | 3 | 1 | 1 | 1 | 1 | 2 | 7 | 0 | 1 | 9 | 1 | 2 | 7 | 1 | 3 | 3 | 5 | 3 |
| 19 | 5 | 3 | 5 | 7 | 3 | 6 | 3 | 1 | 7 | 2 | 5 | 5 | 1 | 4 | 7 | 1 | 6 | 5 | 6 | 5 |
| 20 | 9 | 1 | 1 | 9 | 2 | 8 | 3 | 0 | 3 | 6 | 7 | 7 | 4 | 7 | 5 | 9 | 8 | 1 | 8 | 3 |
| 21 | 9 | 0 | 2 | 9 | 9 | 7 | 4 | 6 | 3 | 6 | 6 | 3 | 7 | 4 | 2 | 7 | 0 | 0 | 1 | 9 |
| 22 | 8 | 1 | 4 | 6 | 4 | 6 | 8 | 2 | 8 | 9 | 5 | 5 | 2 | 9 | 6 | 2 | 5 | 3 | 0 | 3 |
| 23 | 4 | 1 | 1 | 9 | 7 | 0 | 7 | 2 | 9 | 0 | 9 | 7 | 0 | 4 | 6 | 2 | 3 | 1 | 0 | 9 |
| 24 | 9 | 9 | 2 | 7 | 1 | 3 | 2 | 9 | 0 | 3 | 9 | 0 | 7 | 5 | 6 | 7 | 1 | 7 | 8 | 7 |
| 25 | 3 | 4 | 2 | 2 | 9 | 1 | 9 | 0 | 7 | 8 | 1 | 6 | 2 | 5 | 3 | 9 | 0 | 9 | 1 | 0 |
| 26 | 2 | 7 | 3 | 9 | 5 | 9 | 9 | 3 | 2 | 9 | 3 | 9 | 1 | 9 | 0 | 5 | 5 | 1 | 4 | 2 |
| 27 | 0 | 2 | 5 | 4 | 0 | 8 | 1 | 7 | 0 | 7 | 1 | 3 | 0 | 4 | 3 | 0 | 6 | 4 | 4 | 4 |
| 28 | 8 | 6 | 0 | 5 | 4 | 8 | 8 | 2 | 7 | 7 | 0 | 1 | 0 | 1 | 7 | 1 | 3 | 5 | 3 | 4 |
| 29 | 4 | 2 | 6 | 4 | 5 | 2 | 4 | 2 | 6 | 1 | 7 | 5 | 6 | 6 | 4 | 0 | 8 | 4 | 1 | 2 |
| 30 | 4 | 4 | 9 | 8 | 7 | 3 | 4 | 3 | 8 | 2 | 9 | 1 | 5 | 3 | 5 | 9 | 8 | 9 | 2 | 9 |
| 31 | 6 | 4 | 8 | 0 | 0 | 0 | 4 | 2 | 3 | 8 | 1 | 8 | 4 | 0 | 9 | 5 | 0 | 9 | 0 | 4 |
| 32 | 3 | 2 | 3 | 8 | 4 | 8 | 8 | 6 | 2 | 9 | 1 | 0 | 1 | 9 | 9 | 3 | 0 | 7 | 3 | 5 |
| 33 | 6 | 6 | 7 | 2 | 8 | 0 | 0 | 8 | 4 | 0 | 0 | 4 | 6 | 0 | 3 | 2 | 2 | 4 | 6 | 8 |
| 34 | 8 | 0 | 3 | 1 | 1 | 1 | 1 | 2 | 7 | 0 | 1 | 9 | 1 | 2 | 7 | 1 | 3 | 3 | 5 | 3 |
| 35 | 5 | 3 | 5 | 7 | 3 | 6 | 3 | 1 | 7 | 2 | 5 | 5 | 1 | 4 | 7 | 1 | 6 | 5 | 6 | 5 |
| 36 | 9 | 1 | 1 | 9 | 2 | 8 | 3 | 0 | 3 | 6 | 7 | 7 | 4 | 7 | 5 | 9 | 8 | 1 | 8 | 3 |
| 37 | 9 | 0 | 2 | 9 | 9 | 7 | 4 | 6 | 3 | 6 | 6 | 3 | 7 | 4 | 2 | 7 | 0 | 0 | 1 | 9 |
| 38 | 8 | 1 | 4 | 6 | 4 | 6 | 8 | 2 | 8 | 9 | 5 | 5 | 2 | 9 | 6 | 2 | 5 | 3 | 0 | 3 |
| 39 | 4 | 1 | 1 | 9 | 7 | 0 | 7 | 2 | 9 | 0 | 9 | 7 | 0 | 4 | 6 | 2 | 3 | 1 | 0 | 9 |
| 40 | 9 | 9 | 2 | 7 | 1 | 3 | 2 | 9 | 0 | 3 | 9 | 0 | 7 | 5 | 6 | 7 | 1 | 7 | 8 | 7 |

Lesson 20:     Estimating a Population Proportion

| ID | Travel to School | Favorite Season | Allergies | Favorite School Subject | Favorite Music | Superpower |
|---|---|---|---|---|---|---|
| 1 | Car | Spring | Yes | English | Pop | Freeze time |
| 2 | Car | Summer | Yes | Music | Pop | Telepathy |
| 3 | Car | Summer | No | Science | Pop | Fly |
| 4 | Walk | Fall | No | Computers and technology | Pop | Invisibility |
| 5 | Car | Summer | No | Art | Country | Telepathy |
| 6 | Car | Summer | No | Physical education | Rap/Hip hop | Freeze time |
| 7 | Car | Spring | No | Physical education | Pop | Telepathy |
| 8 | Car | Winter | No | Art | Other | Fly |
| 9 | Car | Summer | No | Physical education | Pop | Fly |
| 10 | Car | Spring | No | Mathematics and statistics | Pop | Telepathy |
| 11 | Car | Summer | Yes | History | Rap/Hip hop | Invisibility |
| 12 | Car | Spring | No | Art | Rap/Hip hop | Freeze time |
| 13 | Bus | Winter | No | Computers and technology | Rap/Hip hop | Fly |
| 14 | Car | Winter | Yes | Social studies | Rap/Hip hop | Fly |
| 15 | Car | Summer | No | Art | Pop | Freeze time |
| 16 | Car | Fall | No | Mathematics and statistics | Pop | Fly |
| 17 | Bus | Winter | No | Science | Rap/Hip hop | Freeze time |
| 18 | Car | Spring | Yes | Art | Pop | Telepathy |
| 19 | Car | Fall | Yes | Science | Pop | Telepathy |
| 20 | Car | Summer | Yes | Physical education | Rap/Hip hop | Invisibility |
| 21 | Car | Spring | Yes | Science | Pop | Invisibility |
| 22 | Car | Winter | Yes | Mathematics and statistics | Country | Invisibility |
| 23 | Car | Summer | Yes | Art | Pop | Invisibility |
| 24 | Bus | Winter | Yes | Other | Pop | Telepathy |
| 25 | Bus | Summer | Yes | Science | Other | Fly |
| 26 | Car | Summer | No | Science | Pop | Fly |
| 27 | Car | Summer | Yes | Music | Pop | Telepathy |
| 28 | Car | Summer | No | Physical education | Country | Super strength |
| 29 | Car | Fall | Yes | Mathematics and statistics | Country | Telepathy |
| 30 | Car | Summer | Yes | Physical education | Rap/Hip hop | Telepathy |
| 31 | Boat | Winter | No | Computers and technology | Gospel | Invisibility |
| 32 | Car | Spring | No | Physical education | Pop | Fly |
| 33 | Car | Spring | No | Physical education | Pop | Fly |
| 34 | Car | Summer | No | Mathematics and statistics | Classical | Fly |
| 35 | Car | Fall | Yes | Science | Jazz | Telepathy |
| 36 | Car | Spring | No | Science | Rap/Hip hop | Telepathy |
| 37 | Car | Summer | No | Music | Country | Telepathy |
| 38 | Bus | Winter | No | Mathematics and statistics | Pop | Fly |
| 39 | Car | Spring | No | Art | Classical | Freeze time |
| 40 | Car | Winter | Yes | Art | Pop | Fly |
| 41 | Walk | Summer | Yes | Physical education | Rap/Hip hop | Fly |
| 42 | Bus | Winter | Yes | Physical education | Gospel | Invisibility |

| 43 | Bus | Summer | No | Art | Other | Invisibility |
|----|-----|--------|----|----|-------|--------------|
| 44 | Car | Summer | Yes | Computers and technology | Other | Freeze time |
| 45 | Car | Fall | Yes | Science | Pop | Fly |
| 46 | Car | Summer | Yes | Music | Rap/Hip hop | Fly |
| 47 | Car | Spring | No | Science | Rap/Hip hop | Invisibility |
| 48 | Bus | Spring | No | Music | Pop | Telepathy |
| 49 | Car | Summer | Yes | Social studies | Techno/Electronic | Telepathy |
| 50 | Car | Summer | Yes | Physical education | Pop | Telepathy |
| 51 | Car | Spring | Yes | Other | Other | Telepathy |
| 52 | Car | Summer | No | Art | Pop | Fly |
| 53 | Car | Summer | Yes | Other | Pop | Telepathy |
| 54 | Car | Summer | Yes | Physical education | Rap/Hip hop | Invisibility |
| 55 | Bus | Summer | Yes | Physical education | Other | Super strength |
| 56 | Car | Summer | No | Science | Rap/Hip hop | Invisibility |
| 57 | Car | Winter | No | Languages | Rap/Hip hop | Super strength |
| 58 | Car | Fall | Yes | English | Pop | Fly |
| 59 | Car | Winter | No | Science | Pop | Telepathy |
| 60 | Car | Summer | No | Art | Pop | Invisibility |
| 61 | Car | Summer | Yes | Other | Pop | Freeze time |
| 62 | Bus | Spring | No | Science | Pop | Fly |
| 63 | Car | Winter | Yes | Mathematics and statistics | Other | Freeze time |
| 64 | Car | Summer | No | Social studies | Classical | Fly |
| 65 | Car | Winter | Yes | Science | Pop | Telepathy |
| 66 | Car | Winter | No | Science | Rock and roll | Fly |
| 67 | Car | Summer | No | Mathematics and statistics | Rap/Hip hop | Super strength |
| 68 | Car | Fall | No | Music | Rock and roll | Super strength |
| 69 | Car | Spring | No | Other | Other | Invisibility |
| 70 | Car | Summer | Yes | Mathematics and statistics | Rap/Hip hop | Telepathy |
| 71 | Car | Winter | No | Art | Other | Fly |
| 72 | Car | Spring | Yes | Mathematics and statistics | Pop | Telepathy |
| 73 | Car | Winter | Yes | Computers and technology | Techno/Electronic | Telepathy |
| 74 | Walk | Winter | No | Physical education | Techno/Electronic | Fly |
| 75 | Walk | Summer | No | History | Rock and roll | Fly |
| 76 | Skateboard/Scooter/Rollerblade | Winter | Yes | Computers and technology | Techno/Electronic | Freeze time |
| 77 | Car | Spring | Yes | Science | Other | Telepathy |
| 78 | Car | Summer | No | Music | Rap/Hip hop | Invisibility |
| 79 | Car | Summer | No | Social studies | Pop | Invisibility |
| 80 | Car | Summer | No | Other | Rap/Hip hop | Telepathy |
| 81 |  | Spring | Yes | History | Rap/Hip hop | Invisibility |
| 82 | Car | Summer | No | Art | Pop | Invisibility |

Lesson 20:    Estimating a Population Proportion

| 83 | Walk | Spring | No | Languages | Jazz | Super strength |
| 84 | Car | Fall | No | History | Jazz | Invisibility |
| 85 | Car | Summer | No | Physical education | Rap/Hip hop | Freeze time |
| 86 | Car | Spring | No | Mathematics and statistics | Pop | Freeze time |
| 87 | Bus | Spring | Yes | Art | Pop | Telepathy |
| 88 | Car | Winter | No | Mathematics and statistics | Other | Invisibility |
| 89 | Car | Summer | Yes | Physical education | Country | Telepathy |
| 90 | Bus | Summer | No | Computers and technology | Other | Fly |
| 91 | Car | Winter | No | History | Pop | Telepathy |
| 92 | Walk | Winter | No | Science | Classical | Telepathy |
| 93 | Bicycle | Summer | No | Physical education | Pop | Invisibility |
| 94 | Car | Summer | No | English | Pop | Telepathy |
| 95 | Car | Summer | Yes | Physical education | Pop | Fly |
| 96 | Car | Winter | No | Science | Other | Freeze time |
| 97 | Car | Winter | No | Other | Rap/Hip hop | Super strength |
| 98 | Car | Summer | Yes | Physical education | Rap/Hip hop | Freeze time |
| 99 | Car | Spring | No | Music | Classical | Telepathy |
| 100 | Car | Spring | Yes | Science | Gospel | Telepathy |
| 101 | Car | Summer | Yes | History | Pop | Super strength |
| 102 | Car | Winter | Yes | English | Country | Freeze time |
| 103 | Car | Spring | No | Computers and technology | Other | Telepathy |
| 104 | Car | Winter | No | History | Other | Invisibility |
| 105 | Car | Fall | No | Music | Pop | Telepathy |
| 106 | Car | Fall | No | Science | Pop | Telepathy |
| 107 | Car | Winter | No | Art | Heavy metal | Fly |
| 108 | Car | Spring | Yes | Science | Rock and roll | Fly |
| 109 | Car | Fall | Yes | Music | Other | Fly |
| 110 | Car | Summer | Yes | Social studies | Techno/ Electronic | Telepathy |
| 111 | Car | Spring | No | Physical education | Pop | Fly |
| 112 | Car | Summer | No | Physical education | Pop | Fly |
| 113 | Car | Summer | Yes | Social studies | Pop | Freeze time |
| 114 | Car | Summer | Yes | Computers and technology | Gospel | Freeze time |
| 115 | Car | Winter | Yes | Other | Rap/Hip hop | Telepathy |
| 116 | Car | Summer | Yes | Science | Country | Telepathy |
| 117 | Car | Fall | | Music | Country | Fly |
| 118 | Walk | Summer | No | History | Pop | Telepathy |
| 119 | Car | Spring | Yes | Art | Pop | Freeze time |
| 120 | Car | Fall | Yes | Physical education | Rap/Hip hop | Fly |
| 121 | Car | Spring | No | Music | Rock and roll | Telepathy |
| 122 | Car | Fall | No | Art | Pop | Invisibility |
| 123 | Car | Summer | Yes | Physical education | Rap/Hip hop | Fly |
| 124 | | Summer | No | Computers and technology | Pop | Telepathy |
| 125 | Car | Fall | No | Art | Pop | Fly |

| 126 | Bicycle | Spring | No | Science | Pop | Invisibility |
|---|---|---|---|---|---|---|
| 127 | Car | Summer | No | Social studies | Gospel | Fly |
| 128 | Bicycle | Winter | No | Social studies | Rap/Hip hop | Fly |
| 129 | Car | Summer | Yes | Mathematics and statistics | Pop | Invisibility |
| 130 | Car | Fall | Yes | Mathematics and statistics | Country | Telepathy |
| 131 | Car | Winter | Yes | Music | Gospel | Super strength |
| 132 | Rail (Train/Tram /Subway) | Fall | Yes | Art | Other | Fly |
| 133 | Walk | Summer | No | Social studies | Pop | Invisibility |
| 134 | Car | Summer | Yes | Music | Pop | Freeze time |
| 135 | Car | Winter | No | Mathematics and statistics | Pop | Telepathy |
| 136 | Car | Fall | Yes | Music | Pop | Telepathy |
| 137 | Car | Summer | Yes | Computers and technology | Other | Freeze time |
| 138 | Car | Summer | Yes | Physical education | Pop | Telepathy |
| 139 | Car | Summer | Yes | Social studies | Other | Telepathy |
| 140 | Car | Spring | Yes | Physical education | Other | Freeze time |
| 141 | Car | Fall | Yes | Science | Country | Telepathy |
| 142 | Car | Spring | Yes | Science | Pop | Invisibility |
| 143 | Car | Summer | No | Other | Rap/Hip hop | Freeze time |
| 144 | Car | Summer | No | Other | Other | Fly |
| 145 | Car | Summer | No | Languages | Pop | Freeze time |
| 146 | Car | Summer | Yes | Physical education | Pop | Telepathy |
| 147 | Bus | Winter | No | History | Country | Invisibility |
| 148 | Car | Spring | No | Computers and technology | Other | Telepathy |
| 149 | Bus | Winter | Yes | Science | Pop | Invisibility |
| 150 | Car | Summer | No | Social studies | Rap/Hip hop | Invisibility |
| 151 | Car | Summer | No | Physical education | Pop | Invisibility |
| 152 | Car | Summer | Yes | Physical education | Pop | Super strength |
| 153 | Car | Summer | No | Mathematics and statistics | Pop | Fly |
| 154 | Car | Summer | No | Art | Rap/Hip hop | Freeze time |
| 155 | Car | Winter | Yes | Other | Classical | Freeze time |
| 156 | Car | Summer | Yes | Computers and technology | Other | Telepathy |
| 157 | Car | Spring | No | Other | Pop | Freeze time |
| 158 | Car | Winter | Yes | Music | Country | Fly |
| 159 | Car | Winter | No | History | Jazz | Invisibility |
| 160 | Car | Spring | Yes | History | Pop | Fly |
| 161 | Car | Winter | Yes | Mathematics and statistics | Other | Telepathy |
| 162 | Car | Fall | No | Science | Country | Invisibility |
| 163 | Car | Winter | No | Science | Other | Fly |
| 164 | Car | Summer | No | Science | Pop | Fly |
| 165 | Skateboard /Scooter/ Rollerblade | Spring | Yes | Social studies | Other | Freeze time |
| 166 | Car | Winter | Yes | Art | Rap/Hip hop | Fly |

Lesson 20:    Estimating a Population Proportion

| 167 | Car | Summer | Yes | Other | Pop | Freeze time |
| 168 | Car | Summer | No | English | Pop | Telepathy |
| 169 | Car | Summer | No | Other | Pop | Invisibility |
| 170 | Car | Summer | Yes | Physical education | Techno/Electronic | Freeze time |
| 171 | Car | Summer | No | Art | Pop | Telepathy |
| 172 | Car | Summer | No | Physical education | Rap/Hip hop | Freeze time |
| 173 | Car | Winter | Yes | Mathematics and statistics | Other | Invisibility |
| 174 | Bus | Summer | Yes | Music | Pop | Freeze time |
| 175 | Car | Winter | No | Art | Pop | Fly |
| 176 | Car | Fall | No | Science | Rap/Hip hop | Fly |
| 177 | Car | Winter | Yes | Social studies | Pop | Telepathy |
| 178 | Car | Fall | No | Art | Other | Fly |
| 179 | Bus | Spring | No | Physical education | Country | Fly |
| 180 | Car | Winter | No | Music | Other | Telepathy |
| 181 | Bus | Summer | No | Computers and technology | Rap/Hip hop | Freeze time |
| 182 | Car | Summer | Yes | Physical education | Rap/Hip hop | Invisibility |
| 183 | Car | Summer | Yes | Music | Other | Telepathy |
| 184 | Car | Spring | No | Science | Rap/Hip hop | Invisibility |
| 185 | Rail (Train/Tram/Subway) | Summer | No | Physical education | Other | Freeze time |
| 186 | Car | Summer | Yes | Mathematics and statistics | Rap/Hip hop | Fly |
| 187 | Bus | Winter | Yes | Mathematics and statistics | Other | Super strength |
| 188 | Car | Summer | No | Mathematics and statistics | Other | Freeze time |
| 189 | Rail (Train/Tram/Subway) | Fall | Yes | Music | Jazz | Fly |
| 190 | Car | Summer | Yes | Science | Pop | Super strength |
| 191 | Car | Summer | Yes | Science | Techno/Electronic | Freeze time |
| 192 | Car | Spring | Yes | Physical education | Rap/Hip hop | Freeze time |
| 193 | Car | Summer | Yes | Physical education | Rap/Hip hop | Freeze time |
| 194 | Car | Winter | No | Physical education | Rap/Hip hop | Telepathy |
| 195 | Car | Winter | No | Music | Jazz | Freeze time |
| 196 | Walk | Summer | Yes | History | Country | Freeze time |
| 197 | Car | Spring | No | History | Rap/Hip hop | Freeze time |
| 198 | Car | Fall | Yes | Other | Pop | Freeze time |
| 199 | Car | Spring | Yes | Science | Other | Freeze time |
| 200 | Bicycle | Winter | Yes | Other | Rap/Hip hop | Freeze time |

Name _____     Date_____

# Lesson 21:  Why Worry about Sampling Variability?

Exit Ticket

How is a *meaningful* difference in sample means different from a *non-meaningful* difference in sample means?  You may use what you saw in the dot plots of this lesson to help you answer this question.

**Template for Bags A and B**

| | | | | | | | | | |
|---|---|---|---|---|---|---|---|---|---|
| 5 | 5 | 5 | 5 | 5 | 6 | 6 | 6 | 6 | 6 |
| 7 | 7 | 7 | 7 | 7 | 8 | 8 | 8 | 8 | 8 |
| 9 | 9 | 9 | 9 | 9 | 10 | 10 | 10 | 10 | 10 |
| 11 | 11 | 11 | 11 | 11 | 12 | 12 | 12 | 12 | 12 |
| 13 | 13 | 13 | 13 | 13 | 14 | 14 | 14 | 14 | 14 |
| 15 | 15 | 15 | 15 | 15 | 16 | 16 | 16 | 16 | 16 |
| 17 | 17 | 17 | 17 | 17 | 18 | 18 | 18 | 18 | 18 |
| 19 | 19 | 19 | 19 | 19 | 20 | 20 | 20 | 20 | 20 |
| 21 | 21 | 21 | 21 | 21 | 22 | 22 | 22 | 22 | 22 |
| 23 | 23 | 23 | 23 | 23 | 24 | 24 | 24 | 24 | 24 |

**Template for Bag C**

| | | | | | | | | | |
|---|---|---|---|---|---|---|---|---|---|
| 1 | 1 | 1 | 1 | 1 | 2 | 2 | 2 | 2 | 2 |
| 3 | 3 | 3 | 3 | 3 | 4 | 4 | 4 | 4 | 4 |
| 5 | 5 | 5 | 5 | 5 | 6 | 6 | 6 | 6 | 6 |
| 7 | 7 | 7 | 7 | 7 | 8 | 8 | 8 | 8 | 8 |
| 9 | 9 | 9 | 9 | 9 | 10 | 10 | 10 | 10 | 10 |
| 11 | 11 | 11 | 11 | 11 | 12 | 12 | 12 | 12 | 12 |
| 13 | 13 | 13 | 13 | 13 | 14 | 14 | 14 | 14 | 14 |
| 15 | 15 | 15 | 15 | 15 | 16 | 16 | 16 | 16 | 16 |
| 17 | 17 | 17 | 17 | 17 | 18 | 18 | 18 | 18 | 18 |
| 19 | 19 | 19 | 19 | 19 | 20 | 20 | 20 | 20 | 20 |

Name _____     Date_____

# Lesson 22:  Using Sample Data to Decide if Two Population Means are Different

Exit Ticket

Suppose that Brett randomly sampled 12 tenth grade girls and boys in his school district and asked them for the number of minutes per day that they text.  The data and summary measures follow.

| Gender | Number of Minutes of Texting | | | | | | | | | | | | mean | MAD |
|--------|----|-----|----|-----|----|-----|----|----|----|-----|----|----|------|-----|
| Girls  | 98 | 104 | 95 | 101 | 98 | 107 | 86 | 92 | 96 | 107 | 88 | 95 | 97.3 | 5.3 |
| Boys   | 66 | 72  | 65 | 60  | 78 | 82  | 63 | 56 | 85 | 79  | 68 | 77 | 70.9 | 7.9 |

1.  Draw dot plots for the two data sets using the same numerical scales.  Discuss the amount of overlap between the two dot plots that you drew and what it may mean in the context of the problem.

2.  Compare the variability in the two data sets using the MAD.  Interpret the result in the context of the problem.

3.  From 1 and 2, does the difference in the two means appear to be meaningful?  Explain.

Name _____  Date_____

# Lesson 23:  Using Sample Data to Decide if Two Population Means are Different

1.  Do eleventh grade males text more per day than eleventh grade females do?  To answer this question, two randomly selected samples were obtained from the Excel data file used in this lesson.  Indicate how 20 randomly selected eleventh grade females would be chosen for this study.  Indicate how 20 randomly selected eleventh grade males would be chosen.

2.  Two randomly selected samples (one of eleventh grade females, and one of eleventh grade males) were obtained from the database.  The results are indicated below:

| | Mean number of minutes per day texting | MAD (minutes) |
|---|---|---|
| Eleventh grade females | 102.55 | 1.31 |
| Eleventh grade males | 100.32 | 1.12 |

Is there a meaningful difference in the number of minutes per day that eleventh grade females and males text?  Explain your answer.

**Copy of the Excel student data file:**

| ID Number | Texting | ReacTime | Homework | Sleep |
|---|---|---|---|---|
| 1 | 99 | 0.33 | 9.0 | 8.2 |
| 2 | 69 | 0.39 | 8.6 | 7.5 |
| 3 | 138 | 0.36 | 6.1 | 8.7 |
| 4 | 100 | 0.40 | 7.9 | 7.8 |
| 5 | 116 | 0.28 | 5.1 | 8.8 |
| 6 | 112 | 0.38 | 6.5 | 7.9 |
| 7 | 79 | 0.35 | 6.5 | 8.8 |
| 8 | 111 | 0.41 | 8.8 | 8.5 |
| 9 | 115 | 0.49 | 8.4 | 8.4 |
| 10 | 82 | 0.43 | 8.7 | 8.8 |
| 11 | 136 | 0.46 | 7.2 | 8.4 |
| 12 | 112 | 0.51 | 8.3 | 9.0 |
| 13 | 101 | 0.42 | 7.0 | 8.8 |
| 14 | 89 | 0.38 | 5.6 | 8.3 |
| 15 | 120 | 0.35 | 7.2 | 8.2 |
| 16 | 144 | 0.36 | 3.9 | 8.8 |
| 17 | 131 | 0.26 | 9.0 | 8.9 |
| 18 | 126 | 0.39 | 7.0 | 8.5 |
| 19 | 118 | 0.37 | 9.2 | 8.7 |
| 20 | 83 | 0.34 | 7.4 | 8.6 |
| 21 | 120 | 0.20 | 4.5 | 8.7 |
| 22 | 114 | 0.38 | 6.0 | 8.6 |
| 23 | 90 | 0.25 | 7.0 | 8.4 |
| 24 | 116 | 0.36 | 5.8 | 8.4 |
| 25 | 108 | 0.36 | 8.9 | 8.1 |
| 26 | 89 | 0.31 | 8.4 | 8.8 |
| 27 | 124 | 0.44 | 6.3 | 8.3 |
| 28 | 121 | 0.32 | 5.2 | 8.0 |
| 29 | 104 | 0.30 | 6.7 | 8.1 |
| 30 | 110 | 0.39 | 7.8 | 8.1 |
| 31 | 119 | 0.36 | 8.5 | 8.0 |
| 32 | 113 | 0.40 | 5.3 | 9.4 |
| 33 | 106 | 0.36 | 5.7 | 8.6 |
| 34 | 119 | 0.33 | 5.9 | 8.4 |
| 35 | 129 | 0.38 | 6.2 | 9.0 |
| 36 | 95 | 0.44 | 7.9 | 8.3 |
| 37 | 126 | 0.41 | 7.2 | 8.6 |
| 38 | 106 | 0.26 | 7.1 | 8.5 |

| 39 | 116 | 0.34 | 4.9 | 8.4 |
|----|-----|------|-----|-----|
| 40 | 107 | 0.35 | 9.3 | 8.1 |
| 41 | 108 | 0.48 | 8.1 | 8.6 |
| 42 | 97 | 0.40 | 7.1 | 8.8 |
| 43 | 97 | 0.27 | 4.2 | 8.3 |
| 44 | 100 | 0.24 | 6.2 | 8.9 |
| 45 | 123 | 0.50 | 8.1 | 8.6 |
| 46 | 94 | 0.39 | 5.2 | 8.3 |
| 47 | 87 | 0.37 | 8.0 | 8.3 |
| 48 | 93 | 0.42 | 9.7 | 8.1 |
| 49 | 117 | 0.39 | 7.9 | 8.3 |
| 50 | 94 | 0.36 | 6.9 | 8.9 |
| 51 | 124 | 0.29 | 8.1 | 8.4 |
| 52 | 116 | 0.44 | 4.9 | 8.2 |
| 53 | 137 | 0.25 | 9.1 | 8.3 |
| 54 | 123 | 0.30 | 3.8 | 8.7 |
| 55 | 122 | 0.21 | 6.6 | 8.8 |
| 56 | 92 | 0.41 | 7.6 | 8.6 |
| 57 | 101 | 0.36 | 8.5 | 8.2 |
| 58 | 111 | 0.34 | 6.5 | 8.7 |
| 59 | 126 | 0.31 | 5.6 | 8.6 |
| 60 | 81 | 0.33 | 7.3 | 8.4 |
| 61 | 118 | 0.35 | 8.4 | 8.3 |
| 62 | 113 | 0.37 | 7.7 | 8.4 |
| 63 | 114 | 0.49 | 6.9 | 8.6 |
| 64 | 124 | 0.34 | 8.5 | 8.1 |
| 65 | 90 | 0.48 | 6.6 | 8.7 |
| 66 | 99 | 0.39 | 6.5 | 8.5 |
| 67 | 155 | 0.40 | 6.5 | 9.0 |
| 68 | 77 | 0.45 | 4.9 | 8.3 |
| 69 | 79 | 0.23 | 7.7 | 8.6 |
| 70 | 91 | 0.43 | 6.7 | 8.9 |
| 71 | 107 | 0.42 | 6.9 | 8.6 |
| 72 | 112 | 0.39 | 6.3 | 8.6 |
| 73 | 147 | 0.34 | 7.8 | 8.3 |
| 74 | 126 | 0.52 | 9.2 | 8.9 |
| 75 | 106 | 0.27 | 6.7 | 8.0 |
| 76 | 86 | 0.21 | 6.1 | 8.4 |
| 77 | 111 | 0.29 | 7.4 | 8.0 |
| 78 | 118 | 0.41 | 5.2 | 8.8 |
| 79 | 105 | 0.28 | 5.9 | 8.3 |

| 80 | 108 | 0.40 | 7.0 | 8.6 |
|---|---|---|---|---|
| 81 | 131 | 0.32 | 7.5 | 8.5 |
| 82 | 86 | 0.34 | 7.4 | 8.4 |
| 83 | 87 | 0.27 | 7.7 | 8.6 |
| 84 | 116 | 0.34 | 9.6 | 9.1 |
| 85 | 102 | 0.42 | 5.6 | 8.7 |
| 86 | 105 | 0.41 | 8.2 | 8.2 |
| 87 | 96 | 0.33 | 4.9 | 8.4 |
| 88 | 90 | 0.20 | 7.5 | 8.6 |
| 89 | 109 | 0.39 | 6.0 | 8.5 |
| 90 | 103 | 0.36 | 5.7 | 9.1 |
| 91 | 98 | 0.26 | 5.5 | 8.4 |
| 92 | 103 | 0.41 | 8.2 | 8.0 |
| 93 | 110 | 0.39 | 9.2 | 8.4 |
| 94 | 101 | 0.39 | 8.8 | 8.6 |
| 95 | 101 | 0.40 | 6.1 | 9.0 |
| 96 | 98 | 0.22 | 7.0 | 8.6 |
| 97 | 105 | 0.43 | 8.0 | 8.5 |
| 98 | 110 | 0.33 | 8.0 | 8.8 |
| 99 | 104 | 0.45 | 9.6 | 8.5 |
| 100 | 119 | 0.29 | 7.2 | 9.1 |
| 101 | 120 | 0.48 | 6.7 | 8.4 |
| 102 | 109 | 0.36 | 6.8 | 8.5 |
| 103 | 105 | 0.29 | 8.6 | 8.6 |
| 104 | 110 | 0.29 | 7.1 | 8.9 |
| 105 | 116 | 0.41 | 6.7 | 8.8 |
| 106 | 114 | 0.25 | 9.0 | 8.8 |
| 107 | 111 | 0.35 | 9.9 | 8.6 |
| 108 | 137 | 0.37 | 6.1 | 7.9 |
| 109 | 104 | 0.37 | 7.3 | 9.0 |
| 110 | 95 | 0.33 | 8.3 | 8.0 |
| 111 | 117 | 0.29 | 7.5 | 8.8 |
| 112 | 100 | 0.30 | 2.9 | 9.3 |
| 113 | 105 | 0.24 | 6.1 | 8.5 |
| 114 | 102 | 0.41 | 7.7 | 8.0 |
| 115 | 109 | 0.32 | 5.8 | 8.2 |
| 116 | 108 | 0.36 | 6.6 | 8.3 |
| 117 | 111 | 0.43 | 8.2 | 8.5 |
| 118 | 115 | 0.20 | 5.7 | 8.9 |
| 119 | 89 | 0.34 | 5.1 | 8.1 |
| 120 | 94 | 0.29 | 9.2 | 8.5 |

| 121 | 105 | 0.26 | 7.3 | 8.6 |
|-----|-----|------|-----|-----|
| 122 | 143 | 0.42 | 5.8 | 8.1 |
| 123 | 129 | 0.29 | 6.8 | 8.8 |
| 124 | 118 | 0.31 | 6.2 | 8.7 |
| 125 | 129 | 0.38 | 9.1 | 8.4 |
| 126 | 96 | 0.25 | 6.7 | 8.5 |
| 127 | 95 | 0.27 | 7.7 | 8.3 |
| 128 | 43 | 0.22 | 6.3 | 9.2 |
| 129 | 64 | 0.23 | 3.7 | 8.5 |
| 130 | 38 | 0.41 | 5.2 | 9.1 |
| 131 | 46 | 0.35 | 7.1 | 8.3 |
| 132 | 56 | 0.32 | 4.7 | 8.9 |
| 133 | 41 | 0.45 | 7.2 | 8.8 |
| 134 | 55 | 0.27 | 4.1 | 8.5 |
| 135 | 57 | 0.40 | 7.5 | 8.6 |
| 136 | 66 | 0.41 | 6.9 | 8.0 |
| 137 | 62 | 0.39 | 3.8 | 8.5 |
| 138 | 49 | 0.36 | 6.4 | 8.5 |
| 139 | 33 | 0.21 | 4.8 | 9.0 |
| 140 | 38 | 0.34 | 5.6 | 8.5 |
| 141 | 75 | 0.19 | 5.0 | 8.2 |
| 142 | 68 | 0.25 | 5.4 | 8.4 |
| 143 | 60 | 0.47 | 7.4 | 9.1 |
| 144 | 63 | 0.33 | 4.4 | 8.2 |
| 145 | 48 | 0.28 | 7.1 | 8.2 |
| 146 | 49 | 0.36 | 2.7 | 8.5 |
| 147 | 72 | 0.44 | 5.6 | 7.6 |
| 148 | 54 | 0.51 | 4.3 | 8.7 |
| 149 | 65 | 0.38 | 7.7 | 8.5 |
| 150 | 72 | 0.40 | 3.4 | 9.1 |
| 151 | 51 | 0.16 | 4.9 | 8.4 |
| 152 | 64 | 0.16 | 4.4 | 8.5 |
| 153 | 43 | 0.34 | 0.1 | 8.8 |
| 154 | 57 | 0.38 | 4.4 | 8.2 |
| 155 | 72 | 0.51 | 3.2 | 8.4 |
| 156 | 37 | 0.46 | 5.3 | 8.6 |
| 157 | 50 | 0.33 | 4.1 | 8.2 |
| 158 | 41 | 0.46 | 4.5 | 8.9 |
| 159 | 63 | 0.40 | 5.0 | 8.7 |
| 160 | 51 | 0.33 | 5.4 | 7.9 |
| 161 | 57 | 0.51 | 4.9 | 8.6 |

| 162 | 51 | 0.24 | 1.7 | 8.4 |
| 163 | 73 | 0.32 | 5.6 | 8.6 |
| 164 | 51 | 0.37 | 4.0 | 8.5 |
| 165 | 52 | 0.36 | 5.8 | 8.3 |
| 166 | 52 | 0.34 | 4.6 | 8.1 |
| 167 | 63 | 0.34 | 4.1 | 8.1 |
| 168 | 76 | 0.30 | 6.1 | 8.2 |
| 169 | 56 | 0.40 | 5.7 | 8.5 |
| 170 | 47 | 0.33 | 5.0 | 8.2 |
| 171 | 44 | 0.41 | 5.0 | 8.3 |
| 172 | 60 | 0.33 | 7.7 | 8.4 |
| 173 | 36 | 0.39 | 6.5 | 8.8 |
| 174 | 52 | 0.30 | 5.4 | 8.2 |
| 175 | 53 | 0.27 | 5.6 | 8.2 |
| 176 | 60 | 0.35 | 6.0 | 8.6 |
| 177 | 48 | 0.43 | 3.6 | 8.6 |
| 178 | 63 | 0.49 | 0.2 | 8.2 |
| 179 | 76 | 0.42 | 5.9 | 8.9 |
| 180 | 58 | 0.34 | 7.3 | 8.3 |
| 181 | 51 | 0.43 | 6.4 | 8.7 |
| 182 | 38 | 0.33 | 4.9 | 8.5 |
| 183 | 46 | 0.17 | 4.7 | 8.3 |
| 184 | 53 | 0.34 | 6.4 | 8.7 |
| 185 | 60 | 0.38 | 6.1 | 8.7 |
| 186 | 71 | 0.23 | 6.9 | 8.2 |
| 187 | 54 | 0.41 | 2.9 | 8.3 |
| 188 | 61 | 0.44 | 5.8 | 8.4 |
| 189 | 62 | 0.35 | 3.9 | 8.9 |
| 190 | 55 | 0.15 | 4.8 | 8.0 |
| 191 | 57 | 0.22 | 4.1 | 8.2 |
| 192 | 43 | 0.41 | 7.5 | 8.5 |
| 193 | 51 | 0.34 | 2.4 | 8.6 |
| 194 | 34 | 0.55 | 3.5 | 8.4 |
| 195 | 38 | 0.43 | 7.1 | 8.8 |
| 196 | 49 | 0.38 | 3.5 | 8.3 |
| 197 | 57 | 0.30 | 3.6 | 8.5 |
| 198 | 53 | 0.37 | 5.2 | 9.1 |
| 199 | 51 | 0.36 | 5.1 | 8.2 |
| 200 | 59 | 0.38 | 3.6 | 8.7 |
| 201 | 35 | 0.44 | 4.0 | 8.0 |
| 202 | 73 | 0.32 | 3.0 | 8.3 |

| 203 | 68 | 0.37 | 2.7 | 8.4 |
| 204 | 31 | 0.36 | 4.6 | 8.6 |
| 205 | 40 | 0.33 | 9.0 | 8.3 |
| 206 | 60 | 0.36 | 6.6 | 8.5 |
| 207 | 66 | 0.44 | 4.2 | 8.5 |
| 208 | 47 | 0.22 | 4.5 | 8.7 |
| 209 | 56 | 0.30 | 4.8 | 8.6 |
| 210 | 72 | 0.36 | 2.9 | 8.8 |
| 211 | 68 | 0.50 | 6.6 | 8.3 |
| 212 | 45 | 0.37 | 7.3 | 8.5 |
| 213 | 58 | 0.17 | 4.9 | 9.0 |
| 214 | 64 | 0.34 | 3.2 | 8.6 |
| 215 | 66 | 0.34 | 2.5 | 8.4 |
| 216 | 49 | 0.29 | 5.0 | 8.3 |
| 217 | 83 | 0.39 | 2.5 | 8.8 |
| 218 | 73 | 0.33 | 3.6 | 8.4 |
| 219 | 52 | 0.34 | 3.3 | 8.8 |
| 220 | 56 | 0.28 | 8.8 | 8.7 |
| 221 | 58 | 0.32 | 5.6 | 8.3 |
| 222 | 53 | 0.40 | 5.9 | 8.1 |
| 223 | 50 | 0.23 | 4.4 | 8.4 |
| 224 | 43 | 0.34 | 3.9 | 8.7 |
| 225 | 50 | 0.39 | 4.4 | 8.0 |
| 226 | 44 | 0.31 | 4.4 | 8.4 |
| 227 | 59 | 0.36 | 6.0 | 9.1 |
| 228 | 41 | 0.35 | 3.2 | 8.4 |
| 229 | 53 | 0.29 | 6.6 | 8.7 |
| 230 | 49 | 0.37 | 5.7 | 8.3 |
| 231 | 42 | 0.22 | 8.5 | 8.6 |
| 232 | 48 | 0.34 | 3.9 | 8.2 |
| 233 | 60 | 0.31 | 6.1 | 8.8 |
| 234 | 56 | 0.50 | 2.6 | 8.5 |
| 235 | 43 | 0.25 | 5.3 | 8.9 |
| 236 | 67 | 0.32 | 6.1 | 8.8 |
| 237 | 43 | 0.24 | 8.6 | 8.8 |
| 238 | 41 | 0.46 | 5.1 | 8.7 |
| 239 | 66 | 0.45 | 4.9 | 8.3 |
| 240 | 44 | 0.52 | 4.1 | 8.7 |
| 241 | 70 | 0.43 | 6.6 | 8.8 |
| 242 | 63 | 0.38 | 7.9 | 8.4 |
| 243 | 47 | 0.24 | 3.9 | 8.3 |

| 244 | 52 | 0.38 | 5.4 | 8.8 |
| 245 | 49 | 0.47 | 4.2 | 8.4 |
| 246 | 45 | 0.31 | 8.1 | 8.8 |
| 247 | 46 | 0.37 | 1.9 | 8.3 |
| 248 | 19 | 0.31 | 6.5 | 8.3 |
| 249 | 63 | 0.40 | 6.1 | 8.5 |
| 250 | 64 | 0.35 | 5.8 | 8.1 |
| 251 | 63 | 0.34 | 6.7 | 8.5 |
| 252 | 68 | 0.46 | 6.9 | 8.5 |
| 253 | 48 | 0.43 | 8.6 | 8.7 |
| 254 | 43 | 0.38 | 4.4 | 8.3 |
| 255 | 50 | 0.32 | 4.6 | 8.7 |
| 256 | 76 | 0.31 | 4.0 | 8.3 |
| 257 | 64 | 0.39 | 5.7 | 8.6 |
| 258 | 38 | 0.29 | 6.4 | 8.0 |
| 259 | 90 | 0.30 | 7.0 | 8.6 |
| 260 | 37 | 0.39 | 4.8 | 8.8 |
| 261 | 58 | 0.37 | 6.5 | 8.0 |
| 262 | 42 | 0.27 | 4.5 | 8.6 |
| 263 | 58 | 0.37 | 6.0 | 8.3 |
| 264 | 42 | 0.42 | 7.2 | 8.8 |
| 265 | 66 | 0.33 | 12.6 | 8.8 |
| 266 | 116 | 0.44 | 8.7 | 7.5 |
| 267 | 76 | 0.43 | 9.5 | 6.9 |
| 268 | 125 | 0.46 | 9.9 | 6.8 |
| 269 | 128 | 0.41 | 9.8 | 6.6 |
| 270 | 128 | 0.37 | 11.3 | 7.1 |
| 271 | 125 | 0.44 | 6.7 | 7.9 |
| 272 | 80 | 0.49 | 10.6 | 7.1 |
| 273 | 110 | 0.48 | 9.9 | 7.2 |
| 274 | 135 | 0.41 | 9.8 | 7.8 |
| 275 | 136 | 0.45 | 8.9 | 7.2 |
| 276 | 142 | 0.43 | 10.2 | 8.0 |
| 277 | 120 | 0.48 | 10.2 | 7.5 |
| 278 | 109 | 0.43 | 10.1 | 7.1 |
| 279 | 109 | 0.50 | 10.9 | 7.5 |
| 280 | 111 | 0.35 | 11.8 | 7.4 |
| 281 | 101 | 0.49 | 8.5 | 7.8 |
| 282 | 98 | 0.50 | 11.6 | 7.2 |
| 283 | 91 | 0.56 | 10.0 | 7.3 |
| 284 | 151 | 0.50 | 7.7 | 6.7 |

| 285 | 82 | 0.48 | 14.0 | 7.5 |
|-----|-----|------|------|-----|
| 286 | 107 | 0.48 | 9.5 | 7.5 |
| 287 | 83 | 0.40 | 12.0 | 7.2 |
| 288 | 91 | 0.40 | 9.2 | 7.9 |
| 289 | 127 | 0.40 | 9.1 | 7.6 |
| 290 | 115 | 0.42 | 11.6 | 6.8 |
| 291 | 118 | 0.40 | 9.8 | 7.3 |
| 292 | 89 | 0.42 | 10.8 | 7.0 |
| 293 | 100 | 0.46 | 11.6 | 7.3 |
| 294 | 97 | 0.39 | 8.5 | 7.8 |
| 295 | 110 | 0.36 | 11.1 | 7.7 |
| 296 | 88 | 0.40 | 9.0 | 6.7 |
| 297 | 103 | 0.47 | 11.7 | 6.7 |
| 298 | 82 | 0.49 | 10.7 | 7.5 |
| 299 | 87 | 0.41 | 8.1 | 7.4 |
| 300 | 130 | 0.39 | 9.8 | 7.6 |
| 301 | 116 | 0.42 | 9.6 | 7.6 |
| 302 | 96 | 0.42 | 11.8 | 7.1 |
| 303 | 122 | 0.39 | 7.9 | 7.1 |
| 304 | 70 | 0.38 | 11.1 | 7.4 |
| 305 | 116 | 0.47 | 8.8 | 7.3 |
| 306 | 122 | 0.48 | 9.0 | 7.0 |
| 307 | 109 | 0.45 | 10.0 | 7.3 |
| 308 | 114 | 0.50 | 10.1 | 7.3 |
| 309 | 62 | 0.47 | 11.4 | 7.3 |
| 310 | 120 | 0.51 | 9.5 | 6.3 |
| 311 | 130 | 0.38 | 10.5 | 7.7 |
| 312 | 92 | 0.47 | 11.8 | 7.3 |
| 313 | 81 | 0.55 | 7.9 | 7.3 |
| 314 | 82 | 0.51 | 10.1 | 7.7 |
| 315 | 102 | 0.48 | 10.9 | 6.5 |
| 316 | 113 | 0.43 | 10.2 | 7.9 |
| 317 | 119 | 0.43 | 8.0 | 6.8 |
| 318 | 108 | 0.48 | 8.9 | 7.0 |
| 319 | 130 | 0.53 | 8.3 | 7.3 |
| 320 | 111 | 0.50 | 9.9 | 6.6 |
| 321 | 132 | 0.50 | 11.5 | 6.8 |
| 322 | 110 | 0.47 | 10.8 | 7.1 |
| 323 | 95 | 0.49 | 10.4 | 7.5 |
| 324 | 137 | 0.29 | 9.8 | 7.5 |
| 325 | 98 | 0.53 | 11.5 | 7.0 |

| 326 | 124 | 0.55 | 10.2 | 6.6 |
| 327 | 146 | 0.36 | 10.2 | 7.5 |
| 328 | 126 | 0.51 | 10.6 | 6.5 |
| 329 | 124 | 0.53 | 9.4 | 7.6 |
| 330 | 99 | 0.47 | 8.7 | 7.7 |
| 331 | 100 | 0.51 | 9.5 | 7.9 |
| 332 | 101 | 0.45 | 9.5 | 7.1 |
| 333 | 113 | 0.37 | 9.4 | 7.8 |
| 334 | 139 | 0.42 | 8.9 | 7.1 |
| 335 | 105 | 0.38 | 8.7 | 7.4 |
| 336 | 113 | 0.45 | 10.7 | 7.3 |
| 337 | 104 | 0.45 | 9.6 | 7.2 |
| 338 | 117 | 0.48 | 10.3 | 7.3 |
| 339 | 132 | 0.43 | 10.9 | 7.7 |
| 340 | 100 | 0.44 | 11.8 | 6.8 |
| 341 | 109 | 0.40 | 8.1 | 7.2 |
| 342 | 95 | 0.39 | 9.7 | 7.4 |
| 343 | 139 | 0.39 | 9.8 | 7.7 |
| 344 | 140 | 0.47 | 8.9 | 7.3 |
| 345 | 110 | 0.48 | 12.1 | 7.2 |
| 346 | 97 | 0.56 | 11.5 | 8.2 |
| 347 | 98 | 0.49 | 11.2 | 6.9 |
| 348 | 146 | 0.44 | 10.0 | 7.2 |
| 349 | 92 | 0.47 | 12.0 | 6.5 |
| 350 | 128 | 0.43 | 10.8 | 7.7 |
| 351 | 156 | 0.50 | 11.4 | 6.3 |
| 352 | 134 | 0.39 | 9.1 | 8.2 |
| 353 | 110 | 0.44 | 7.6 | 6.6 |
| 354 | 104 | 0.45 | 12.4 | 7.5 |
| 355 | 98 | 0.54 | 11.0 | 7.1 |
| 356 | 120 | 0.50 | 10.5 | 7.3 |
| 357 | 140 | 0.50 | 10.6 | 6.9 |
| 358 | 130 | 0.53 | 10.7 | 7.4 |
| 359 | 115 | 0.45 | 10.1 | 7.1 |
| 360 | 159 | 0.41 | 10.7 | 7.5 |
| 361 | 114 | 0.43 | 9.9 | 6.9 |
| 362 | 128 | 0.46 | 9.3 | 7.0 |
| 363 | 96 | 0.49 | 7.6 | 7.5 |
| 364 | 61 | 0.49 | 12.0 | 6.7 |
| 365 | 60 | 0.46 | 8.2 | 7.6 |
| 366 | 51 | 0.50 | 7.8 | 7.6 |

| | | | | |
|---|---|---|---|---|
| 367 | 61 | 0.49 | 7.9 | 7.1 |
| 368 | 46 | 0.57 | 7.5 | 6.5 |
| 369 | 60 | 0.44 | 8.0 | 7.7 |
| 370 | 53 | 0.36 | 12.5 | 7.1 |
| 371 | 55 | 0.45 | 10.7 | 7.3 |
| 372 | 59 | 0.38 | 9.0 | 7.0 |
| 373 | 61 | 0.38 | 9.3 | 6.9 |
| 374 | 69 | 0.57 | 10.4 | 7.4 |
| 375 | 63 | 0.51 | 10.6 | 7.0 |
| 376 | 62 | 0.48 | 12.8 | 7.1 |
| 377 | 57 | 0.49 | 7.8 | 7.4 |
| 378 | 70 | 0.40 | 11.2 | 6.9 |
| 379 | 31 | 0.46 | 9.2 | 6.6 |
| 380 | 70 | 0.41 | 10.8 | 6.8 |
| 381 | 66 | 0.39 | 10.9 | 7.8 |
| 382 | 62 | 0.51 | 9.8 | 6.3 |
| 383 | 75 | 0.50 | 9.6 | 7.2 |
| 384 | 58 | 0.34 | 9.1 | 7.2 |
| 385 | 50 | 0.47 | 11.3 | 7.3 |
| 386 | 73 | 0.44 | 9.1 | 7.4 |
| 387 | 61 | 0.37 | 10.8 | 7.2 |
| 388 | 48 | 0.48 | 8.1 | 6.8 |
| 389 | 54 | 0.52 | 12.4 | 7.6 |
| 390 | 63 | 0.52 | 9.4 | 7.1 |
| 391 | 69 | 0.35 | 13.2 | 7.1 |
| 392 | 71 | 0.39 | 12.0 | 7.6 |
| 393 | 44 | 0.40 | 10.6 | 7.4 |
| 394 | 60 | 0.42 | 11.8 | 6.8 |
| 395 | 79 | 0.37 | 9.4 | 7.9 |
| 396 | 38 | 0.50 | 9.9 | 7.3 |
| 397 | 80 | 0.57 | 11.0 | 7.0 |
| 398 | 54 | 0.46 | 8.8 | 6.9 |
| 399 | 74 | 0.44 | 10.8 | 7.6 |
| 400 | 37 | 0.40 | 9.3 | 7.8 |
| 401 | 69 | 0.47 | 9.8 | 7.1 |
| 402 | 54 | 0.47 | 9.6 | 7.3 |
| 403 | 68 | 0.42 | 10.1 | 8.1 |
| 404 | 49 | 0.56 | 8.9 | 7.2 |
| 405 | 55 | 0.45 | 6.1 | 7.2 |
| 406 | 64 | 0.43 | 10.2 | 6.9 |
| 407 | 83 | 0.41 | 7.3 | 6.6 |

Lesson 23:  Using Sample Data to Decide if Two Population Means Are Different

| 408 | 36 | 0.46 | 11.5 | 7.3 |
| 409 | 44 | 0.43 | 11.0 | 6.7 |
| 410 | 65 | 0.44 | 11.1 | 7.0 |
| 411 | 77 | 0.39 | 12.1 | 7.7 |
| 412 | 33 | 0.44 | 6.9 | 7.1 |
| 413 | 45 | 0.47 | 8.2 | 6.9 |
| 414 | 70 | 0.53 | 7.3 | 7.2 |
| 415 | 77 | 0.44 | 8.9 | 7.2 |
| 416 | 53 | 0.46 | 9.2 | 6.8 |
| 417 | 60 | 0.49 | 11.0 | 7.4 |
| 418 | 86 | 0.44 | 5.8 | 7.8 |
| 419 | 49 | 0.55 | 8.4 | 7.2 |
| 420 | 50 | 0.45 | 12.3 | 6.5 |
| 421 | 64 | 0.41 | 10.7 | 7.2 |
| 422 | 57 | 0.45 | 7.0 | 7.1 |
| 423 | 56 | 0.40 | 12.1 | 6.5 |
| 424 | 41 | 0.45 | 12.7 | 7.2 |
| 425 | 50 | 0.50 | 8.3 | 6.8 |
| 426 | 63 | 0.45 | 11.6 | 7.4 |
| 427 | 44 | 0.43 | 7.7 | 7.1 |
| 428 | 51 | 0.42 | 10.3 | 7.5 |
| 429 | 51 | 0.50 | 8.7 | 7.1 |
| 430 | 54 | 0.43 | 8.4 | 7.2 |
| 431 | 45 | 0.44 | 7.0 | 6.9 |
| 432 | 65 | 0.46 | 10.5 | 7.5 |
| 433 | 60 | 0.45 | 7.4 | 7.1 |
| 434 | 52 | 0.42 | 4.1 | 7.1 |
| 435 | 50 | 0.49 | 11.2 | 6.9 |
| 436 | 61 | 0.52 | 10.7 | 6.7 |
| 437 | 42 | 0.43 | 9.2 | 6.8 |
| 438 | 42 | 0.50 | 11.4 | 6.8 |
| 439 | 66 | 0.47 | 7.9 | 7.2 |
| 440 | 65 | 0.43 | 9.2 | 7.1 |

Name _____ Date _____

Round all decimal answers to the nearest hundredth.

1. You and a friend decide to conduct a survey at your school to see whether students are in favor of a new dress code policy. Your friend stands at the school entrance and asks the opinions of the first 100 students who come to campus on Monday. You obtain a list of all students at the school and randomly select 60 to survey.

   a. Your friend finds 34% of his sample in favor of the new dress code policy, but you find only 16%. Which do you believe is more likely to be representative of the school population? Explain your choice.

   b. Suppose 25% of the students at the school are in favor of the new dress code policy. Below is a dot plot of the proportion of students who favor the new dress code for each of 100 different random samples of 50 students at the school.

   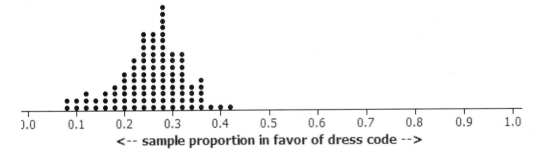

   If you were to select a random sample of 50 students and ask them if they favor the new dress code, do you think that your sample proportion will be within 0.05 of the population proportion? Explain.

c.  Suppose ten people each take a simple random sample of 100 students from the school and calculate the proportion in the sample who favors the new dress code. On the dot plot axis below, place 10 values that you think are most believable for the proportions you could obtain.

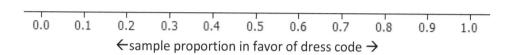

0.0   0.1   0.2   0.3   0.4   0.5   0.6   0.7   0.8   0.9   1.0
←sample proportion in favor of dress code →

Explain your reasoning.

2. Students in a random sample of 57 students were asked to measure their hand-spans (distance from outside of thumb to outside of little finger when the hand is stretched out as far as possible). The graphs below show the results for the males and females.

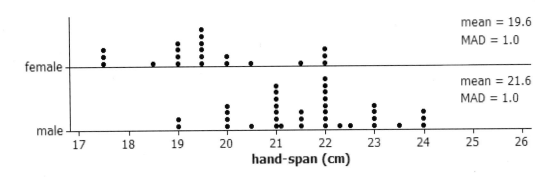

a. Based on these data, do you think there is a difference between the population mean hand-span for males and the population mean hand-span for females? Justify your answer.

b. The same students were asked to measure their heights, with the results shown below.

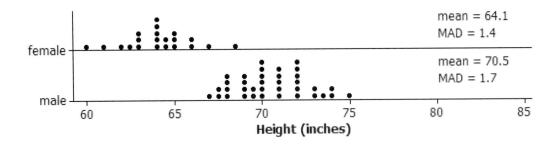

Are these height data more or less convincing of a difference in the population mean height than the hand span data are of a difference in population mean handspan? Explain.

3.  A student purchases a bag of "mini" chocolate chip cookies, and, after opening the bag, finds one cookie that does not contain any chocolate chips! The student then wonders how unlikely it is to randomly find a cookie with no chocolate chips for this brand.

   a.  Based on the bag of 30 cookies, estimate the probability of this company producing a cookie with no chocolate chips.

   b.  Suppose the cookie company claims that 90% of all cookies it produces contain chocolate chips. Explain how you could simulate randomly selecting 30 cookies (one bag) from such a population to determine how many of the sampled cookies do not contain chocolate chips. Explain the details of your method so it could be carried out by another person.

   c.  Now explain how you could use simulation to estimate the probability of obtaining a bag of 30 cookies with exactly one cookie with no chocolate chips.

d.  If 90% of the cookies made by this company contain chocolate chips, then the actual probability of obtaining a bag of 30 cookies with one chipless cookie equals 0.143.  Based on this result, would you advise this student to complain to the company about finding one cookie with no chocolate chips in her bag of 30?  Explain.

**Eureka Math: *A Story of Ratios* Contributors**

Michael Allwood, Curriculum Writer

Tiah Alphonso, Program Manager – Curriculum Production

Catriona Anderson, Program Manager – Implementation Support

Beau Bailey, Curriculum Writer

Scott Baldridge, Lead Mathematician and Lead Curriculum Writer

Bonnie Bergstresser, Math Auditor

Gail Burrill, Curriculum Writer

Beth Chance, Statistician

Joanne Choi, Curriculum Writer

Jill Diniz, Program Director

Lori Fanning, Curriculum Writer

Ellen Fort, Math Auditor

Kathy Fritz, Curriculum Writer

Glenn Gebhard, Curriculum Writer

Krysta Gibbs, Curriculum Writer

Winnie Gilbert, Lead Curriculum Writer / Editor, Grade 8

Pam Goodner, Math Auditor

Debby Grawn, Curriculum Writer

Bonnie Hart, Curriculum Writer

Stefanie Hassan, Lead Curriculum Writer / Editor, Grade 8

Sherri Hernandez, Math Auditor

Patrick Hopfensperger, Curriculum Writer

Sunil Koswatta, Mathematician, Grade 8

Brian Kotz, Curriculum Writer

Henry Kranendonk, Statistics Lead Curriculum Writer / Editor

Connie Laughlin, Math Auditor

Jennifer Loftin, Program Manager – Professional Development

Abby Mattern, Math Auditor

Nell McAnelly, Project Director

Saki Milton, Curriculum Writer

Pia Mohsen, Curriculum Writer

Jerry Moreno, Statistician

Ann Netter, Lead Curriculum Writer / Editor, Grades 6-7

Roxy Peck, Statistician, Statistics Lead Curriculum Writer / Editor

Terrie Poehl, Math Auditor

Spencer Roby, Math Auditor

Kathleen Scholand, Math Auditor

Erika Silva, Lead Curriculum Writer / Editor, Grade 6-7

Hester Sutton, Advisor / Reviewer Grades 6-7

Shannon Vinson, Statistics Lead Curriculum Writer / Editor

Julie Wortmann, Lead Curriculum Writer / Editor, Grade 7

David Wright, Mathematician, Lead Curriculum Writer / Editor, Grades 6-7

Kristen Zimmerman, Document Production Manager